Synthese Library

Studies in Epistemology, Logic, Methodology, and Philosophy of Science

Volume 415

The aim of *Synthese Library* is to provide a forum for the best current work in the methodology and philosophy of science and in epistemology. A wide variety of different approaches have traditionally been represented in the Library, and every effort is made to maintain this variety, not for its own sake, but because we believe that there are many fruitful and illuminating approaches to the philosophy of science and related disciplines.

Special attention is paid to methodological studies which illustrate the interplay of empirical and philosophical viewpoints and to contributions to the formal (logical, set-theoretical, mathematical, information-theoretical, decision-theoretical, etc.) methodology of empirical sciences. Likewise, the applications of logical methods to epistemology as well as philosophically and methodologically relevant studies in logic are strongly encouraged. The emphasis on logic will be tempered by interest in the psychological, historical, and sociological aspects of science.

Besides monographs *Synthese Library* publishes thematically unified anthologies and edited volumes with a well-defined topical focus inside the aim and scope of the book series. The contributions in the volumes are expected to be focused and structurally organized in accordance with the central theme(s), and should be tied together by an extensive editorial introduction or set of introductions if the volume is divided into parts. An extensive bibliography and index are mandatory.

More information about this series at http://www.springer.com/series/6607

The Sylvan Jungle

This book is part of a collection of four books that present the work of the iconic and iconoclastic Australian philosopher Richard Routley (né Sylvan).

The four books are:

- Exploring Meinong's Jungle and Beyond
- Noneist Explorations I
- Noneist Explorations II
- Ultralogic as Universal?

All books are published in the Synthese Library.

Richard Routley and Val Routley *Authors*
Dominic Hyde *Editor*

Noneist Explorations I

The Sylvan Jungle — Volume 2

with Supplementary Essays

Springer

Richard Routley and Val Routley (deceased)

Editor
Dominic Hyde
Philosophy, HPRC
University of Queensland
St. Lucia, QLD, Australia

Synthese Library
ISBN 978-3-030-26307-2 ISBN 978-3-030-26309-6 (eBook)
https://doi.org/10.1007/978-3-030-26309-6

This Springer imprint is published by the registered company Springer Nature Switzerland AG.
The registered company address is: Gewerbestrasse 11, 6330 Cham, Switzerland

Editors' Preface

This volume continues the reprinting of a new edition of Richard Routley's 1980 monograph *Exploring Meinong's Jungle and Beyond: An investigation of noneism and the theory of items.*

Some of the work reprinted in this volume contains material co-authored with Val Plumwood (formerly Val Routley), in particular, chapter 4. Other material was developed with Val, for example parts of chapter 7 (cf. the footnote beginning §7.7(2)). For this reason, she is listed as co-author.

We have sought to correct obvious errors, and have updated some references. As noted in the preface to Vol. 1, attempts were made to locate and reproduce the photographs that Sylvan included in the original publication (see the introductory essay by Hyde, Vol. 1), but the passage of time has erased their tracks. We have, however, been able to find further photographs of the same area as that pictured in the original monograph by one of the original photographers, Henry Gold, and with his kind permission reproduce another of his photos, *Vines and epiphytes*, in this volume.

For help in the production of this second volume, thanks go to: Louise Sylvan for her permission to reprint Richard's original material; Elizabeth Smart for permission to reprint the essay by J.J.C. Smart; the ever-helpful archivists at the University of Queensland Fryer Library (where Sylvan's extensive archive is housed); Michael Kebrt and the Word-to-LaTeX crew; and funding from the University of Adelaide.

*　　*　　*　　*　　*

A note on referencing: the original 1980 edition is cited as "*EMJB*$_1$" (available at: http://hdl.handle.net/11375/14805), and this four-volume reprint is cited as "*EMJB*$_2$". Page, section and chapter references are to *EMJB*$_2$.

Contents

The original version of this book was revised. Due to a mistake by the publisher, the editor's name 'Dominic Hyde' was mistakenly appeared as a co-author in all the chapters on Springerlink and third-party websites. The correction to this chapter is available at https://doi.org/10.1007/978-3-030-26309-6_7.

Contributors

Graham Priest, Philosophy, CUNY Graduate Center, New York, USA

J.J.C. Smart (deceased), Philosophy, Research School of Social Sciences, Australian National University, Canberra, Australia

Maria Elisabeth Reicher, Philosophy, RWTH Aachen University, Aachen, Germany

Fred Kroon, Philosophy, University of Auckland, New Zealand

Introduction: some personal reflections

Graham Priest

With the perspective afforded by time, Richard Sylvan, *nè* Routley (1935-1996), will, I believe, come to be seen as the most important Australasian philosopher of the 20th Century. This is not at all to denigrate his currently better known compatriot contemporaries: a number of these made highly significant contributions to philosophy. But what set Richard apart was the originality he deployed and the scope of his vision. He made original contributions to logic, metaphysics, the philosophy of language, value theory, environmental philosophy, political philosophy. Moreover, though he never wrote anything that integrated all of these, it is clear to those who know his work that his views on all these matters formed part of an overall and systematic philosophical picture.

Richard's work is not as well known as it could be for a number of reasons. One is that he paid no heed to contemporary philosophical sensibilities. He delighted in taking unpopular views, articulating and defending them. He was in his element demolishing the views of his opponents. In other words, he was asking to be ignored, particularly since he was working in an outpost of the philosophical world. He did not have the prestige of an Oxford, Princeton, or Paris, behind him to force people to take note. Next, Richard's work was not always user-friendly. It was exemplary in clarity and argumentation, but he did not have the literary style of a Quine or a Ryle. His work could often appear heavy and laboured. Third, the work was often not easily accessible. He distrusted professional publishers, their connection with "the establishment", and their profit-driven motives. Hence, much of his work appeared in typescript form in in-house publications and pre-prints that never got beyond the *pre* stage. (His work predates the current era, when computers and programs such as Latex can produce elegant documents in-house.)

Exploring Meinong's Jungle and Beyond – or *The Jungle Book*, as it became known to Richard and his friends – fits neatly into this picture. When Richard started to write on the topic, the view that some objects do not exist was about as unpopular as it is possible to be – just a shade short of insanity. As Ryle said of Meinong's view in an article in the *Revue de Metaphysique* in 1973: "*Gegenstandstheorie* is dead, buried and not going to be resurrected". Richard coined the term *noneism* for the view that Meinong – and incidentally, all the great medieval logicians – held, that some objects do not exist. Richard was one of the small band of philosophers (and to my mind the most thoroughgoing of them) who worked to bring noneism back from the dead; and he set about it with the intellectual power of a steam-roller.[1]

The Jungle Book collects the many papers he wrote on the topic, pulling them together from a number of places, of various degrees of accessibility. The result is the thousand-and-some page book – termed coyly, "Interim Edition" – which appeared originally in 1980. It was published in-house and distributed from Richard's office at the Australian National University (ANU), by himself and his research assistants. Both the length of the book and the fact that it is in typewriter font make it particularly hard to read. (I suspect that the number of people who have read the whole thing can still be counted on the fingers and toes of a normal human being.) The manuscript never went through the mill of professional editing. It would certainly have benefitted from this, by being cut down to get rid of the repetitions, being reorganised so as to structure the material better, and having some of the weaker parts removed. Still, what it gave us was Richard's uncensored iconoclastic and creative brilliance.

The book was highly influential amongst Richard's friends and colleagues, but the circumstances around its accessibility mean that it never achieved the distribution and uptake it deserved. So I'm delighted that it has been possible to have the book reappear in the current multi-volume Latexed form. It will make the work much better known, as is its desert – though I have no doubt that were Richard to come back from the grave, since the work is being published by Springer, he would veto the project.

The first of the four volumes contains the core of Richard's noneism: what eventually became his Princeton PhD thesis – submitted many years after he left prematurely

[1] By the way, I often meet philosophers – especially non-native-English speakers – who mis-pronounce the word or, at least, who pronounce it in a way that Richard did not intend. Say the English word *none*, and just stick *ism* on the end. In other words, don't pronounce the middle *e*.

because he didn't like the place. The fourth volume contains an appendix of the original book, "Ultralogic as Universal?". This is an important essay in its own right, and well worth the republication, though it has little connection with the rest of the book. I think that the reason Richard appended it was simply his somewhat optimistic desire to make it more accessible. The other two volumes, of which this is the first, contain the other chapters. These are on a variety of themes with nothing much in common except noneism. There was no way to cut these chapters up in a systematic fashion to produce volumes with greater integrity, so the contents of each are patently diverse.

That does not mean, of course, that the essays in the volumes lack interest. They certainly do not. But it does mean that it is hard to write an editorial introduction to the essays of the usual kind. So I decided not to try. Just read the essays! What I have decided to do instead in the rest of this introduction is to say a bit more about Richard himself.

I well recall the first time I met him. This was at the first conference I ever attended after I moved to Australia from the UK in 1976. It was a meeting of the Australasian Association for Logic in Canberra at the ANU. Richard and the other ANU logicians of course attended the meeting. I gave the paper which was to be published a few years later as "The Logic of Paradox". As I was leaving the session, I recall hearing someone saying (not to me), "What a seditious paper". That was not Richard's reaction. As we were walking up the stairs to the Coombs Building tea room for a break, Richard turned to me and said "So you're a dialectician, are you?" "Dialectician" was the word he was using for what is now called "dialetheism", a word that we junked soon afterwards because of its heavy intellectual baggage. At any rate, so started our close friendship and collaboration which was terminated only 20 years later by Richard's untimely death. Richard was not a dialetheist at the time I met him, but he had been playing with the idea. Targetting the very big apple of the Principle of Non-Contradiction greatly appealed to his iconoclasm.

Richard was already a noneist when I met him, though. I was outraged by the idea. My Quinean orthodoxy told me that this really was beyond the pale – much more so than dialetheism! So although we saw many things in very much the same way in our collaboration, noneism was not one of them; we argued about it a lot. In the end, I had to agree that all the Quinean objections that I thought were so devastating were just lame. I didn't become a noneist at that time, though. There was still the question of how to address the characterisation problem is a satisfactory way. The characterisation problem is this: under some condition or other, a thing has the properties it is characterised as having (the Characterisation Principle). Everyone, noneist or otherwise, accepts this, but no one can accept it in full generality. It leads in a two-line argument to triviality. The problem was how best to accommodate the Principle in a noneist context. The various suggestions for doing so mark the crucial difference between current Neo-Meinongians. Richard struggled with the problem in the *Jungle Book*, though he never really solved it to his satisfaction. Indeed, he was still wresting with it in his final essay on noneism, "Re-exploring Item Theory" (pp. 546-81 of Volume 1 of this edition of the *Jungle Book*).

I finally became a noneist when I found a solution to the problem which satisfied me, and which resulted in my own *Towards Non-Being.* Unfortunately I did not find this till some years after Richard's death, so we never had the pleasure of arguing about it.

Richard and I argued a lot; but there was never anything confrontational about the arguments. We were both interested in the other's ideas, and intent on getting to the bottom of things in a collegial fashion. This was the way that Richard argued with people with whom he was intellectually sympathetic. Arguments could be quite different with people with whom he was not.

When it came to disagreement, Richard was no shrinking violet. He could be blunt in saying that an idea didn't stack up. This could be, and sometimes was, interpreted as personal hostility. And indeed, Richard didn't have much time for people whom he thought were intellectually closed-minded. So bad blood could easily develop. And at the ANU, it eventually did. In the heyday of the Canberra Logic Group – the group of logicians which formed around Richard and Bob Meyer – the chair of the Department was Jack (J.J.C.) Smart. Jack, being a committed Quinean, disagreed with Richard's views deeply, but the two always got on well together. On his retirement in 1985, Jack was replaced by Frank Jackson, and for reasons that are not part of the story here, Richard and Frank did not get on well together. Matters went from bad to worse. And after all the other logicians in the Department left it to found the Automated Reasoning Project, Richard found himself completely isolated in the Department. (Why Richard didn't leave with the other logicians was never entirely clear to me.) By the end of his life he had become entirely alienated from the Department. After his death, when someone else moved into his office, the copies of the *Jungle Book*, which filled its long back wall, were simply disposed of.

Richard's tensions in his personal relationships with other people could be exacerbated by the fact that he did not have the mannerisms which put people at their ease. Most people, when they chat with others, do things with their words, tone of voice, and body language, which serve to put them at their ease. When these are not forthcoming, it is naturally interpreted as hostility. Richard had none of these mannerisms. This was not normally due to the fact that he felt hostile, however. He was in fact, a rather shy person, and I think that social graces did not come easily to him. I must confess that it took me a few years to realise all this about Richard; and though our relationship was never anything but friendly, it was not till I finally did so that I felt completely at ease in his company. Indeed, once you came to understand all of this about Richard, you came to see a very different side of him. Under his rather tough exterior, Richard was a genuinely warm and caring person. This bred loyalty and affection in his friends, close colleagues, and students. Indeed, as a supervisor of graduate students – there were no undergraduate students in the Department – Richard was everything one could wish for: conscientious, supportive, friendly. Students or groups of them would often go out to his home on Plumwood Mountain, where they would work with him on the land whilst discussing philosophy, before they adjourned to the house for something to eat and a few bottles of good Australian red wine.

Richard loved the land. He was a committed environmentalist, and he and his first wife, Val Routley, made important contributions to both environmental ethics and policy, such as in their book *Fight for the Forests*. He cared greatly for the rain forest around his house on the mountain, as he cared for his graduate students, and for philosophy. It is no accident that the cover or the original jungle book was a photograph of Australian rainforest. The term "Meinong's jungle" was coined, as far as I'm aware, by William Kneale in his book *Probability and Induction*. It was meant, I am sure, as a put-down: jungle = tangled mess. Richard, however, adopted it as a term of love: jungle = complex and integrated eco-system. He was greatly at home in sylvan environments, both those of existent trees and those of non-existent objects.

ORIGINAL MATERIAL

First Edition Front Matter [Abridged]

Acknowledgements

My main historical debt is of course to the work of Alexius Meinong. But, as will become apparent, I am also indebted to the work of precursors of Meinong, in particular Thomas Reid. I have been much helped by critical expositions of Meinong's work, especially Findlay 1963, and, in making recent redraftings of older material, by Roderick Chisholm's articles. I have been encouraged to elaborate earlier essays and much stimulated by recent attempts to work out a more satisfactory theory of objects than Meinong's mature theory, in particular the (reductionist) theories of Terence Parsons. That I am, or try to be, severely critical of much other work on theories of objects in no way lessens my debt to some of it.

Among my modern creditors I owe most to Val Routley, who jointly authored some of the chapters (chapters 4, 8 and 9 [the latter two reprinted in Vol. 3 —*Eds.*]) and who contributed much to many sections not explicitly acknowledged as joint. For example, the idea that the Reference Theory underlay alternatives to the theory of objects and generated very many philosophical problems, was the result of joint work and discussion.

I have profited – as acknowledgements at relevant points in the text will to *some* extent reveal – from constructive criticism directed at earlier exposure of this work, in particular extended presentations in seminar series at the University of Illinois, Chicago Circle, in 1969, at the State University of Campinas in 1976, and at the Australian National University in 1978.

* * * * * *

Some of the essays which follow are redraftings, mostly with substantial changes and additions, of previous essays, which they supersede. Main details are as follows:

Chapter 2 – which has not been subject to nearly as much revision as it deserves – incorporates virtually all of "Existence and identity when times change", a 69 page typescript from 1968. The paper was subsequently re-entitled "Exploring Meinong's Jungle. II. Existence and identity when times change". Professor Sobocinski kindly offered in 1969 to publish both parts, I and II, of "Exploring Meinong's Jungle" in the *Notre Dame Journal of Formal Logic.* Perhaps fortunately for other contributors to the Journal, part II was never submitted in final form, and part I has recently been withdrawn.

Parts of several of the essays [included in this Volume $_{-Eds.}$] have been published elsewhere; *Chapter 3* in *Philosophy and Phenomenological Research*; *Chapter 6* in *Grazer Philosophische Studien*; and *Chapter 7* in *Poetics*; while some of *Chapter 4* has previously appeared in *Revue Internationale de Philosophie*, the remainder of the paper involved (referred to as Routley^2 1973) being largely taken up in Chapter 1 [see Vol. 1 $_{-Eds.}$].

Excerpts from earlier articles on the logic and semantics of nonexistence and intensionality and on universal semantics have also been included in the text; these are drawn from the following periodicals: *Notre Dame Journal of Formal Logic* (papers referred to as EI, SE, NE), *Philosophica* (MTD), *Journal of Philosophical Logic* (US), *Communication and Cognition* (Routley^2 1975), *Inquiry* (Routley 1976), and *Philosophical Studies* (Routley 1974). Permission to reproduce material has been sought from editors of all the journals cited, and I am indebted to most editors for replies granting permission.

Parts of many of the essays have been read at conferences and seminars in various parts of the world since 1965 and some of the material has as a result (and gratifyingly) worked its way into the literature. It is pleasant to record that much of the material is now regarded as far less crazy and disreputable than it was in the mid-sixties, when it was taken as a sign of early mental deterioration and of philosophical irresponsibility.

* * * * * *

References, notation, etc. Two forms of reference to other work are used. Publications which are referred to frequently are usually assigned special abbreviations (e.g., SE, Slog); otherwise works are cited by giving the author's name and the year of publication.

In case an author has published more than one paper in the one year the papers are ordered alphabetically.

The bibliography records only items that are actually cited in the text. Delays in production made feasible – what was always thought desirable (as even the *authors* of Slog have repeatedly found) – the addition of an index: this too was compiled by Jean Norman.

In quoting other authors the following minor liberties have been taken: notation has been changed to conform with that of the text, and occasionally passages have been rearranged (hopefully without distortion of content). Occasionally too citations have been drawn from unfinished or unpublished work (in particular Parsons 1978 and Tooley 1978) or even from lecture notes (Kripke 1973): sources of these sorts are recorded in the bibliography, and due allowance should be made.

Standard abbreviations, such as "iff" for "if and only if" and "wrt" for "with respect to", are adopted. The metalanguage is logicians' ordinary English enriched by a few symbols, most notably "$\rightarrow$" read "if ... then ..." or "that ... implies that ...", "&" for "and", "$\vee$" for "or", "$\sim$" for "not", "P" for "some" and "U" for "every". These abbreviations are not always used however, and often expressions are written out in English.

Cross references are made in obvious ways, e.g. "see 3.3" means "see chapter 3, section 3" and "in §4" means "in section 4 (of the same chapter)". The labelling of theorems and lemmata is also chapter relativised. Notation, bracketing conventions, labelling of systems is as explained in companion volume RLR; but in fact where these things are not familiar from the literature or self-explanatory they are explained as they are introduced.

*　　*　　*　　*　　*　　*

Notes for prospective readers. By and large the chapters (and even sections) can be read in any order, e.g. a reader can proceed directly to chapter 3 or to chapter 9, or even to section 12.3. Occasionally some backward reference may be called for (e.g. to explain central principles, such as the Ontological Assumption), but it will never require much backtracking.

In places the text becomes heavily loaded with logical symbolism. The reader should not be intimidated. Everything said can be expressed in English, and commonly is so expressed, and always a recipe is given for unscrambling symbolic notation into

English. However the symbolism is intended as an aid to understanding and argument and to exact formulation of the theory, not as an obstacle. Should the reader become bogged down in such logical material or discouraged by it, I suggest it be skipped over or otherwise bypassed.

Richard Routley
Plumwood Mountain
Box 37
Braidwood
Australia 2622.

Chapter 2

Exploring Meinong's jungle and beyond
II. Existence and identity when times change

§1. Existence is existence now.

Not all items that have existed or will exist currently exist: some, like Aristotle and Queen Hatshepsut, have ceased to exist, others, like the greatest philosopher born in the 21st century, do not yet exist. The fact that most of us really want to claim that purely past and purely future items do not exist, that Aristotle does not exist, is part of the case for the thesis:

E0. existence is existence now.

For if an item does not exist now then either it never exists or it is purely past or purely future. But if it never exists it does not exist, and if it is purely past or future it does not, by the former points, exist. Therefore, contraposing, what does exist exists now. The converse of this, that what exists now does exist, is fairly unproblematical. For if any item exists now it satisfies whatever criterion of existence is adopted, and so exists. Again, if an item does exist *now* then it *is* existent (transferring the now into the tense), so it exists. The converse, that only what exists now exists, can likewise be presented as a grammatical transformation; if an item exists then it *is* existent, so it exists now. The thesis reflects a rationally-based determination to use "exists" as a present-tensed verb, and not in some other way. But that is not all.

Purely past and purely future items are, *like merely possible items*, not (now) determinate in all extensional respects: hence (applying the results of 1.19 [Vol. 1 $_{\text{—}Eds.}$]) they do not exist. Compare the items Aristotle and Polonius, and remember Peirce's question as to how long before Polonius died had he had a hair cut and Russell's as to

The original version of this chapter was revised. Due to a mistake by the publisher, the editor's name 'Dominic Hyde' was mistakenly appeared as a co-author in all the chapters on Springerlink and third-party websites. The correction to this chapter is available at https://doi.org/10.1007/978-3-030-26309-6_7.

D. Hyde (ed.), *Noneist Explorations I*, Synthese Library 415,
https://doi.org/10.1007/978-3-030-26309-6_1

the baldness of the present king of France. Well, is Aristotle bald now? If he is, how long has he been bald? If not, how long since he had a haircut and how long is his hair? Since Aristotle has ceased to exist, it is false that Aristotle is now bald and false that he is not now bald, even on Russell's theory of descriptions naturally (i.e. temporally) construed. Thus Aristotle is indeterminate in respect of the extensional property of (present) baldness. Hence he does not exist now; hence he does not exist. Likewise the future sea battle is indeterminate in various respects, so even it if will exist it does not exist. Of course there are substantial differences between the various sorts of nonentities alluded to, between Aristotle (a past object) who did exist but does not now, the future sea battle (a future object) which will exist but does not now, and Polonius (a possibilium) who never existed.

Criteria for existence, which are not biassed by the Ontological Assumption, converge with common sense and natural language in the main claims made, in particular on E0. For consider what we *do* say and how we argue (before philosophers and positivistically-inclined scientists got at us).[1] We say, and say freely, such things as "Aristotle does not exist, he's dead", and we are prepared to argue, correctly, "Aristotle is dead. Dead men do not exist. So Aristotle does not exist". The same results emerge from other criteria for existence. Consider, for instance, the spatial test. Aristotle, like Polonius, does not have an actual location, a postal address, or identity card etc.; he is not to be encountered in Greece or anywhere else in the world.

[1] Compare Putnam (1967b, p. 240, my rearrangement):

> The main principle underlying the "man in the street's" view of the nature of time [is] as follows: (1) All (and only) things that exist *now* are real. Future things (which do not already exist) are not real...; although, of course they *will* be real when the appropriate time has come to be the present time. Similarly past things (which have ceased to exist) are not real, although they *were* real in the past.
>
> Obviously, we shall have to make some assumptions about the concept *real* if we are to discuss the "man in the street's" view at all.

If we make, not Putnam's (relativistically-) loaded assumptions for which no motivation is offered (especially his assumption III), but the commonplace assumption that what is real is what exists – which certainly coincides with one central sense of "real", cf. OED – then what results from (1) is not Putnam's conclusion that future things are (already) real and, a conclusion Putnam does not draw, the inconsistency and incoherence of the man-in-the-street's view. (For "future things (which do not already exist)" are real and also not real.) What results instead is, firstly, E0, that all and only things that exist exist now, and secondly a coherent view of the nature of time.

Putnam's position, which eventually vanishes into a tautology, is further examined below when the question of the reality of the future is considered.

Or consider in more detail the criterion tentatively adopted in 1.19 [Vol. 1 —*Eds.*], according to which existent items, entities, are completely determinate in all extensional respects. Since

(1) $xE \Leftrightarrow .\ (U\,\mathrm{ext}\, f)(x{\sim}f \equiv {\sim}xf),^{\alpha}$

i.e. that item x exists logically coimplies that for all extensional predicates of x predicate negation and sentence negation coincide, and since

(2) x is $\sim$-indeterminate $\Leftrightarrow .\ (P\,\mathrm{ext}\, f)(xf \,\&\, x{\sim}f),^{\beta}$

i.e. that x is (negation-) indeterminate logically coimplies that for some extensional property x both has and lacks that property, it follows that if an item is indeterminate then it does not exist.[1] Hence again, since purely past and future items are indeterminate, are incomplete in many extensional respects (e.g. where they are lodging, what they ate for breakfast, what clothes they are wearing), they do not exist. Nor is the key principle (1) that is used a consequence merely of a stipulative definition of existence. Rather it derives from an explicative definition of existence; for there are independent arguments for the correctness of the coimplication (see 1.19 [Vol. 1 —*Eds.*]). Equally important here, (1) cannot be thrown over by classical logicians since the special case of (1) needed in the argument holds classically. When confronted by names naming items not of acquaintance the classical logician typically eliminates these in favour of definite descriptions, in the way Russell has explained. Most names, including all names of purely future and purely past items, should strictly be replaced by definite descriptions since the items (purportedly) referred to are not items of acquaintance. Many later workers in the classical paradigm go further and insist upon eliminating all names in favour of definite descriptions. But for definite descriptions (as noted) a classical version of (1) holds. This is a consequence of PM, *14.32. After modification of scope conventions and predicate relettering and reordering, *14.32 reads:

$$(\iota x\ xf)E \equiv .\ (\iota x\ xf){\sim}g \equiv {\sim}(\iota x\ xf)g$$

Then since g is not free on the left-hand side, and since all predicates of described items are supposed to be extensional (see PM, p. 182),

[α] Where 'U' is the ontologically neutral universal quantifier, 'for all'; see Vol. 1, §1.9. —*Eds.*
[β] Where 'P' is the ontologically neutral particular quantifier, 'for some'; see Vol. 1, §1.9. —*Eds.*
[1] Almost all strict implications and coimplications can be replaced by entailments and coentailments, and in a more polished theory would be so replaced.

(1′) $(\imath x\ xf)E \equiv .\ (U\,\mathrm{ext}\,g)((\imath x\ xf){\sim}g \equiv {\sim}(\imath x\ xf)g)$

follows. To argue the same point differently, if the x which has f exists then $(\imath x\ xf){\sim}g \equiv {\sim}(\imath x\ xf)g$ holds for all g, and so, in particular, for all extensional g. Conversely, if the x which is f does not exist then $(\imath x\ xf){\sim}g$ is always false and ${\sim}(\imath x\ xf)g$ is always true, by PM, *14.21; hence ${\sim}(U\,\mathrm{ext}\,g)((\imath x\ xf){\sim}g \equiv {\sim}(\imath x\ xf)g)$. It follows from (1′) that $(P\,\mathrm{ext}\,g)({\sim}(\imath x\ xf)g\ \&\ {\sim}(\imath x\ xf){\sim}g) \supset .\ {\sim}(\imath x\ xf)E$, i.e. an item which neither has nor lacks some extensional feature does not exist. Therefore, since the Stagyrite neither has nor lacks various extensional properties, he does not exist; and likewise for all other purely past and purely future objects.

Similar arguments to those that support E0 support the deductively equivalent thesis, *nonexistence is nonexistence now.* Adoption of E0 need entail *no loss* in what can be said or argued. On the contrary, a great advantage accrues from the adoption of the thesis within the theory of items to be developed, that without loss of expressive power ontological commitments can be cut back to a correct and reasonable level. In particular, adoption of E0 does not exclude continuity of times or a temporal continuum, as Williams mistakenly claims (Gale 1967, p. 114): continuity is however defined in a less ontologically obnoxious way than usual, using neutral quantification over times.

That past and future items do not exist is not the only corollary of E0. It also follows (with the help of minor lemmas) that past and future items, which are not present, do not exist, and that the past and the future do not exist. So much of what we talk of does not exist. Immediately, acute problems are raised for the Reference Theory and for its associated logic, classical logic.[α] For according to the Reference Theory what does not exist lacks an unproblematic meaning and cannot be truly spoken of, and according to a dogma of classical logic what does not exist has no true (primary) properties. But we constantly speak of purely past and purely future items, of the past and the future, and ascribe true properties to them.

Because purely past and future items both do not exist and yet have quite definite properties *classical logic is inadequate to formalise discourse and arguments concerning past and future items.* For it follows from classical logic (PM, *14.21) that no item which does not exist has any true properties. Similarly on other theories of descriptions they

[α] Recall the Reference Theory, Vol. 1, §1.6, one version of which is: "the meaning of a subject expression is given by, or is a function of, its reference [an existing item]". —*Eds.*

have no (or but few) distinctive properties. Accordingly, to accommodate logical discourse about purely past and purely future items a nonclassical chronological logic is essential. The advantages of neutral logic are really highlighted when chronological extensions are made, much as the advantages of an alternative theory to the Reference Theory really emerge when becoming and ceasing to exist are to be explained. Because the reference of "a" may change over time, and because the reference of "a" may even cease to exist and will eventually do so, a more satisfactory theory has to take account of change of reference, and of identity and existence, over lapse of time. To formalise such a theory properly, a neutral chronological logic is needed.

But not any old chronological logic will serve. Any suitable logic should satisfy certain conditions of adequacy (including those cited at the beginning of §4), and given these the inadequacy of all current theories can be seen.[1]

§2. Enlarging on some of the chronological inadequacies of classical logic and its metaphysical basis, the Reference Theory.

Though Aristotle does not exist we can continue to talk about and signify Aristotle; though he does not exist now it is not nonsense (non-reference), but true, that Aristotle *was* born in Stagira, and taught Alexander. Aristotle has true properties – including the property of not existing, the property of having ceased to exist, etc.

According to classical doctrine on purely past and purely future items, names or descriptions such as "Aristotle" and "Man first born in the 21st century", (hereafter "Manfibo"), since not signifying existing ostensible items, must be eliminated through descriptions. And *all* primary statements containing the resulting descriptions are false; for (short of an obnoxious and implausible platonism) Aristotle no longer exists and so does not exist, and Man first born in the 21st century does not yet exist, and so does not exist. The doctrine strictly construed is preposterous: it is true, not false, that Aristotle, who is a dead philosopher, was born at Stagira, and that Man first born in the

[1] This is not to disparage recent work in chronological logic; the researches of Prior and co-workers (reported in Prior 1967) have notably advanced the subject. The present essay is especially indebted to the work of Prior and of Rescher.

21st century is first born in that century.[1] On the strict doctrine even Caesar's classic statement on the division of Gaul is false.

Nor can Aristotle and Alexander be *values* of subject (or individual) variables of standard quantification theories with identity. So it is not true on these theories that Aristotle is identical with Aristotle or was a Greek. Only by attributing to these items what they do not have – existence – could they be handled by the theories in some of the expected ways; and then other anomalies would appear. For example, it would follow that Aristotle is determinate in all respects, including present baldness, whereabouts, etc. For similar reasons moves to save the position by introducing some secondary sense of existence in which past items still exist are classically inadmissible; for these moves are inconsistent with such outcomes of the theory of descriptions as the indeterminacy of items which do not (primarily) exist. In any case the shuffle is much less inviting with respect to future items, since future items do not yet exist and never have.

On classical theory one cannot even make any true (primary) predictions about future items.[2] Classical theory raises intolerable problems for non-platonic architects, engineers and planners. Consider, for example, the hotel that will be built on site s, a hotel which has definite and approved specifications and plans but which does not yet exist. Among other things, the hotel will be built on site s. But since the hotel does not exist it is *false*, according to classical logical theory, that the hotel will be built on site s.[3] All primary predictions about future buildings are false.

---

[1] The correct classical response, made in fact by Prior, is that such statements as "Aristotle was born in Stagira" "are not 'primary' statements containing the description in question, but contain 'secondary' occurrences of them, the form of the propositions being 'it was the case that (Aristotle is born in Stagira)', so there's no ... reason ... why they *shouldn't* come out true...". Note that:

1) the response does take it for granted that "Aristotle" is not, what it appears to be, a name, but a (disguised) description; indeed it must be such, since names have (classically) no scope. "According to Prior, 'Bucephalus' 'can no longer count as a proper name' because Alexander's horse has ceased to exist". However, as Hintikka goes on to remark (1959, p. 136, n. 15), "this implies the eminently unsatisfactory conclusion that the *logical status* of a name changes when its bearer dies". The thesis that purely past objects cannot be named is considered further in §5.

2) the response blocks legitimate inferences, such as that to "Someone was born at Stagira".

3) the response fails against present tense ascriptions to past objects, as in "Aristotle is dead", "Aristotle is the author of *Nichomachean Ethics*", etc. All such ascriptions have to be paraphrased away; "dead" declared like "exists" not a predicate; etc. Canonical forms are fixed to fit the (failing) theory.

[2] Thus Prior: "*no* prediction is a 'primary' prediction". The severe difficulties in the development of this position are like those set out in the previous footnote.

[3] On the platonistic alternative now favoured classically, the hotel exists. Then why bother to design it or built it?

Similar problems for the classical theory are raised by events that do not exist, for example Hitler's invading England in 1940. That such events do not exist should follow, even classically, from their indeterminacy: after all which day did Hitler invade, how many troops did he have? Yet just such events are important for counterfactual conditionals.

Not surprisingly classical practice does not square with the theory; there is a big discrepancy, and a good deal of double talk to try to close it. Clear examples as to the discrepancy are provided by Russell's and Carnap's work. In PM (e.g. pp. 66–7), "Socrates" and "Scott" are treated as individual names, Scott and Socrates as individuals. But it is apparent from later work that this should be regarded as a temporary expositional expedient. For Socrates and Scott are not ostensively indicable now to any one; they are introduced rather by some sort of description or source book account. Socrates is in the same position as Homer (see PM, p. 175); "Socrates" is not a logically proper name and should really be eliminated by way of a description. For one thing we can say of Socrates is that he does not now exist though he did exist a very long time ago. But according to Russell, of a logically properly-named individual we cannot significantly say that it exists, or that it does not. Thus "Socrates" is not a logically proper name; and nor is "Scott". Socrates and Scott are not really individuals of Russellian logic. On the theory of Russellian logic indeed *no purely past individuals are individuals of the logic.* But classical practice appears quite different, and in applications of Russellian logic in ever-multiplying numbers of textbooks and in Russell himself (e.g. PM) we find purely past individuals taken as values of the variables, and thereby also licence to quantify over some of the past assumed. This practice is not warranted by the theory, and it cannot be correct. For, to illustrate, from the truth "Socrates is a citizen of ancient Athens" it would follow what is false – that there exists a citizen of ancient Athens.[1]

Taking advantage of the discrepancy between classical practice and theory we can present this dilemma: if the practice is followed we are obliged to accept as true many

[1] As always there is a classical response, made again by Prior: "It *isn't* a truth that 'Socrates is a citizen of ancient Athens' (in fact no one is) – he *was* one". Similarly it is false that Socrates is a philosopher, false that Socrates is smarter than Rescher and also false that Rescher is at least as smart as Socrates, etc. To arrive at the truths such "philosophically-misleading" statements are intended to express, two stages of regimentation are prescribed, retensing with the insertion of extra verbs where required, and removal of pseudo-names, such as "Socrates".

false existential statements; but if the theory is strictly pursued we are obliged to reject as false many historical truths.

The problem once again for classical theory is that familiar, often truthful, discourse, such as talk about the history or the future of the universe or items in it, does include quantification over items which do not exist. Not unexpectedly the same escape routes that are taken in dealing with difficulties raised by nonentities are again raised in the case of purely past or future items.

First, the platonistic strategies reappear.[1] Purely past items do in some sense exist. What sense? One piece of evasive action sometimes favoured is to introduce *timeless existence*, and a related ploy is the introduction of *omnitemporal existence*. According to the timeless existence move, even if Aristotle doesn't now exist, he timelessly exists. Where? How? This shuffle runs into the difficulties that face *any* escape according to which Aristotle still, in some sense, exists. For Aristotle, if he exists, is a concrete individual: it is significant to ask of him, if he exists, where he is to be found, what his current properties are, and so forth. But as we know, Aristotle isn't now to be found anywhere. We can't say what he now has for breakfast, or what he thinks of Prior's tense logics. In these respects Aristotle is indeterminate, like a nonentity. Nor does the timeless entity Aristotle satisfy usual requirements for existence. Against this universal item Berkelian jibes can be made. What height does he have; the height Aristotle had when 10 years old, or the maximum height Aristotle attained, or the height he had when he died? The actual Aristotle died, the timeless Aristotle presumably did not since otherwise he would not be timeless and would have ceased to exist. Was there a time at which the actual changing Aristotle and the timeless Aristotle coincided? If so, when? If not, what has the so-called "timeless Aristotle" got to do with Aristotle? The problems regarding purely future items are even worse. Enough has been said to make it plain that talk of timeless existence of individuals so far from being a viable escape route leads to considerable difficulties and soon involves heady metaphysics.

In fact the timeless Aristotle, since *indeterminate* in so many directions, does not exist, even classically; hence the timeless Aristotle, whatever he is like and however he differs from the timeless Plato, cannot with his timeless mates uphold the classical theory

[1] These have enjoyed more plausibility than they might otherwise have had because of the Minkowski pictures of relativistic theory. Some of the philosophical shortcomings of Minkowski representations are indicated below.

(as strictly construed). For if the timeless Aristotle does not exist, then Aristotle does not exist timelessly:[1] so timeless existence does not reinstate Aristotle as a value of classical variables or as an item having true properties.

Furthermore, what is correct about timeless existence can be replaced by discourse either about existence at all times or existence at some time. For example "a past time timelessly exists" can be paraphrased by "a past time exists at some time". But nothing is correct about existence out of time or existence without time. For d exists out of time if, by E0, d exists now out of time – a conceptual muddle. Nontemporal (mostly Humpty Dumpty) senses of existence can be defined in terms of omnitemporal existence, or sometime-existence, or by abstraction from times. No such manoeuvres work as escape devices: they all, in virtue of PM, *14.21, ascribe either too many or too few properties to Aristotle. Consider what happens with omnitemporal existence.

Many of the difficulties that afflict the attribution of timeless existence to Aristotle likewise render the ascription of omnitemporal existence to Aristotle quite implausible. For if Aristotle exists omnitemporally, he exists at all times, and so exists now. Well, then, what sort of features does this unlocatable but currently existing Aristotle have? Is he bald, at least 2,000 years old and something of a medical wonder? Like the timeless Aristotle, the current Aristotle of the omnitemporal subterfuge does not exist because he is radically indeterminate; for the same reason this Aristotle is of no more avail to the classical theory than Phoebus, the bringer of light.

Nor does "sometime-existence" afford an escape. For sometime-existence, existence at some time, is not a kind of existence; sometime-existence does not entail existence. While *sometime-existence* is an admissible notion for noneism it is not for other theories without a heavy injection of platonism. For an object that has sometime-existence, such as Aristotle, at most times does not exist. Hence it is not a value of classical bound variables, and not open to classical logical assessment, etc.

The sempiternal hypothesis, that if an item exists at some time it exists always – symbolised $(Pt)(xE, t) \dashv (Ut)(xE, t)$ – is sometimes assumed in an attempt to rectify the classical theory. But since the sempiternal hypothesis entails omnitemporal existence it does not rectify matters at all. Because it is inconsistent with the E0 thesis, and because it entails omnitemporal existence, the sempiternal hypothesis is false.

[1] The logical transformation used at this point of the argument is usually conceded.

Finally, as is well-known (and as Prior 1967 nicely explains, p. 139), platonistic salvage operations lead to many unpalatable theses inconsistent with the thesis that existence is existence now and with commonsense, for instance Barcan-type formulae such as $F(\exists x)xf \dashv (\exists x)Fxf$ (e.g. if it will be the case that there exists a man who proves Fermat's last theorem, then there exists a man who will prove Fermat's last theorem!) and $(\exists x)xf \dashv P(\exists x)Fxf$ (e.g. if there exist a breeder reactor then it was the case that there exists an object that will be a breeder reactor!).

In characteristic opposition to platonism, reductionist moves reappear. According to nominalistic moves discourse concerning purely past items can somehow be reduced to or paraphrased by talk about present entities. That such a move has little to commend it and encounters serious difficulties has been explained at some length by Ayer 1956.[1] The idealist strategy which attempts some reduction through memory and remembered items has likewise been implicitly criticised by Ayer. Neither reduction resolves the problem of quantification over past items unless past items are also eliminated. And this is what would have to happen. On nominalistic reductions Aristotle is typically eliminated by way of the name "Aristotle", and on idealistic reductions through the idea or concept of Aristotle. It is easy to make difficulties for both these sorts of elimination; for the first by variations on "It is contingently true that 'Aristotle' names Aristotle"; and for the second by variations on "Their idea of Aristotle bears no resemblance to (is different from) Aristotle". The reductions normally make no effort to cater for, what are even more recalcitrant, purely future items.

The Reference Theory is at least as vulnerable to the foregoing criticisms as classical logic which it metaphysically underpins. For the Reference Theory leads to the principles, worked into classical logic, which result in the chronological inadequacies of classical logic. But a larger charge is laid against the Reference Theory, which subsequent sections do something to substantiate. It is that very many, indeed most,[2] problems in the philosophy of time are a product of the Reference Theory. Indeed the Reference Theory has done an immense amount of damage in the philosophy of time in diverting

[1] Actually Ayer treats the related problem of verification of statements about the past. But the verification difficulties arise of course from the non-existence, and therefore nonlocatability, of past items; and the attempted analyses to guarantee verification coincide with those designed to secure talk of past items.

[2] Most, but not all. The theory of relativity, at least in the hands of empiricists, is second only to the Reference Theory and classical logic in generating such spurious problems.

attention away from the obvious commonsense answers to puzzles over time and change.

§3. Change and identity over time; Heracleitean and Parmenidean problems for chronological logics.

The traditional problems and solutions, both western and eastern, to the problem of change, and especially of the wheel of life, of birth and extinction, are set within and shaped by the Reference Theory. The basic assumption is that things cannot really go in and out of existence, since we can go on thinking, talking and conceptualising about them. The classical western position is well summed up (though in a slightly different context) by Reid (1895, p. 370; see also p. 372, p. 374):

> The universe must be made of something. Every workman must have materials to work upon. That the world should be made out of nothing seemed to them absurd, because everything that is made must be made of something.
> Nullam rem e nihilo gigni divinitus unquam.[α] – LUCR.
> De nihilo nihil, in nihilum nil posse reverti.[β] – PERS.
> This maxim never was brought into doubt; even in Cicero's time it continued to be held by all philosophers. What natural philosopher (says that author in his second book of Divination) ever asserted that anything could take its rise from nothing, or be reducted to nothing?

The characteristically western metaphysical solution was through an underlying substance (such as Aristotle's prima materia) which persisted: the typical eastern metaphysical solution was through a life cycle theory. On both metaphysics the result was that things do not really come into and go out of existence but are simply transformed into something else, as a brass pig may be wrought into a brass monkey.

Except insofar as the maxim reflects an early formulation of a conservation principle, it is mistaken. The problem is engendered by a mistaken insistence, in accord with the Reference Theory, in existentially-loaded quantifiers. Otherwise, and unproblemat-

[α] Nothing is ever gotten out of nothing by divine power. —*Eds.*
[β] Nothing is generated from nothing, nothing returns to nothing. —*Eds.*

ically, something existent can come into existence from nothing existent but something nonexistent.

The Reference Theory also engenders serious problems concerning change and identity over time, problems going back to Parmenides and Heracleitus. Two important connected conditions of adequacy on a philosophical theory of time and on chronological logics are these:

First, it should be possible to assert truly sometimes that x at t_1, is the same as y (or x) at a different time t_2, and yet that the item has changed over time. *Identity and change over time should be compatible.* For instance, we should be able to say truly that Russell in 1911 is the same man as Russell in 1968 though Russell has changed over the years. But both Heracleitus and Parmenides have arguments to show that this is impossible.

Second, it should be possible to assert truly that an item has come into existence or has ceased to exist. But Parmenides has an argument to show that this is impossible, and this argument can be developed so as to vitiate chronological logics based on classical logic.

Consider first Heracleitus's thesis that everything is in constant flux or change: that *no* individual *thing persists.* Suppose for a reductio argument that some item a does persist, say from time t_1 to time t_2. But then a at t_1 would have a different property from a at t_2, for example existing at t_1 (using E0 and $t_1 \neq t_2$), or being in a certain state or position p (see Aristotle's comments on Heracleitus). Therefore, by Leibnitz's identity principle a at t_1 differs from a at t_2 since a at t_1 has different properties from a at t_2. Therefore a changes from time t_1 to t_2 since its properties vary, and therefore a does not persist. For a is not the same at t_1 as at t_2 entails that a has changed from t_1 to t_2; and conversely a has changed from t_1 to t_2 entails that a at t_1 is distinct from a at t_2. Finally if a at t_1 is distinct from a at t_2, a does not persist from t_1 to t_2 (cf. the river fragments of Heracleitus, in Freeman 1947).

Now (classical) Leibnitzean identity just does lead straight to Heracleitus's outrageous thesis. Classical logic provides no ready-made escape. However, it is not hard to find a way in which the classical logician can escape the Heracleitan flux argument. For Leibnitz identity only requires coincidence of all *properties*, and the classical logician can simply extend his dogma according to which existence is not a property to the dogma that *existence at a time t* is not a property.[1] In this way he avoids the argument presented on behalf of Heracleitus. Actually the escape is not quite so simple: first the Russellian is obliged to classify as nonproperties various other features that may be used to distinguish a at t_1 from a at t_2 ; second he is at pains to develop a satisfactory account of a's change (or to explain how something can come to exist or cease to exist). For as a's change is not in respect of properties, classical logic cannot be applied to yield implications about change (e.g. we are not strictly entitled to argue: if a changed from t_1 to t_2 then something changed from t_1 to t_2).

Yet the usual requirement on change seems, at first glance, correct: a changes from t_1 to t_2 if a at t_1 has some property that a at t_2 does not have, or conversely.[2] A logically equivalent account of change is offered initially by Russell:

> Change is the difference, in respect of truth or falsehood, between a proposition concerning an entity and a time T and a proposition concerning the *same* entity and another time T', provided that the two propositions differ only by the fact that T occurs in the one where T' occurs in the other (1937, p. 469, my italics).

There is a serious tension in this account; namely that under conditions where change occurs the entity cannot be the *same* according to Leibnitz identity. This difficulty Russell does not meet; what he does say is that his definition requires emendation if it is to accord with usage, since "usage does not permit us to speak of change except where what changes is existent throughout". This is false. Orlando changed over time; when someone is killed he changes; when a plant dies or is burnt it changes. The fact is that classical logic is simply unable to handle adequately changes in which some items come to be or cease to exist. Classical logic even when extended chronologically, in the way Russell and Carnap for example have extended it, not only does not provide any solution to the problem of identity with change over time, but by its adherence to a Leibnitzian identity principle and its legislation on properties precludes itself from easily providing a solution.

Part of the trouble is due to the complete unsatisfactoriness of the Leibnitz theory of identity. The Leibnitz theory of identity depends for its support on the Reference

[1] Naturally on the theory of items, existence at time t, like existence, is a property.

[2] Occurring at t_2, for example, will not do as a distinguishing property; for if the items are the same they may both have this property.

Theory and good reasons for adopting it vanish once the Reference Theory is abandoned (see 1.11 [Vol. 1 —*Eds.*]). But for identity of changing items at different times extensional identity too requires elaboration. Extensional identity though satisfactory for contingent identity of items *at the one time*, has to be replaced by *extensional identity over time*, since extensional identity is inconsistent with change. For extensional identity over time the class of preserved properties must be further restricted to what will be called (sufficiently) *dated properties.* For example the boy of 1958 is extensionally identical with the man of 1968 because they both have the dated extensional properties of being $5'$ tall in 1958 and $6'$ tall in 1968, 10 years old in 1958 and so on. But the individual has changed for the boy of 1958 has the undated extensional property of being $5'$ tall but the man of 1968 does not have this property. Briefly $(x, t_1) = (y, t_2) \leftrightarrow . \ (U \, \text{ext dated} \ f) . \ (x, t_1)f \equiv (y, t_2)f$, i.e. x at t_1 is extensionally identical with y at t_2 iff they share all sufficiently dated extensional predicates. These details indicate the way in which the theory of items can solve the logical problem of identity throughout change.[1]

The usual requirement on change needs, like Leibnitz identity, amendment; for a does not change from t_1 to t_2 just because of someone's belief or knowledge of a changes from t_1 to t_2. An improved account runs thus: a changes from t_1 to t_2 iff $(P \, \text{ext} \ f) . \ (a, t_1)f \ \& \ (a, t_2){\sim}f$. On the theory of items an item which ceases to exist or comes to exist thereby changes. For if an item ceases to exist between t_1 and t_2 say, it will be indeterminate wrt some extensional property f at t_1 and determinate wrt this property at t_2. Hence a changes.

Parmenides' argument for his thesis that change is impossible, that everything steadfastly is, is, from a referential standpoint, at least as formidable. The argument for the impossibility of change splits into subarguments of independent interest:

(1) Coming to be (exist) is impossible.

For suppose that a came to be. Then a must come to be either from what is or from what is not (does not exist). But a cannot come to *be* from what exists already; for this contradicts "coming to *be*". Nor can a come to be from what is not, for from what is not nothing can come to be. For (by the OA[α]) what does not exist has no features, and

[1] The matter of distinguishing dated properties remains, but the preanalytic notion is clear enough for present purposes.

[α] OA is the Ontological Assumption (see Vol. 1, §1.3): "a non-denoting expression cannot be the proper subject of a true statement". —*Eds.*

so cannot sustain the performance of coming to be. Thus Aristotle writes "from what is not nothing could have come to be (because something must be present as substratum)". An existent substratum would of course guarantee properties (hence later substratum theories[1]); but nonexistence likewise excludes a present substratum.

A similar argument tells against starting to exist (contrary to Prior 1967, p. 139). For suppose a starts to exist at t_1. Then, firstly, a has the property of starting to exist at t_1 and, secondly, at t_1 - δt, a does not exist. Hence, at t_1 - δt, a has no (true) properties, by the Ontological Assumption. Hence a does not have any properties associated with starting to exist at t_1, and in particular does not have the property of starting to exist at t_1 – a contradiction.[2]

(2) Ceasing to be (exist) is impossible, by an argument similar to (1).

(3) Change entails coming to exist or ceasing to exist.

Suppose item a changes from time t_1 to t_2, where t_1 is earlier than t_2. There will be a time (t_2 itself) when t_2 is present time and t_1 is in the past: consider a's change from this perspective. By the requirements on change a has a property now which it did not have formerly. Hence the past a is strictly different from the present a, so the existence of one does not guarantee the existence of the other. Since existence is existence now the different past a does not exist whereas the present a does exist. Thus, in changing, a comes to exist.

From (1), (2) and (3) it follows

(4) Change is impossible.

(5) What Is Is, Being Is.

Since Being is the whole of what is, this thesis may be interpreted:

The sum total (fusion) of all entities exists and never alters.

[1] As noted, the power of these referential considerations helps explain the popularity of substratum theories which guarantee properties and of rearrangement theories, got by weakening Leibnitz or settling upon resemblance instead of identity. As Wittgenstein correctly emphasized (in 1953), substratum and substance theories are based on the Reference Theory (of meaning). With the noneist abandonment of the Reference Theory the point of problem-making rearrangement and substratum theories disappears.

[2] Given the Reference Theory, items which start and stop existing gain and lose all their properties! [Recall that a stated consequence of the Theory (cf. Vol. 1, p. 30) is "the following formulation of the Ontological Assumption: *it is not true that nonentities ever have properties.*" —Eds.]

The argument is this: The sum exists since each part exists. The sum never alters for if it did some part would come to exist or cease to exist, which by (1) and (2) is impossible.

Parmenides' argument has sometimes been represented as an early piece of science supporting a physical conservation principle to the effect that creation of matter *ex nihilo* is impossible. This is an error: it would not then exclude the impossibility of movement or change of position or alteration of colour, all of which Parmenides thought he had established. No, Parmenides' argument is a metaphysical argument, based on the Reference Theory. Hence the relevance of Parmenides' pronouncements that we cannot speak or think about what is not or think that what is not exists.

To make it clear that Parmenides' arguments strike at the heart of modern chronological logics the arguments have been reconstructed using only weaker assumptions, consequences of the Reference Theory and established parts of classical logic and the calculus of individuals, together with the E0 thesis. It has been argued that rejection of E0 leads to grave difficulties; so rejection of this thesis is not a viable way of avoiding (3) and (4) above. Moreover its rejection does not provide an escape from (1), (2) and (5).

Hence the further thesis:

Classical logicians are stuck not just without a satisfactory chronological logic; they *are stuck with Parmenides' changeless universe.*

Is there an escape? Of course: there are always escape routes of *some* sort. But the main escape – apart from the already repeated chronological platonism which has it that everything that at some time exists exists timelessly, and so yields its own changeless universe, paradigmatically Minkowski's modernisation of Parmenides – is methodologically unsound and at a serious cost. The escape involves further curtailing the class of predicates which express properties. It is not just that existence is said not to be a property, and likewise existence at t for any time t. Becoming, coming into existence, dying, ceasing to exist, perishing, persisting, being killed, being conceived, spontaneously combusting, and so on: none of these are properties, indeed *no change feature* with relevant connections with existence *is really a property.* For by the Parmenidean argument (3) every change feature entails coming into existence or ceasing to exist at some time. Hence these implausible theses into which the classical logician is forced:

(I) No change or existence feature is a property. Consequently any theory based on *classical logic is inadequate to treat of change fully.* For as change features are not properties predicate logic cannot be applied forthwith to change features.

(II) All features that nonentities truly have are nonproperties (unless these features can be re-represented as secondary properties). (II) has been considered before (ad nauseam).

Classical logic incorporates, at the interpretational stage, a property filter, which filters out features, which apparently falsify the theory, as nonproperties or not traits. Plainly this property-filter is a theory-saving device, and is methodologically unsatisfactory. A serious deficiency of the classical escape from Parmenides is that the distinction between predicates which express properties and those which do not gets increasingly hazy. Classical doctrine needs to be underpinned by a technical distinction between properties and nonproperties. Until this is done part of the theory remains insufficiently assessible and falsifiable. Yet it is not easy to see exactly how the distinction should be drawn or can be satisfactorily drawn.

With the theory of items, however, Parmenides' arguments are easily halted. The argument under heading (3) depends on taking identity of items at different times as at least extensional coincidence on all properties. As this requirement has been rejected, for reasons already advanced, the argument fails. The argument itself provides a further reason for abandoning the usual account of (extensional) identity as applying to identity over time. The remaining arguments are premissed on the Ontological Assumption. As against these arguments, nonentities do have definite properties, and sometimes the properties of being about to exist or of starting to exist within time δt. Hence becoming and ceasing to exist are perfectly possible, and illustrative examples can be consistently added to the logic. Becoming is E-coming, coming into existence (see OED for this sense), and relevant logical notions are easily defined in terms of terminology soon to be introduced; for example:

x first comes into existence at $t_1 =_{\mathrm{Df}} (Ut)(t < t_1 \supset . \ (x{\sim}E, t)) \ \& \ (xE, t_1)$.

x (simply) comes into existence between t_1 and $t_2 =_{\mathrm{Df}} (x{\sim}E, t_1) \ \& \ (xE, t_2)$.

This entails that qu(x) has a referent at t_2 but not at t_1.[α]

[α] Where "qu" is the quotation function of Slog. —*Eds.*

Everything existent first comes into existence at $t =_{\text{Df}} (Ux).[(xE, 0) \supset (Ut^1)(t^1 < t \supset . \sim(xE, t^1))] \;\&. \; (xE, t)$.[α]

Etc.[1]

These definitions of neutral logic refute incidentally the thesis (advanced, e.g. by Godfrey-Smith 1976, p. 401) that such definitions require tensing.

§4. Developing a nonmetrical neutral chronological logic.

The key to neutral chronological logic, and therewith to the solution of several hoary traditional philosophical problems, is neutral quantification over times without commitment to the existence of past and future times and objects, temporal instants, etc. Thus neutral chronological logic builds on neutral logic enriched by specific temporal variables.[2] To extend neutral nonchronological logic to a full chronological logic an at-relation is introduced. Let us add to the logic of items (of chapter 1 [Vol. 1 —*Eds.*]) the primitive @ – read "at", or sometimes "in", "on" or "of" – which conforms to the formation rule:

> If A is a wff and s is a singular subject then $(A \;@\; s)$ is a wff – written alternatively (A, s) where s is a temporal term.

@ can be viewed as projecting A onto a scale, coordinate or map system of which s specifies a part. Some examples are: He was killed at 10 o'clock on highway 9; H weighed in at 10 stone; J cornered at 100 m.p.h. at Horseshoe Bend; It is hot in Uganda in summer. It is easy to wind up with nonsense by taking inappropriate subjects as in "H weighed in at the author of Waverley", "Uganda is hot at David Hume": this feature will be exploited in defining *Time*.

The preposition *at* gives a fundamental logical operation (from wff and subject to wff) for the development of formal theories of time, space, motion and matter (see Russell 1937, e.g. p. 465; also Quine WO, pp. 172–3, Łoś as reported in Prior 1967, p. 212, Rescher in Gale 1967). Whereas Russell observes (at least in 1937) that "at" is indefinable, Quine claims (WO, p. 172 and p. 104) that, where x is a spatio-temporal

[α] Where "0" indicates the present time, now. —*Eds.*

[1] Thus the Hegelian doctrine of Becoming as a transition from Nothing into Being is correct, and the "Principle of Sufficient Reason", that Nothing comes from Nothing, false, under the intended (existential) construal of quantifiers: under other construals it is different.

[2] It does *not* exclude a tensed approach, but combines with such.

object and t is a time, x at t can be construed as the common part of x and t. Quine's proposal is defective. Were it right at-ness like the common part and interaction relations would be symmetric. But as Russell, nearer right, remarks (1937, p. 465) the at-relation is (significance features neglected) asymmetrical and intransitive. Quine's construal of at-ness in terms of juxtaposition is similarly defective. Comparisons like the following reveal the inadequacy of Quine's constructions. Compare "the moon at 2 o'clock" with "2 o'clock at the moon"; the difference, and the non-symmetry, can be emphasized by coupling each sentence in turn with predicates like "is photographed" and "is round". Or consider "10 o'clock at Eastern Australia (at E.A. time) is 12 o'clock at New Zealand (at N.Z. time)"; if the relation were symmetrical it would follow "Eastern Australia at 10 o'clock is New Zealand at 12 o'clock" or what is just as bad "E.A. time at 10 o'clock is N.Z. time at 12 o'clock". Goodman (in 1977) does define at-connexions in terms of W, read "with", but even in the calculus of individuals W is taken as a new primitive. Also Russell (in 1936) defines the special "at" he requires in terms of membership "$\in$"; but the interconnexion does not hold generally. For the logical properties of @ are different from those of $\in$; many of the generally accepted properties of $\in$, as given by set theory postulates, fail for @. Despite some similarities to predication, @ cannot be reduced to predication any more than to membership, and conversely predication cannot be eliminated in favour of at-ness, for at least an "is at" copula must remain.

Temporal qualification is a special sort of at-projection, namely qualification with respect to times. To distinguish such qualification and to bypass the complications of a significance filter in the main subsequent developments, time variables $t, t^1 \ldots, t_1 \ldots$ are introduced. Time variables have as substitution-range such expressions as "now", "10 o'clock", "last week", "1964", "when Caesar died", "before the end of the last week of 1968", "between 2 and 3 o'clock Greenwich mean time", "10 years ago". In some cases an *at* must be elided (in English) in forming A @ t. These new variables, time variables, are eliminable in suitable set-ups by way of restricted variables and the predicate "is a time". For example, within a significance logic the expression (A, t), i.e. A @ t, can be eliminated in context by $T\, x$time $\rightharpoonup A$ @ x.[1] Thus $(Ut)(A, t)$ is replaced by $(Ux)(T\, x\text{time} \rightharpoonup A$ @ $x)$.

[1] The notation from significance theory in what follows is that of Slog. [I.e. "T $x$time" reads "it is truly the case that x is a time", i.e. "x is truly a time"; "$\rightharpoonup$" reads "provided that ... then ..." (Slog, p. 294), used to delimit significance ranges for what appears to its right; "q" (see next paragraph) is the particular objectual quantifier (Slog, pp. 148, 471). —*Eds.*

Can the predicate "is a time", used in the elimination, be defined? Given the relation $<$ of *temporal* precedence the following definition can be tried: $x\text{time} =_{\text{Df}} (\mathcal{q}y)S(x < y)$, i.e. *$x$ is a time* is defined: for some y it is significant that x temporally precedes y. Then 10 o'clock is a time, 1964 is a time, now is a time. But the definition is too liberal; for under it events such as Clay's knocking down Patterson also count as times, and Locke and the model T Ford are times, for example Locke because he was temporally earlier than Hume. An improved definition which appears to escape these difficulties is this:

$$T\ x\text{time} =_{\text{Df}} (Up)(Sp \rightharpoonup S(p \mathbin{@} x))\ \&\ S(\mathcal{q}y)(x < y).$$

Since this definition in effect assumes that

D. $St\ \&\ Sp \dashv S(p \mathbin{@} t)$,

i.e. whenever both qu(t) and qu(p) are significant qu(p @ t) is significant, and so in particular supposes that truths of logic and mathematics are significant *at* times, it invites objections. For example, Smart's theory of time leans heavily on the contrary thesis to be criticised shortly, that truths of mathematics and logic are not significantly asserted when temporally qualified (see Smart 1963, pp. 133–141). In D, $St =_{\text{Df}} (\mathcal{q}q)\ S\ (q \mathbin{@} t)$, i.e. qu($t$) is significant iff for some sentence qu(q), qu(q @ t) is significant. To defend D, first make the familiar separation of English declarative sentences into those that are *temporal*, like "Socrates is smoking", "Tom beat Bill" and "The next glacial period is in the remote future", and those like, "2+2 = 4", "Scarlet is a determinate form of red", "Murder is wrong", "Numbers don't exist" and "The circle cannot be squared (by ruler and compass techniques)", that are *nontemporal* (often misleadingly called *tenseless* sentences). This distinction can be given some precision along lines suggested by Broad (in Gale 1967, p. 123), namely temporal sentences are those such that their truth or falsity depends on their (possible) time of assertion. On this account compound sentences like "Tom either beat Bill or did not" are nontemporal. Subsequently the characterisation is taken up through the implication (for significant sentences),

$$A \text{ is nontemporal } \dashv .\ (\pi t)\ .\ (A, t) \leftrightarrow A,$$

which in turn coentails a generalisation of Broad's account,

$$A \text{ is nontemporal } \leftrightarrow .\ (\pi t_1, t_2)\ .\ (A, t_1) \leftrightarrow (A, t_2);$$

but for the time being such a characterisation assumes the point at issue.[α]

In arguing for D, there is little (not no) loss in generality in concentrating on grammatically simple sentences; for if D holds for such sentences then D can be proved inductively for compound sentences using such significance compounding principles as $S(p \mathbin{\&} q) \leftrightarrows Sp \mathbin{\&} Sq$, $S(a$ believes $p) \dashv Sa \mathbin{\&} Sp$, etc. Since there is little doubt but that D holds for grammatically simple temporal sentences it remains to haggle the case for simple nontemporal sentences. Such sentences as "$2+2=4$ now (at midnight)", "The circle cannot ever be squared" are ordinarily treated as significant, so it seems from a limited sampling, and it is important that they should be so treated, for several reasons.

First, ruling such sentences out as nonsignificant would preclude normal application of the statements, for example to empirical subject matter. Simplified situations can be envisaged where the grocer, say, argues in this sort of way: I have two flagons of claret here, and I've two more out the back; two and two are four now; so I now have four flagons of claret in stock. Similarly the student applying classical mechanics to solve a standard problem involving cylinders rolling on one another needs to know that the mechanical laws, as well as the arithmetical, apply at the times in question, and needs to apply them at the relevant times. We remonstrate against the man trying to square the circle, or to devise a decision procedure for full quantification theory: Don't you understand the circle can't be squared; therefore it can't be squared today – so you might as well pack up – or tomorrow – so it's a waste of your time to start trying again then, or ever! Second, empiricist theories of arithmetic and idealist theories like Kant's and neo-intuitionistic theories according to which arithmetical statements involve a reference to time presumably cannot be refuted simply on significance grounds. Third, in virtue of logical transformations such as substitutivity of identity, temporal qualifications of many necessary statements can be derived. Consider, to illustrate, the following argument: The number of planets is 9 at present; the number of planets $= 9$ (or $4+5$); therefore, by substitutivity of identity, 9 is 9 at present, and $9 = 4+5$ now. By a similar argument, starting from the fact that Jupiter has 4 moons at midnight tonight, it follows that $2+2=4$ at midnight tonight. A different set of transformations with the same outcome goes as follows: $2+2=4$ is true at midnight tonight, since it is always true; so at midnight

[α] Where "π" reads "for all possible" – in contrast to "U" which, recall, reads "for all (possible or otherwise)". —*Eds.*

tonight $2+2=4$ is true; so at midnight tonight $2+2=4$; hence $2+2$ will equal 4 at midnight tonight.

One reason why some philosophers are reluctant to concede that necessary truths such as $2+2=4$ can significantly be temporally qualified, is that they think that such a concession will commit them to assertions like 2 at midnight + 2 at midnight = 4 at midnight, and thereby to the significance of phrases like "2 at midnight". But such commitment would follow only given the correctness of transformations like $(xf, t) \simeq (x, t)f$; and such transformations are not generally correct, as the example under discussion already indicates (and is explained below).[α]

Once the predicate "is a time" is satisfactorily defined, Augustine's famous question "What, then, is Time?" can be answered, though the answer may not appear very illuminating. Time is a property – of times – of 1984, midnight, doomsday, Spring, and when the lark sings in the meadow – much as Number is a property of numbers. Accordingly, Time $\equiv \lambda x$ $x$time, i.e. Time is strictly the property of times; or, more specifically, Time $\equiv (\imath f)(Ux)(xf \leftrightarrow T\ x\text{time})$, i.e. Time is the property of all and only those items that are truly times.

The familiar circularity objection will be fired at this definition. To define "T $x$time" the primitive notion of temporal precedence was isolated; but temporal precedence is precedence *in time*. To meet this objection the requirement that $<$ be a *temporal* precedence relation is abandoned: let "$<$" simply read "precedes". The adequacy of the definition of "is a time" to exclude spatial locations as times when the temporal requirement on precedence is dropped depends on facts such as the unidimensionality of the time sequence as opposed to even 2-dimensional space (compare Broad, in Gale 1967, pp. 119–20), and the non-locatibility spatially of certain mental phenomena which are temporally ordered.

In what follows sortal variables $t, t^1 \ldots$ for times are used (they can be eliminated as restricted variables through the predicate "is a time"). The basic logic is an extension of that in chapter 1 [Vol. 1 —*Eds.*] – without significance. The logic is extended to embrace new wff containing @ and to include neutral quantification over times.

Since expressions of the form (A, t), B @ t^1 are wff, the logic includes as wff $((Ux)xf, t)$ & $((Px){\sim}xf, t^1)$, $(p$ & $q, t) \dashv (p, t)$, etc. Alternatively (p, t) could have been

[α] "$\simeq$" is the relation of strong extensional equivalence (Slog, p. 91): the expressions flanking it take the same value: true, or false or non-significant or incomplete. —*Eds.*

introduced as a primitive wff, just for sentential variables, and truth-functions introduced recursively by definitions like: $\sim(A,t) =_{\mathrm{Df}} (\sim A,t)$, (A,t) & $(B,t) =_{\mathrm{Df}} (A$ & $B,t)$. (Compare PM, *9.) But this method lacks generality: no means of introducing (A,t) & (B,t^1) is available and there is no direct way of introducing intensional sentence predicates as in $\Diamond(B,t)$. With the method adopted such wff are already available, but their logical interrelations are less specific. The initial assumption of the chronological logic is designed to rectify this deficiency.

Postulate 0 (P0) (i)

$$\mathrm{ext}\,\Phi^1 \dashv .\ \Phi^1(p,t) \leftrightarrows (\Phi^1 p,t),$$

$$\mathrm{ext}\,\Phi^2 \dashv .\ (p,t)\Phi^2(q,t) \leftrightarrows (p\Phi^2 q,t); \quad \text{etc.}$$

The assumption, adopted generally for n-place extensional predicate Φ^2, asserts that extensional predicates distribute within @t. It follows, for example, $\vdash \sim(p,t)$ & $(q,t) \leftrightarrows (\sim p$ & $q,t)$. P0 does not extend to intensional sentence predicates as these counter-examples (easily generalised) to an extension show. (Smart believes that $2+2=4$, midnight) is true, but Smart believes that ($2+2=4$, midnight) is not true. Consider too the different non-equivalent construals of "Alcoholism was being discussed at a conference yesterday (in Canberra)", and "It is possible that John is running when he is not running". In the case of modalities the (equivalent) implications $(\Diamond p,t) \dashv \Diamond(p,t)$ and $\Box(p,t) \dashv (\Box p,t)$, fail: the converse implications are however correct for consistent times. For $\Diamond(p,t) \dashv \Diamond t$: the general principle $(\Box p,t) \dashv \Box(p,t)$ appears however to conflict with the admission of impossible times, such as the time when a squared the circle. Accordingly only the following qualified postulate is adopted in the case of modalities (the qualification is explained in the discussion of P1)

Postulate 0 (P0) (ii)

> $(\pi t)\,.\,(\Box p,t) \dashv \Box(p,t)$, i.e. for all consistent times t, necessarily p at t entails it is necessary that p at t.

Compounding of times by connectives such as "and" and "or" cannot be added to the system without some scoping precautions. For example, the definition $(A, t_1 \vee t_2) =_{\mathrm{Df}} (A,t_1) \vee (A,t_2)$ would lead to inconsistency: consider alternative expansions of (It was hot and it was wet, Monday $\vee$ Tuesday).

Quantification, neutral quantification, can be based on the (indefinite) description operator ξ, or introduced independently. The time term $\xi t(A, t)$ for example, reads "a (time) t such that A at t". The particular quantifier P is connected with ξ by the expected coimplication (a consequence of the extension of the logic), $(Pt)(B, t) \leftrightarrow (B, \xi t(B, t))$, i.e. that, for some time t, B is true at t coimplies that B is the case at a time t such that B at t. The coimplication reveals that descriptions are values of time variables, a fact that is a consequence of the extended ξ-postulate: $B(N) \rightarrow B(tB)$, where N is a time expression, i.e. a time variable or a time term, ξtB for some B.

Definite descriptions are defined as in 1.22 [Vol. 1 *—Eds.*]: thus $\imath tB =_{\mathrm{Df}} \xi t(B\ \&\ (Ut')(B \supset .\ t = t'))$ i.e. the time such that (of) B is a contingently unique time such that B. Times are *extensionally* (contingently) identical (or simultaneous) if and only if they share all extensional properties.

Postulate 0 (P0) (iii)

$$(Uu)(A, T) \leftrightarrow ((Uu)A, T), \text{ provided } u \text{ is not free in time term } T.$$

Hence too $\vdash (Pu)(A, T) \leftrightarrow ((Pu)A, T)$ under the same proviso. P0 (iii) can in fact be asserted for all extensional quantifiers, where a quantifier Q is extensional iff it distributes across an equivalence $A \equiv B$ whenever the universal closure of $A \equiv B$ holds.

Use of neutral (i.e. nonontological) quantifiers [i.e. "U", "P" *—Eds.*] eliminates many puzzles; for example the apparently valid argument (considered in Prior 1967, p. 144) – Nothing that has perished exists, and some house has perished, therefore some house does not exist – is valid when formulated with neutral quantifiers. Similarly "Alexander rode Bucephalus" *does* entail "Alexander rode something", though it does not entail "Alexander rode an actual horse".

Since logical laws are necessarily true they are true in all possible worlds, and so at all possible times – in *this* sense independently of time. Thus it would be expected that some principle like

J. $\Box A \rightarrow (Ut)(A, t)$

can be used to link achronological with chronological logics. But J itself is too strong. First, since impossible times are included in the range of quantifiers – because expressions like "the time at which Hobbes squared the circle" belong to the substitution-range of time variables – J would assert that truths which are necessary are true even at impossible times or in impossible worlds. Impossible times are however just times at which even necessary truths are not all true: to require that they should be true would be to over-restrict the class of impossible times. Secondly, J would lead to inconsistency in a logic which extended the characterisation postulates of chapter 1 [Vol. 1 *—Eds.*]. Suppose to illustrate the point, the theory includes the truth that it is impossible that Hobbes squared the circle, i.e. abbreviated $\sim\Diamond h$. Now given that the characterisation postulate applies to $\xi t(h,t)$ it follows $(h, \xi t(h,t))$, and so also $(Pt)(h,t)$. But using J it follows $(Pt)(p,t) \dashv p$, thus:

$$\begin{aligned}\sim\Diamond p &\dashv \Box\sim p\\ &\dashv (Ut)(\sim p, t) \quad \text{by J.}\\ &\dashv \sim(Pt)(p,t).\end{aligned}$$

[Thus from $(Pt)(h,t)$ it follows that h. *—Eds.*] Hence $\Diamond h$, and inconsistency.

By weakening J as follows these troubles at least are avoided:

Postulate 1 (P1)

$$\Box A \dashv (\pi t)(A,t),$$

i.e. necessarily A entails for all consistent (possible) t, A at t. Possibility-restricted temporal quantifiers "π" and "Σ", and consistent times, are explained as follows: $t\Diamond =_{\text{Df}} \sim(Pp)(\sim\Diamond p \ \&\ (p,t))$, i.e. a time t is possible iff no impossible statement is true at t.[α] $\Sigma t\ B =_{\text{Df}} \xi t(t\Diamond \ \&\ B)$,[1] i.e. a consistent time t such that B is a time which is consistent such that B. $(\Sigma t)B(t) =_{\text{Df}} B(\Sigma t\ B(t))$; $(\pi t)B(t) =_{\text{Df}} \sim(\Sigma t)\sim B(t)$. A stronger, but less satisfactory, definition of "Σ" is this: $(\Sigma_1 t)B(t) =_{\text{Df}} (Pt)(t\Diamond \ \&\ B(t))$. $\vdash (\Sigma_1 t)B(t) \dashv (\Sigma t)B(t)$; $\vdash \check{S}^t_N t\Diamond \ \&\ B(t)| \dashv (\Sigma_1 t)B(t)$; $\vdash \Box A \dashv (\pi_1 t)(A,t)$, where $(\pi_1 t)B(t) =_{\text{Df}} \sim(\Sigma_1 t)\sim B(t)$. Since times which exist, have existed or will exist are consistent times (see below), it follows using P1, since $\Box(2+2=4)$, that $2+2=4$ next week as well as at 1000 B.C. But though the general principle $\Box A \dashv (\pi x)A \ @\ x$ should produce significance

[α] "$\Diamond$" is the logical predicate "is possible"; cf. Vol. 1, 1.19. *—Eds.*

[1] Should entailment be used in place of necessary material implication the following Σ-*postulate* seems to be needed (in lieu again of a more satisfactory theory of restricted variables: $\check{S}^t_T B| \Rightarrow B(\Sigma t\ B)$, where time term T is consistent. Hence $\vdash (\pi t)(A,t) \Rightarrow \check{S}^t_T(A,t)|$, provided $T\Diamond$.

qualms since it leads to results such as "$2 + 2 = 4$ at 10 stone", P1 need not for reasons already given. It follows, using P1,

$$\Box A, t\Diamonddot \rightarrow (A, t),$$

but the weaker rule

$$A, t\Diamonddot \rightarrow (A, t)$$

is incorrect. For applying the rule – given the fact, which we can easily add to the logic, that Bertrand Russell exists – we should wind up with the falsehood that Bertrand Russell existed 1000 years ago. Indeed the E0 thesis would be falsified.

Although P1 is true, since necessary truths are true in all possible worlds and each possible time generates a possible world via the statements true at the time, the converse of P1,

$\breve{1}$. $$(\pi t)(A, t) \strictif \Box A$$

is false. However it is sometimes supposed that $\breve{1}$ is true, for instance by theories according to which the past is necessary. Thus Diodorus apparently adopted an account of necessity which implies $\breve{1}$. A first counterexample to $\breve{1}$ is provided by $(\pi t)((Px)xE, t) \strictif \Box(Px)xE$. For it is perfectly consistent to assert that at every consistent time some item exists, but that it is not logically necessary that some item exists. Further counterexamples to $\breve{1}$ (or better to a version of $\breve{1}$ which strengthens the antecedent to all physically realizable times) are got on the false but consistent supposition that consistent times have a last term (or a first term). If consistent time has a last term then $(\pi t)((\Sigma t')(t < t'), t)$, but it is not logically necessary that $(\Sigma t')(t < t')$.

The interpretation of wff has not yet been made sufficiently explicit. Although expressions like $((A, t), t)$ are well-formed according to the logic, their interpretation is not unambiguously fixed; in particular the interpretation of A and (A, t) should be further clarified. Before the introduction of (A, t) in addition to A, the logic was inadequate to formalise and assess a wide class of sentences and arguments containing temporal expressions, e.g. sentences like "Whatever is future will be present", "It is possible that in the future all speech will end and all sentient organisms will die", "If a person is always happy, he is happy when he ceases to exist".[1] This is not to say that the nonchronological logic is not interpreted over tensed expressions: it is. Thus both "Socrates is happy" and "Socrates was happy" are admissible substitution-instances of xf; though the subjects of these sentences are the same, the predicates "is happy" and "was happy" are different, and "was happy" is not amenable to further analysis in terms of "is happy". Furthermore all the sentences of the (underlying) nonchronological logic are tensed sentences; for they each include a verbal component and every verb carries a tense. That the sentences of the underlying theory are tensed sentences encourages construal of indicative sentences as true iff they are true now. In fact a treatment of this sort is already forced in special cases by the E0 thesis: $xE \leftrightarrow (xE, 0)$. A generalisation of this thesis, adopted by Prior in the intended interpretations of his tense logics, is the next assumption:

Postulate 2 (P2): $A \leftrightarrow (A, 0)$

where the new primitive "0", read "now" or "the now time", is a substitution-value of time variables. Thus, for example, $2 + 2 = 4$ coentails $2 + 2 = 4$ now. Hence too, $\vdash (\pi t)(A, t) \rightarrow A$, and $\vdash$ E0. P2 reflects the tensed present-oriented feature of discourse. It is of course possible to construct *different* tense logics which are not present-oriented, for example by taking another orientation point, such as last week or 1984 at Plumwood Mountain for such discourse.

The logic is inconsistent with that relational theory of time according to which it is impossible to have time unless something exists at each time. On this relational theory, $(\pi t) .\ (p \supset p, t) \rightarrow ((Px)xE, t)$, since it is not logically possible that t is a time unless something exists at t. Hence

	$\Box(\pi t)(p \supset p, t) \rightarrow (\pi t)((Px)xE, t)$	
	$\rightarrow ((Px)xE, 0)$	
	$\rightarrow (Px)xE$	by P2
But	$\Box(p \supset p) \rightarrow (\pi t)(p \supset p, t)$	by P1
So	$\Box\Box(p \supset p) \rightarrow \Box(\pi t)(p \supset p, t)$.	
Detaching	$\Box(Px)xE$;	

[1] This point is enough to cast serious doubt on the dogma that chronological logics are philosophically unimportant and that there is nothing special about time as a qualifier or quantifier or property. For the dogma see Goodman 1977 and Quine WO. The philosophically relevant sections of Prior 1967 provide a good base for a case against the dogma.

but this is inconsistent with Meinong's theorem (previously defended and adopted), $\sim\Box(Px)xE$, i.e. it is not logically necessary that something exists. Such relational theories are condemned independently in that they rest on the Reference Theory: otherwise why should some item have to exist at a time in order for a consistent role to be ascribed to that time?

It is a consequence of P2 that statements like "Socrates exists" and "Meinong is happy" – unlike noncontingent statements such as $2 + 2 = 4$ – are true at some times and false at others. Though such an outcome accords with the Aristotelian and scholastic account of statements it conflicts with a standard theory of propositions according to which only temporally invariant sentences express propositions (these points are explained in detail in Prior 1967). But there is nothing to stop us having *both* temporally variable statements, such as "Socrates exists (now)", which are true at one time and false at another, and temporally definite statements, like "Socrates exists at 400 B.C."; and the chronological logic, with its temporal qualification, like the universal theory of 1.24 [Vol. 1 —*Eds.*], in fact caters for both. A logic which allows both for temporally variable statements and for temporally more definite statements has the immediate advantage of greater generality; it is more comprehensive than logics which only admit temporally definite statements, unless temporally variable statements can be suitably reduced to temporally definite ones. Because of the comprehensiveness of the logic, P2 does not, despite appearances, legislate against temporally definite or nontemporal discourse; such discourse is included within the range of the logic.

A first stab at marking out nontemporal discourse is summed up in the following characterisation:

$$A \text{ is nontemporal}_1 =_{\mathrm{Df}} A \leftrightarrow (\pi t)(A, t).$$

It follows that: $\vdash \Box A \rightarrow A$ is nontemporal$_1$; $\vdash \sim\Diamond A \rightarrow .\ A$ is nontemporal$_1$; and that $\vdash \sim(A \text{ is nontemporal}_1) \rightarrow .\ \nabla A$. Hence "$2 + 2 = 4$" is a nontemporal statement. But although all statements of logic and mathematics emerge, as desired, as nontemporal, given that they are necessary, it does not follow (what used to be taken as a desideratum) that statements of physical laws are nontemporal. To rectify the matter the notion of physical possibility, symbolised Ⓟ, needs to be introduced somehow or other. Here it is taken as a primitive modal connective. (Alternatively ⓅA could be defined, in a familiar if defective[1] way, as "A is compatible with all physical laws".) Then *physically realisable* times may be defined: $\langle\mathrm{P}\rangle t =_{\mathrm{Df}} \sim(Pp)(\sim\langle\mathrm{P}\rangle p \,\&\, (p,t))$. A similar development to that based on logical possibility "$\Diamond$" is made for "$\langle\mathrm{P}\rangle$". Thus, for example, universal and particular quantifiers "π_p" and "Σ_p" are introduced. With this development an improved definition of nontemporality can be provided, namely:

$$A \text{ is nontemporal} =_{\mathrm{Df}} A \Leftrightarrow (\pi_p t)(A,t).$$

Then $\vdash \sim\nabla A \Rightarrow .\ A$ is nontemporal, $\vdash \boxed{\mathrm{P}} A \Rightarrow .\ A$ is nontemporal (where $\boxed{\mathrm{P}} A =_{\mathrm{Df}} \sim\langle\mathrm{P}\rangle\sim A$), $\vdash \sim\langle\mathrm{P}\rangle A \Rightarrow .\ A$ is nontemporal. Hence if $\boxed{\mathrm{P}}(E = mc^2)$, then "$E = mc^2$" is nontemporal. As an outcome of the definition of nontemporality, nontemporal statements are relatively time-independent; and insofar as they are time independent time can be abstracted from their assertion. Thus nontemporality provides a formal analogue for the widely deployed but ill-explained notion of temporal assertion. The myth of tenseless assertion should however be exploded. The notion of tenseless sentences is one infected with difficulties; for instance, standard examples of tenseless sentences of English are present-tensed and accordingly not tenseless. Nor can this tense be eliminated (even by abstraction feats) because of the tensed character of all English verbs.

Physical and logical non-contingencies do not provide the only cases where temporal qualifications may be omitted and time accordingly abstracted from. A further class is provided by the next assumption:

Postulate 3 (P3): $(\pi t)(A,t) \Leftrightarrow A$, provided A is not free for temporal qualification.

P3 does not hold generally under the interpretation adopted; for otherwise $(xE,t) \Leftrightarrow xE \Leftrightarrow (xE,0)$, and disaster. A wff A is *not* free for temporal qualification iff every atomic subwff of A is eventually qualified by a bound time variable. For example $(Pt)(xE,t)\ \&\ (Pt')(p \vee q,t')$ is not free for temporal qualification. Applying P3,

$$((Pt)(\text{It snows in Melbourne}, t), t') \Leftrightarrow (Pt)(\text{It snows in Melbourne}, t)$$
$$\Leftrightarrow ((Pt)(\text{It snows in Melbourne}, t), 0).$$

Among the class of time expressions dates can be distinguished. Roughly a date is a temporal specification, not involving a present or egocentric reference point, defined within some (recognised) calendar system, for example within the Christian or Moslem

[1] See the discussion in Bressan 1980.

calendars. A calendar provides a time metric oriented on a standard event. Thus "on July 4, 1968" and "at the last hour of 1900" are dates, but "now", "10 years ago" and "at the time when the dodo disappeared" are not. A predicate expression is *dated* if it contains a date specification and no temporal qualifiers which are not dates. A sentence is dated if all its predicate-expressions are dated. Aren't dated sentences temporally definite? Though they are not temporally variable, they are very often not as temporally definite as they could be! Consider the successive consistent qualifications indicated in (((The war stopped, in 1958), 10 Jan.), 10 p.m.). This says more than (The war stopped, 1958); it coentails what we can represent thus: (The war stopped in 1958 on Jan. 10 at 10 p.m.) or schematically (The war stopped, 10 p.m. Jan. 10, 1958). Generalising this connection of times leads to:

Postulate 4 (P4): $((A,t),t') \leftrightarrows (A, t \oplus t')$.

A new class of time expressions, concatenated times, formed using the new primitive $\oplus$, has thus been admitted. $\oplus$ conforms to the formation rule: If t and t' are time expressions then $(t \oplus t')$ is a time expression.

Not all time qualifications are consistent; consider for example "The Blue Nile Falls are dry in 1957 in 1958", and "Socrates died in 399 B.C. 10 years ago". What is consistent is the very different "The Blue Nile Falls are dry in 1957 and 1958", not "The Blue Nile Falls are dry @ (1957 $\oplus$ 1958)". Concatenation of times is very like addition of times, only for proper addition times must be rationalised and made consistent, for example the same units must be used and the datings made consistent. Without a metric on times the usual recursive axiomatisation of addition cannot be adopted since no units of time are available for the recursion clause. When a uniform metric is imposed $\oplus$ can serve for addition of metrized time expressions. But even without metrisation a mathematical treatment is available; for:

Postulate 5 (P5). Times form a commutative group under $\oplus$, where the identity determinate is that of extensional identity.

Spelling this out is simply a matter of setting down some set of axioms for the group. The closure property already follows from the formation rule for $\oplus$. So the postulates P5 entails are these:

5.1 $t \oplus 0 = t$

5.2 $t \oplus t' = t' \oplus t$

5.3 $t \oplus (t' \oplus t'') = (t \oplus t') \oplus t''$

5.4 $(Pt') \,.\, t \oplus t' = 0$

Extensional identity for times is defined in terms of predicates of times thus: $t = t' =_{\mathrm{Df}} (U \operatorname{ext} f)(tf \equiv t'f)$, where extensionality is characterised as before. Finally it is assumed

5.5 $\operatorname{ext}(\oplus T)$, for any time term T whose further predicates, if any, are extensional.

The effect of 5.5 is to give the effect of Leibnitz identity for a large class of time terms, including all those needed for proving familiar group properties. It follows, for instance, that $t \oplus t = t$, that the group identity 0 is unique and the only idempotent element of the group, and that each element of the group has a unique inverse. The inverse of t is written, as usual in a module, $-t$, i.e. $-t =_{\mathrm{Df}} \imath t'(t \oplus t' = 0)$

Admitting inverse elements practically amounts to admitting, as has been done, expressions like "10 years ago" and "last week" as substitution-values of time variables. So, for example, we have ((Russell is happy, 10 years ago), 10 years in the future) $\leftrightarrow$ (Russell is happy, 10 years ago $\oplus$ 10 years hence) $\equiv$ (Russell is happy, 0). In the last step the fact that $\operatorname{ext} f$ and $\operatorname{ext}(f, t)$ are interderivable is used. The inverse of dates such as 1958 are expressions relating times such as 10 years later; the inverses of dates vary with change of time. Inverses though unique are not unique up to synonymy or even up to strict identity: otherwise 1958 would (in 1968) be strictly identical with 10 years ago.

A temporal precedence relation, "earlier than", can be introduced into the theory in various ways. For example, if the notion of positive time from any given origin is introduced precedence can be defined just as the *less than* relation is defined in Peano arithmetic. A less devious route is simply to introduce the relation $<$, read "is earlier than" or "wholly precedes", as a new primitive satisfying the formation rule: If t_1 and t_2 are time terms then $(t_1 < t_2)$ is a wff. Temporal precedence, precedence of times, should, it seems, apply initially to times, only derivatively to events and other items that mark times, even though it may eventuate that times are only abstractions from features of these other items. Some times do precede others: 1958 precedes 1959, the time when Brutus killed Caesar precedes the time of Cleopatra's death. But since times other than now do not exist, quantification over times is neutral quantification, and the times that are ordered by $<$ do not in general exist.

In terms of $<$ other ordering relations are defined:

$$t_1 \leq t_2 =_{\mathrm{Df}} \ t_1 < t_2 \vee t_1 = t_2;$$
$$t_1 \sim t_2 =_{\mathrm{Df}} \ {\sim}(t_1 < t_2) \ \& \ {\sim}(t_2 < t_1).$$

$<$ may be compared with Russell's primitive P [i.e. the relation (on events) "wholly precedes" —*Eds.*], $\sim$ with Russell's S [i.e. the relation (on events) "neither wholly precedes nor wholly follows" —*Eds.*] (OT, pp. 347–8): but the relations differ in their ranges, $<$ being defined on times whereas P is defined over (token) events. The relations can however be extended (given certain assumptions), so that their ranges coincide. For example, given $<$ and an event predicate H, read "happens", P can be introduced thus:

$$e_1 P e_2 \leftrightarrows (e_1 H, t_1) \ \& \ (e_2 H, t_2) \ \& \ t_1 < t_2,$$

i.e. that event e_1 precedes event e_2 coentails that e_1 happens at t_1, and e_2 happens at t_2 and t_1 precedes t_2.

Relation $\sim$ reads, like Russell's S, "temporally overlaps" or "is partially simultaneous with" or "is not temporally connected with". Much as in Russell (OT, p. 348), further relations may be defined; e.g.

$$t_1 \sim\!/\!< t_2 =_{\mathrm{Df}} (Pt)(t_1 \sim t \ \& \ t < t_2),$$
$$t_1 \sim\!/\!\breve{<} t_2 =_{\mathrm{Df}} (Pt)(t_1 \sim t \ \& \ t_2 < t).$$

Russell in effect adopts the Newtonian reading "t_1 begins before t_2" for $t_1 \sim\!/\!< t_2$; relativistically this reading is inadequate. Further Russell's assumption (OT, p. 348) that $\sim\!/\!<$ is transitive, though correct for absolute and proper time, appears to fail for relativistic time. Accordingly only a weaker relativistically admissible assumption is made.

Postulate 6 (P6). $<$ is an extensional partial ordering on times, i.e. $<$ is a transitive and irreflexive relation on times, and ext $(<)$ [i.e. ext_1 $(< t)$ and ext_2 $(t <)$] for all time terms t which contain no non-extensional predicates.

It follows at once that $<$ is an asymmetrical relation on times; and also $\vdash$ Reflex $(\leq)$. It also follows, as is proved:

$\vdash$ Trans $(\leq)$

Proof. $t_1 \leq t_2 \,\&\, t_2 \leq t_3 \;\text{-3}\,.\; (t_1 < t_2 \vee t_1 = t_2) \,\&\, (t_2 < t_3 \vee t_2 = t_3)$

$$\text{-3}\,.\; (t_1 < t_2 \,\&\, t_2 < t_3) \vee (t_1 < t_2 \,\&\, t_2 = t_3)$$
$$\vee\, (t_1 = t_2 \,\&\, t_2 < t_3) \vee (t_1 = t_2 \,\&\, t_2 = t_3)$$
$$\text{-3}\,.\; t_1 < t_3 \vee t_1 < t_3 \vee t_1 < t_3 \vee t_1 = t_3$$
$$\text{-3}\,.\; t_1 \leq t_3.$$

$\vdash t_1 \leq t_2 \,\&\, t_2 \leq t_1 \;\text{E-3}\,.\; t_1 = t_2$

Proof. $t_1 = t_2 \;\text{-3}\; t_1 \leq t_2 \,\&\, t_2 \leq t_1$. Also,

$$t_1 \leq t_2 \,\&\, t_2 \leq t_1 \;\text{-3}\,.\; (t_1 < t_2 \,\&\, t_2 < t_1) \vee (t_1 = t_2 \,\&\, t_2 < t_1)$$
$$\vee\, (t_1 < t_2 \,\&\, t_1 = t_2) \vee (t_1 = t_2 \,\&\, t_1 = t_2)$$
$$\text{-3}\,.\; t_1 = t_2 \vee t_1 < t_1$$
$$\text{-3}\,.\; t_1 = t_2, \text{ since Irreflex } (<)$$

$\vdash t_1 = t_2 \;\text{-3}\,.\; t_1 \sim t_2$

Proof. $t_1 < t_2 \;\text{-3}\; {\sim}(t_1 = t_2)$,

since $t_1 < t_2 \,\&\, {\sim}(t_1 < t_1) \;\text{-3}\,.\; (P\,\text{ext}\, f)(t_1 f \,\&\, {\sim} t_2 f)$.

Hence $t_1 < t_2 \vee t_2 < t_1 \;\text{-3}\; {\sim}(t_1 = t_2)$.

That the converse $t_1 \sim t_2 \;\text{-3}\; t_1 = t_2$ does not hold generally is an important feature of relativity theory. Equally important the relation $\sim$ is not transitive, despite Trans $(=)$. The relation $\sim$ behaves logically like the spacelike separation relation of relativity. Not only is it not generally transitive: also $\vdash$ Reflex $(\sim)$ and $\vdash$ Sym $(\sim)$, just as for the spacelike separation relation of relativity. There seems no crushing objection, when elaborating the theory, to identifying (contingently) $\sim$ with the spacelike separation relation of relativity theory. Given such a connexion, absoluteness of time (in one important sense) can be defined:

$$\text{Time is absolute} =_{\text{Df}} (\pi_p t_1, t_2)(t_1 \sim t_2 \;\text{-3}\; t_1 = t_2).$$

Familiar presentations of relativity theory, which adopt $\sim$ as the simultaneity relation, are definitely misleading. For a simultaneity relation is a sameness-of-time relation, and hence an identity relation; and the transitivity of an identity relation is a logical feature of such a relation. Thus, as a matter of logic, the relation $\sim$ is not an identity relation, and therefore not a simultaneity relation. Contrary to Grünbaum (1964, p. 351) then, this usage restriction is not simply an ordinary language requirement, and its retention has nothing to do with retention of Newtonian beliefs.

In order to obtain usual tense distinctions terrestrial proper time (or an analogue) should be introduced, for example through a further time ordering relation $\lessdot$.

P6′: $\lessdot$ is a connected extensional partial ordering which coincides with $<$ for the cases where $t_1 < t_2$ holds (i.e. $t_1 < t_2 \prec t_1 \lessdot t_2$).

In the preliminary discussion of tenses which follows, however, the distinction between $\lessdot$ and $<$ is neglected: terrestrial tenses, as distinct from universal tenses relative to 0, are obtained by replacing $<$ by $\lessdot$. Tenses other than the present are reduced to the present tense of the logic using Prior's tense logic functors (for details of this reduction of tenses see Prior 1967; the feasibility of such a reduction has long been realised). Prior's connectives are defined as follows:

$$PA =_{\text{Df}} (Pt)(t < 0 \ \& \ (A, t)),$$

i.e. it has been the case that A iff for some time earlier than now A is the case at t.

$$HA = (Ut)(t < 0 \supset (A, t)) \ \& \ (Pt)(t < 0)$$
$$FA = (Pt)(0 < t \ \& \ (A, t))$$
$$GA = (Ut)(0 < t \supset (A, t)) \ \& \ (Pt)(0 < t).$$

H, F and G read respectively "It has always been the case that", "it will be the case that" and "it will always be the case that". This representation of Prior's connectives is not perfect, as regards the recovery of postulates of (basic) tense logic, but suffers from the same sort of defects as the syntactical representation in classical quantification logic of traditional syllogistic (discussed in 1.16 [Vol. 1 —*Eds.*]). Thus for instance, the particular clauses tacked on to the definitions of H and G to ensure that $\vdash HA \prec PA$ and $\vdash GA \prec FA$, parallelling tacking clauses to resolve problems of existential import in representing the syllogism. Once again (as in 1.16) given a superior analysis of "such that" and restricted variables to the classical extensional analysis, an improved syntactical representation could be provided (and therewith strict implication, which ties with the extensional analysis, superseded by entailment).[1] Alternatively, by switching to a semantical representation, imperfections of fit can be removed, in a rather classical way. Adequate extensional modellings *can* be given for all the main tense logics, with quantificational-style representation (like that of modal functors in 1.17 [Vol. 1 —*Eds.*]) of Prior's functors in terms of quantification over possible worlds or times. The perfection of fit is achieved however by the imposition of semantical modelling conditions, which serve as semantical postulates to bridge gaps. There is with the quantificational-relational representation of tenses and Prior's functors, as in the case of context (discussed in Slog, 7.2), the question of whether representation should be overtly syntactical or remain semantical. The guidance of discourse is clear enough however: *an adequate theory of time should include in its syntax* (as primitive or derived) *both Prior's functors and tenses and also the representational apparatus, interrelations and orderings of and quantifications over times.*

Once past, present and future times and the existence of times are characterised many results on McTaggart's A-series (the series: past, present, future) follow:

t is present @ $t =_{\mathrm{Df}} t = t$; $\vdash t$ is present $\leftrightarrow$ $t = 0$;
the present $=_{\mathrm{Df}} \imath t(t = 0)$;
t is past @ $t' =_{\mathrm{Df}} t < t'$;
the past $=_{\mathrm{Df}} \hat{t}(t < 0)$;
t is future @ $t' =_{\mathrm{Df}} t' < t$;
the future $=_{\mathrm{Df}} \hat{t}(0 < t)$;
x is a present entity @ $t =_{\mathrm{Df}} (xE, t)$; $\vdash x$ is a present entity $\leftrightarrow$ xE;
x is a past entity @ $t =_{\mathrm{Df}} (Pt')(t' < t \;\&\; (xE, t'))$;
x is a future entity @ $t =_{\mathrm{Df}} (Pt')(t < t' \;\&\; (xE, t'))$;
x is a future entity $\leftrightarrow$ x will exist; $\vdash$ Future things are things that will exist.
$\vdash x$ is a present entity $\leftrightarrow$ entity x was future & is present & will be past.

Re-expressions of Findlay's theses (presented in Gale 1967, p. 160) likewise follow; namely:

$\vdash x$ is present $\leftrightarrow$ x is present at (the) present;
$\vdash x$ is future $\leftrightarrow$ x is future at present; etc.

[1] One of the main factors preventing immediate retooling with relevant connections concerns the loss of expected inferential power when restricted variables are analysed classically.

A definition of the existence of times can be based on the pervading thesis, existence is existence now; namely

$tE =_{\mathrm{Df}} t = \imath t(t = 0)$. Then,

$\vdash \imath t(t = 0)E$, i.e. the present exists.

$\vdash 0 = \imath t(t = 0)$ i.e. Now = the present.

Proof. Since $\imath t(t = 0)E$, $\imath t(t = 0)$ is reliable by the ECP.[α]
$\vdash tE = .\ t = 0$; $\vdash 0E$.

$\vdash$ Past times do not exist; $\vdash$ the past does not exist. $\vdash$ Past times have existed, i.e. $(Ut)(t$ is past $\supset t$ has existed). $\vdash$ Future times do not exist; $\vdash$ the future does not exist. $\vdash$ Whatever is a past entity has existed. $\vdash$ The past and future do not exist, now; nor do they exist timelessly. $\vdash$ The past has existed but does not now: in *this* sense the past is not real. $\vdash$ The future will exist but does not yet: it too is not real in *this* sense. To say that the past is unreal is not of course to say that history is mythology or bunk or that the past is prefabricated;[1] for past times have existed. But though each past time has existed, at no time does or has the past existed.

But definition of tE or not, there is a real question as to whether any such objects as times, time instances, time slices, even present times, exist. For times do not have, at least in a straightforward way, locations – unless we say, very artificially, that they are everywhere, or, relativistically, everywhere in the local fuzzily-bounded region. Nor do they have entire physical relations, e.g. they cannot (significantly) be hit, caused, irradiated, etc. Thus time instances appear to fail tests for existence; they appear to fail the unified criteria for existence subsequently adduced (in chapter 9); they appear not to exist.

Whether Time itself exists depends – once the Ontological Assumption is got past – on many things, especially, given that Time is a universal, is a property of times, on the criterion of property existence assumed. If an instantial criterion were adopted then Time would exist, assuming further that the present (instance) exists. However, if a different

[α] ECP: $(\xi x A(x))E \dashv A(\xi x A(x))$, i.e. "existence is sufficient for reliability in the way an object present itself (as $\tau x A$) as to how it is (whether it is indeed A)" – cf. Vol. 1, §1.21.4, rearranged. —*Eds.*

[1] History and the past are genuine enough. And history has many interesting features yet to be touched upon, e.g. the intensionality of so much of it, especially the theoretical elements (see chapter 10 [Vol. 3 —*Eds.*]).

criterion is preferred, according to which such universals as properties do not exist – a criterion ultimately demanded by the indeterminacy of properties should properties be objects of a sort – then Time does not exist. Subsequent arguments (given in chapter 9 [Vol. 3 —*Eds.*]) will lead to just such a conclusion: that Time does not exist. Again that does *not* imply that it is a fabrication or fiction; that things don't genuinely happen at different times, that some things do not happen after others, that none are truly past, that nothing will happen, that some things do not exist now, etc.

The logic so far elaborated caters for the symbolisation of sentences, which reveal the breakdown of Smart's theory of tense elimination (of 1963, p. 134), and for the formalization of valid arguments which Smart's theory excludes. Consider for example "Some entity was future and is present and will be past" which indeed entails "Some entity is present"; or (after Prior 1967, p. 12) "Eventually all speech will have come to an end", or "At some future time everything actual will have ceased to exist". Indeed the logic permits the formalisation of many sentences, about nonentities and what will not exist, which Prior's theory (in 1957 and 1967) so far provides no way of formalising.

A cross-classification of times is provided by the following inclusive series: times, possible times, physically realisable times, sometimes actual times, historical times, and now. Sometime actual times are tentatively defined thus: $tSE =_{\mathrm{Df}} (tE, t)$, and historical times are defined as past or present sometimes actual times. It is tempting, but probably a mistake, to introduce a bridging assumption, namely,

$$(tE, t) ⥽ \langle P \rangle t;$$

from which it will follow that times that exist, have existed or will exist are consistent times. Even if the result is right, the route is doubtful.

An extension of the preliminary definition of item existence also follows, namely $\vdash (xE, t) ⥼⥽ (U \,\mathrm{ext}\, f)((x{\sim}f, t) \equiv {\sim}(xf, t))$. This characterisation of existence at time t provides a basis for accounts already sketched (in §3) of becoming and ceasing to exist. Then too *sometime-existence* – often misleadingly called atemporal or tenseless existence – can be defined: $x\underline{E} =_{\mathrm{Df}} (Pt)(xE, t)$. Remember that sometime-existence is not strictly a kind of existence.

If 0 can be defined within the logic the A-series of McTaggart can be reduced to the B-series (of earlier and later than relations) together with the tensings of the logic.

A promising way of defining 0, as suggested by the E0 thesis and P2, is in terms of the present tense. A present tensed expression available in the theory is $(Px)xE$, and hence an initially promising definition is simply $0 =_{\text{Df}} \imath t(Px)xE$: the definition uses a vacuous iota term. An alternative definition, based on the nonexistence is nonexistence now thesis, which avoids difficulties about times, if any, when nothing exists, is this: $0 =_{\text{Df}} \imath t(Px){\sim}xE$.

Objections are bound to be raised to the introduction of Now into the logic – objections which are independent of the adequacy of the proposed definitions or others. A first objection is that it destroys the purely topological approach to time by introduction of an origin. If P2 and the definition of 0 are abandoned then a topological "chronological" theory remains. But many of the distinctive features of time, as distinct from space say, are lost. And without P2 the interpretation of the logic is obscured. Moreover the topological approach is not just inadequate for the treatment of existence, since existence is existence now, of other scientifically relevant properties, and of anisotropy; it also excludes the treatment of tenses and of McTaggart's A-series and its features. In sum, the strictly topological approach is simply inadequate to the subject matter.

A second objection, based on the theory of relativity, is that the theory "does not make any allowance for the transient Now of common sense" (Grünbaum 1964, p. 318). So far this only indicates – as observed by Reichenbach (see also Grünbaum 1964, p. 318) – the incompleteness of the 4-dimensional picture used by the theory of relativity: it simply omits treatment of A-series features. To reinforce the objection very sweeping (but no doubt false and certainly anthropocentric) claims are sometimes made, such as that the concept of Now cannot be assigned a legitimate place in physics, and is not part of physics, but that Now is a purely psychological mode introduced by man (see, e.g., [Hugo] Bergmann[α], and Grünbaum in 1964, pp. 323–4). According to Grünbaum (p. 324),

> the coming into being or becoming of an event, as distinct from its merely being, is thus no more than the entry of its effect(s) into the immediate awareness of a sentient organism (man);

[α] Cf. Bergmann, H. 1929, *Der Kampf um das Kausalgesetz in der Jüngsten Physik*, Vieweg, Braunschweig. —*Eds.*

and he attributes to Weyl his own thesis that coming into being, happening, as contrasted with being (that is *existence*: p. 324), is only coming into the present awareness of a sentient organism.

None of these dogmas are consequences of the theory of relativity at all, in particular the platonic claim about existence. For *the theory of relativity can, and should, be formulated using nonexistential quantifiers.* It can be so formulated because formalisations of (fragments of) the theory using existential quantifiers can be replaced by formalisations using nonexistential quantifiers; it should be so formulated because what exists is not relativistically invariant (see also below), and because much of which the theory of relativity speaks (especially when applied) does not exist. Since, furthermore, becoming and coming into existence can be defined in a quite nonpsychological way using "0", the remaining issue comes down to the question of the correctness of the following thesis:

(Q). Now is a psychological mode which has no legitimate place in physics.

But Q is false. 0 has been defined without appeal to psychological matters. McTaggart's A-series would not somehow vanish with the extermination of all sentient beings. For now is the time at which something fails to exist, or exists; and what exists, or does not, is not a psychological, or anthropocentric, matter. Also, as Black has emphasised (see Grünbaum 1964, p. 327), the concepts of McTaggart's A-series and of change are inseparable components of the commonsense and initial scientific outlook, and are a standard part of scientists' laboratory concepts. What, then, is the case for Q? Part of the case is built on verification principles; for example, the only way one can verify features of Now is by my own experience and awareness. This is clearly a poor argument, quite apart from the well-known deficiencies of verification principles: because this is, in the relevant sense, the only way one can verify any empirical matter whether of physics or elsewhere.

Similarly objectionable is the connected claim (Grünbaum 1964, p. 325), that "the *instantaneous awareness of succession* ... is an essential ingredient of the meaning of 'now'." But reflection on the sense of "now", reinforced by consultation of dictionaries, does little or nothing to support this deliberate attempt to psychologize the meaning of "now". Awareness, by some conscious item, is not entailed by the sense of "now". Moreover objections to the old idea that Now marks out an instant of time, as distinct from a neighbourhood, are nowadays well known (instants of time themselves are suspect in some quarters, because as ideal limits or infinitesimals they do not exist).

A further common argument for Q is this: Now cannot be defined absolutely because, given relativity theory, the notion of absolute simultaneity cannot be introduced in a convention-free way. This does nothing to show that Now is a psychological matter – relativity theory does not show that space is a psychological matter – but it may be thought that it supports the thesis that Now has no legitimate place in physics. It doesn't, because many notions which are not relativistically invariant and are not entirely "convention-free" have legitimate places in physics. This undermines the first part of Grünbaum's argument, namely:

1. Any definition of "now" which makes use of absolute simultaneity is inadmissable. For it is only inadmissable if the notion defined is supposed also to be invariant. If, however, we define "now" in our usual terrestrial framework, why shouldn't we then, if we want to, introduce "absolute" simultaneity, even if it rests on an assumption which is *alleged* to be, as a matter of physical necessity, untestable? And even if it depends on a local framework?

2. Any definition of "now" in terms of the class of events not causally or signal connectible with a particular (Here-)Now "yields a conception of the present which differs from the Now of conscious experience" essentially (Grünbaum 1964, p. 319). For "a given event E_2 at point P_2 in space will remain simultaneous for an observer at a distant point P_1 *throughout* [...] a continuum of events having a spacelike separation from E_2". On the contrary, similar features are already involved in everyday and scientific discourse about the present. It is perfectly satisfactory to talk of Now, referring to today or to this year; in such cases a continuum of events is included given real number metrisation of time intervals. Moreover the velocity of light is so great that few discrepancies with ordinary usage occur over terrestrial distances.

If, however, the Now "of conscious experience" is *extrapolated* to apply to astronomical distances and to allow for really high-speed travel, then discrepancies will appear. Then, given the received theory, differences between 0 defined in the logic and "now" as defined in 2 have to be recognised. The unsubscripted 0 is defined for our standard terrestrial system. 0_S for a system S moving at a high relative velocity, may coincide terrestrially with 0 but differ from 0 at astronomical distances. Not only is Now system relative; *given relativity, what really exists is likewise system relative.* A computerised mechanical system passing the earth at a high velocity may report that a star, which still exists according to terrestrial astronomers, has ceased to exist. There are discrepancies not just over what exists; there may be differences over a wide range of extensional properties, for example what colour a distant star is, how old and how heavy a man is (compare the clock paradox), and so on. Because of discrepancies over what exists, the predicate "E" and likewise normal existential quantifiers have to be system relativized, given the theory of relativity: the unsubscripted ones adopted have been those of the terrestrial framework. Because of the system relativity of existential quantifiers, neutral quantifiers are important for the formulation of the invariants of relativity theory. For quantifiers of the form "There exists timelessly..." are excluded through the E0 thesis and because of the total unsatisfactoriness of the notion of timeless existence; and quantifiers of the form "There exist sometimes..." and "There exist always...", which can be freed from a particular system, are either not adequate to the task or else unnecessarily platonize the theory of relativity. But reconstructions of the confused classical notion of timeless existence may, of course, be fitted into the theory; for example $(\underline{\exists}x)xf =_{\text{Df}} (Pt)(Px)(xE \mathbin{\&} xf, t)$. "$(\underline{\exists}x)xf$" reads "there sometimes exists an x which is f". Then $\vdash (\underline{\exists}x)xf \leftrightarrows ((\underline{\exists}x)xf, t)$, i.e. $(\underline{\exists}x)xf$ is independent of time.

The foregoing criticisms of timeless and tenseless expression and existence are bound to encounter heavy opposition. For it is often assumed that tense can be eliminated by "paraphrasing tensed sentences into terms of eternal relations of things to times" (Quine, WO, p. 172). The same assumption underlies the widespread view that ordinary language tensed discourse can be equivalently represented tenselessly in terms of the Minkowski 4-dimensional representation of space-time (see, e.g., in Gale 1967, both Williams and Smart). However these assumptions depend on the correctness of the following transformation:

T. $(x, t)f = (xf, t)$, i.e. x at t has f iff x has f at t,
where (x, t) is a stage or time-slice.

One of the more explicit attempts to so eliminate tenses appears in Quine's work, and in his sample eliminations the transformation T is repeatedly applied (see, e.g., WO, pp. 172–4); for example "x is eating y at a" transforms to "x at a is eating y". The same transformation is applied in linking individuals with their stages or time-slice; for example, a Cäyster-stage connects with the Cäyster river as illustrated by the following application of T: "The Cäyster (flooded the lower Cäyster valley) @ the first day of the year 400 BC" is equivalent to "The Cäyster @ the first day of the year 400 BC (flooded the lower Cäyster valley)." (For background see the introduction of stages, FLP, p. 65ff.). On Smart's projected elimination, a further element is added to T. Smart, a strong advocate of 4-dimensional tenseless discourse, claims that the temporal facts that are ordinarily represented through tensed discourse can be represented within a 4-dimensional (Minkowski) representation (in Gale 1967, pp. 164–5). This thesis requires at least a truth-functional equivalence between ordinary tensed statements and non-temporal 4-dimensional statements; for otherwise facts would not be preserved under change of representation. If only truth is preserved the transformation will fail for sentences within the scope of intensional functors; and it does then fail as simple examples using functors like "Strawson believes that ..." reveal. But, apart from this, how is the truth-functional equivalence between the alternative representations established? It hasn't been established: as so often we are expected to rely on a few sketchy examples. Smart re-expresses the sentences "The spherical cricket-ball becomes ellipsoidal" in the 4-dimensional representation by saying that the 3-dimensional cross-section of cricket-ball$_4$ at $t = t_1$ is spherical and that the 3-dimensional cross-section at $t = t_2$ is ellipsoidal. How is the equivalence shown? By inserting, it appears, the following connections: The spherical cricket-ball becomes ellipsoidal $\equiv$ the cricket ball is spherical and becomes ellipsoidal $\equiv$ for some times t_1 and t_2, t_1 is earlier than t_2 and the cricket ball is spherical at t_1 and the cricket ball is ellipsoidal at $t_2 \equiv$ for some times t_1 and t_2, t_1 is earlier than t_2 and the cricket ball at t_1 is spherical and the cricket ball at t_2 is ellipsoidal $\equiv$ for some times t_1 and t_2, t_1 is earlier than t_2 and the 3-dimensional cross-section of cricket-ball$_4$ at t_1 is spherical and the 3-dimensional cross-section of cricket ball$_4$ at t_2 is ellipsoidal. Smart's theory, then, applies both T and the further identity:

A. x @ t = the 3-dimensional cross-section for t of x_4, where x_4 is a certain 4-dimensional item.

Thus Smart's theory adds to Quine's account of tense elimination a further important ingredient, that generalised in A which links tense elimination with the 4-dimensional picture. But Minkowski's *geometric* representation of the Lorentz transformations, e.g. as rotations, which is not in dispute, should be distinguished from the *general* Minkowski representation, which is intended to provide an alternative representation of tensed discourse. It is the general representation that Smart, Williams and others recommend (see Gale 1967).

Because T and its immediate improvements are false and A raises serious difficulties, *the attempted elimination of tense fails, and the general Minkowski representation is defective.* Counterexamples to T abound. First T breaks down for intensional predicates: compare "It is necessary that John is running when he is running" with "When he is running it is necessary that John is running", "Alcoholism was discussed in Canberra yesterday" with "Alcoholism yesterday was discussed in Canberra", "The number 2 was thought of at 10 o'clock this morning" with "The number 2 at 10 o'clock this morning was thought of", "Aristotle was portrayed (visualised) as a young man yesterday" with "Aristotle yesterday was portrayed (visualised) as a young man". To the objection that expressions have not been put into proper logical form before T is applied – for example "by some people" should be inserted in the alcoholism example, and this insertion should bear the temporal qualification – it should be said: no recipe is given as to what are the appropriate starting logical forms or as to how, or onto which individual expression, the temporal qualification is transferred, and until all this is done, tense elimination using T is just not general and not effective. To the objection that T only holds for extensional predicates it should be replied that the eliminations and re-representations have been put up quite generally and accordingly fail if they fail anywhere; moreover, as some succeeding counterexamples will show T does not hold generally for extensional predicates. Second, T fails for various dispositional, state and performative properties and more generally for long-term properties: consider transformations of "Jack refused (promised) to sleep on Friday", "Jack only eats fish on Fridays", "Joan was married at 10am sharp on Wednesday" (it wasn't the stage of Joan that got married: it was Joan), "The widow of the former prime minister married on Saturday", "John walked down the street at about 10 o'clock". Third, T breaks down in the following sorts of cases: for certain relations of stages – thus Quine today is older than Quine last year, but it is false that last year Quine is older than Quine – and for the predicates "is (not) a time-slice (stage)".[1] It may be contended that only a principle weaker than T is wanted for tense elimination, for example:

T′. Every temporally qualified or tensed sentence can be transformed to a form (xf, t) such that T then holds.

[1] These particular counterexamples were suggested by M.K. Rennie.

So, for example, "Jack promised to sleep on Friday" is first replaced by "Jack promised that Jack sleep on Friday" and then, using T, by "Jack promised that Jack on Friday sleep". A still weaker principle may be proposed, such as:

T″. For every tensed sentence there is an equivalent untensed sentence.

But T″ only restates the tense elimination thesis, without giving any of the details that make the thesis good. Nor has T′ been made good generally; worse, as it stands it is non-effective.

Several of the counterexamples to T destroy T′ as well, in particular examples which employ intensional and long-term predicates, but also examples such as "Smart is not a time-slice at 10 a.m. May 24, 1968". The claim made is not that it is impossible to eliminate tense: in a weak sense the theory so far developed (like Reichenbach's original theory) eliminates tenses – other than the present. What is contended is that it is impossible to eliminate tensed temporal qualifications of sentences in favour of equivalent sentences stating tenseless properties and relations of stages or time-slices, in favour of timeless features of slices. The reason for this is the same as the main reason for the failure of T′. For the elimination of tenses it is essential that stages have all their properties at the one time. For in order that stages or slices be the medium for the elimination of tenses all their predicates must be present tensed, so that they can be reconstrued as tenseless. Hence the stages or slices required for the elimination can have no past or future properties and no properties stretching over time or held at different times. Since, however, it is not true of individual objects, even at a momentary time, that their properties are all held at one time, at the momentary time, the proposed elimination and T and T′ must fail. Most of the counter examples are based on this very feature: that objects considered at a time have various properties that stretch beyond that time. The feature suggests further counter examples, using for example identity and continuity. The same feature explains why the tense elimination project strikes serious difficulties with doubly temporally qualified sentences such as "The New Zealander who climbed Everest in 1952 is alive (is in Nepal) in 1968"; for the 1952 time-slice of Hilary does not have 1968-properties. The 1952 Hilary slice cannot be identified with the 1968 Hilary slice, given the slices required for tense elimination. Nor can resort be made to the persisting item, Hilary, to save the day; for the persisting Hilary reintroduces a heap of tensed predicates and lets back temporal qualification.

At such a point appeal may be made to the 4-dimensional Hilary$_4$, to glue the different time-slices timelessly together. However theory-saving 4-dimensional items such as cricket-ball$_4$ and Aristotle$_4$ create difficulties quite as acute as those raised by the timeless Aristotle. For what properties does cricket-ball$_4$ have? How is cricket-ball$_4$ related to the cricket ball? Smart writes (in Gale 1967, p. 164) "Let 'cricket-ball$_4$' be the expression which in our 4-dimensional representation refers to the cricket-ball through its entire history". But on this account one would expect cricket-ball$_4$ to have all the properties that the cricket ball has at each stage of its history. Then it would be an impossibilium, unless the universe is Parmenidean and never changes: otherwise cricket-ball$_4$ is both spherical and not spherical, both has and lacks change properties, and so on. Perhaps instead cricket-ball$_4$, like the average cricket-ball, has at least the properties the cricket-ball always has, for these properties the cricket-ball has through its entire history. But whereas the cricket-ball is always a cricket ball and always 3-dimensional, cricket-ball$_4$ has a shape like a distorted 4-dimensional cylinder and is accordingly neither a cricket-ball nor 3-dimensional. It would be ludicrous to suppose that cricket-ball$_4$ be bowled, in the way the cricket ball always may, or have anything much to do with a cricket-match. Nor does cricket-ball$_4$ have the common properties of its cross-sections; for cricket-ball$_4$ is not a slice or cross-section; Smart's explanation of cricket-ball$_4$ is defective. How, to take another example, does "Aristotle$_4$" refer to Aristotle through his entire history? Some features of cricket-ball$_4$ are almost as obscure as those of cricket ball noumenon. The sole properties of cricket-ball$_4$ appear to be its space-time location (worm), which may intersect with the space-time locations of other 4-dimensional items, and that it has temporal cross-sections which coincide with instantaneous (localised) stages of the persisting cricket-ball from which it is constructed. Otherwise, however, cricket-ball$_4$ is radically indeterminate: hence it does not exist, and should not be existentially quantified over. As to the indeterminacy, cricket-ball$_4$ neither stays the same nor fails to stay the same by changing, is neither hard nor lacking in hardness, is neither red nor some other colour; for it has no tensed properties. It follows that cricket-ball$_4$ is not an entity. Because they connect only with instantaneous stages of the persisting objects on which they are based, 4-dimensional items provide no reflection of intensional, dispositional and long-term properties of the persisting items they are supposed to replace. Thus it is just not true that one can switch over to 4-dimensional discourse without heavy loss: how does one re-express dispositionals such as that H used to be an alcoholic and a smoker and is becoming an alcoholic again, or counterfactuals; and statements about nonentities must, like so much else, be simply written off. Finally, because of its featurelessness, cricket-ball$_4$ fails its appointed job as *independent* amalgam of stages or cross-sections; for how do we decide from which stages to manufacture the fictitious cricket-ball$_4$: only by appeal back to the persisting object, the ordinary cricket-ball, which has these stages. Thus tense-elimination, if it is to avoid circularity, will have to be accomplished without appeal to 4-dimensional items. The unsatisfactoriness of cricket-ball$_4$ and Hilary$_4$ transfers, of course, to the principle A.

Despite the troubles of time-slice and stages a distinct but related notion, that of *item-sections*, is important for characterizing identity of items over time, and for treating everyday sentences such as "Albert last year ran faster than John now does" and "Russell in 1920 still thought atomism was correct". The 1928-section of Russell, henceforth symbolised "(Russell, 1928)" and read "Russell of (in) 1928", is understood as a section of the persisting object Russell, as an individual at a time, not as a slice of Russell which begins and ends with 1928. The 1928 slice of Russell is more like Russell just at 1928, and better symbolised (Russell $\upharpoonleft$ 1928), Russell restricted to 1928; for it is restricted to 1928 property-wise.[α] The Δt time-slice of x is that item $(x \upharpoonleft \Delta t)$ which has all and only the *present* properties that x has during interval Δt. Time-slice items do not exist because they are indeterminate with respect to times outside their time base Δt. Quine's stages are momentary time-slices, i.e. time-slices where Δt is momentary. Different time-slices or stages of Russell are different items; for example (Russell $\upharpoonleft$ 1928) is quite distinct from (Russell $\upharpoonleft$ 1940–1950). But sections may be identified; for instance (Russell, 1928) = (Russell, 1948).

Persisting or stable items are among the (object-)values of item variables; it is normally sections of these stable items that are identified. Things, not slices and slice abstractions such as sense data and stages provide the interpretational backdrop of the logic. The piecing together of slice items to obtain stable individuals, without somehow smuggling in stable individuals, raises insuperable difficulties. Introduction of sections, as distinct from slices, avoids these problems; for the properties of temporally distinct sections may coincide.

[α] Cf. PM, *35.01. —*Eds.*

The logic is extended by adding section couples of the form (x, t) satisfying the formation rule: if x is an item term and t a time term, then (x, t) is an item (section) term, and meeting the postulate

P7 $(xf, t) = (x, t)f$, provided ext f.

Thus a section has at *least* the extensional properties that the item, of which the section is made, has at that time. Individuals may be viewed as special cases of sections, for example as sections over their own life-span; thus (Russell is not a section, 1928) is false, so disposing of certain counterexamples. Sections, unlike slices, have past, future and long-term properties. (Russell, 1928) has the property of being a future octogenarian, of being prepared to go on philosophizing, and of having been born many years before. Sections, unlike stages, have all tensed predicates; consequently they do not provide a device for eliminating tense. Sections do not eliminate individuals; they are individuals considered at a time. Sections have a history and a future; slices do not. Yet in spite of the differences *the confusion between sections and slices is central to the tense elimination enterprise.* For sections T holds for a large class of predicates just because sections are items at a time and have properties beyond that time. Yet just because they have these features which guarantee T, they fail to provide a medium for tense eliminations. In contrast slices, through their abstraction from non-present tensed properties, do provide a medium for elimination; but because of their depletion of properties T frequently fails. It is easy then to see how a conflation of sections and slices would make the tense elimination enterprise seem plausible; and through this confusion the enterprise has gained plausibility.

The solution offered to the problem of identity over time depends on considering sections, not slices. Temporally distinct slices are never identical. But sections are identical iff they coincide on all extensional dated properties. To illustrate: the man of 1968 has changed from the boy of 1958, because the man of 1968 has the undated property of being 6 feet tall but the boy of 1958 does not have this property. However both sections have the dated properties of being about 5 feet tall in 1958 and being 6 feet tall in 1968.

In rounding out the solution the theory stands in need of a fuller account of dated predicates than that suggested earlier or those indications to be found in the literature. Although the distinction has a good intuitive base, it is not sufficiently sharply drawn. There are other important respects in which the theory remains sketchy or unsatisfactory:

the formalism is incomplete in ways that matter, no semantical theory is developed (other than the universal one, which is not applied), etc. In short, much remains to be done in working out noneist chronological logic.

Even so, the noneist theory developed, sketchy though it is, has a number of corollaries, not so far elaborated or not so far adumbrated at all, for the philosophy of time. Such an account applies both to entities and to nonentities. For nonentities like entities may change. Orlando and Macbeth both changed over time. Change is not confined to $d(\underline{\underline{G}})$ but is a feature of many objects in $d(\underline{\underline{T}}) - d(\underline{\underline{G}})$.[α] Nor are change and time restricted to the (evolving) factual-world $\underline{\underline{T}}$; other worlds may change over time, some other worlds are Newtonian, and so on. Such facts have an important bearing not only on philosophical problems concerning time, but also on the conceptual foundations of physical theories, especially relativity where several of the underlying verificationist assumptions require semantical overhaul.

§5. Further corollaries of noneism for the philosophy of time.

Noneism casts a new and different light on many of the perennial problems in the philosophy of time, and on some of the newer problems thrown up by modern physics as well. Almost all the "problems", if not directly generated by the Reference Theory, are influenced and distorted by it; and so are resolved or seen differently with its removal. For example, several problems in the philosophy of time arise from a steadfast referential resistance to consideration of alternative worlds, e.g. problems of the connection of time with change, problems of fatalism and determinism. Only a selection from among the problems to which noneism can be profitably applied are considered in what follows.

1. *Reality questions: the reality of time?*. Even if reality is taken (as the OED takes it) to be the property of being real, "real" itself has several different roles, e.g. what is real may be (actually) existing or occurring in fact or objective (in some sense) or genuine or natural or sincere or not hypothetical or not pretended or ....[1] It makes a big difference

[α] For a discussion of these domains, see Vol. 1, §1.17: $d(\underline{\underline{T}})$ is the domain of the factual (actual) world, the actual world or factual model, including nonexistent items; $d(\underline{\underline{G}})$ is the (sub)domain of the existent items of factual (actual) world. —*Eds.*

[1] The importance of the differences for philosophical reality questions has been brought out particularly sharply by Austin 1962.

which role is intended. For example, though time does not exist it is genuine (not a pretence or human construction) and objective (as opposed to being a subjective, e.g. a mental feature of subjects of certain sorts, or a property of observers) and natural (not artificial) – so at least it will be maintained. When philosophers assert or deny or argue about the reality of time or of the future, they are usually, but not always, exercised about the existence issue. Let us consider this issue first.

Time is not itself a notion free of ambiguity. Even as an abstract object, it can figure both as a class, e.g. of tensed events (as in "future in time"), and as a property, an abstraction from such predicates as "is a time" (e.g. "10 o'clock is a time"). It also has roles as a particular, as for instance with particular times such as the time of the Black Death. But all these roles can be recovered (more or less) from the property of time – using variables restricted by "is a time" and classes collected with certain tensed predicates, as the logic reveals – so let us concentrate on the property (which is distinguished, where expedient, by capitalisation).

No universals [properties] exist. Time is a universal [a property]. Therefore Time does not exist. The argument is valid. The major premiss is established in chapter 9 [Vol. 3 —*Eds.*]. The minor premiss has already been obtained. Hence Time does not exist. So in one sense Time is unreal: it does *not* follow that it is an illusion, or that Time is an inconsistent object.[1]

It is a corollary that the thesis that Time exists if [or iff] there are temporal facts is false; for there are temporal facts such as that Moore is now dead. The thesis was heavily exploited by Moore and forms the basis of his famous argument for the reality of Time (see, e.g., Gale 1967, p. 69). The thesis is not a commonsense one at all (despite one of the contexts, the paper "A defence of common sense", in which Moore presented it), but an instance of the Ontological Assumption, namely of a form much favoured by Prior, that x exists if there are facts about x. Since if x has (extensional) properties there are facts about x, the earlier elaborate case against the OA transfers intact to refute this form of it. There are however more relevant counterexamples, e.g. it is a (temporal) fact about Moore that he presented an argument that Time is real, another (temporal) fact that Moore is now dead, so he doesn't exist.[2]

[1] F.H. Bradley's arguments for the unreality of Time should be rejected as logically defective. They purport to show that Time is inconsistent.

[2] Moore's thesis is in effect rejected by relativity buffs, for quite different reasons. On relativity theory there are temporal facts, e.g. some processes are contemporaneous, in the sense that, for any two, part

In fact all the usual philosophical arguments for the reality and existence of Time rely at bottom on the Reference Theory, and can accordingly be dismissed.

2. *Against the subjectivity of time: initial points.* A notable lacuna appears in a well-known argument for the subjectivity of time. According to Augustine, for example,

> Neither the past or the future but only the present really *is*; the present is only a moment, and time can only be measured while it is passing. Nevertheless there really is time past and future. We seem here to be lead into contradictions (Russell 1946, p. 373).

Escaping these contradictions leads to Augustine's subjectivist theory of time. But existential and non-existential verbs and quantifiers have not been distinguished. The present exists, the past does not; but the past can still be spoken of, past items can be quantified over non-existentially, and so on; and in this way time can be measured. Thus the past can be signified, discoursed about, and its times metrised, though it does not exist. But not according to the Reference Theory; the Reference Theory blots out all these requisite distinctions. And it is really the Reference Theory, clearly exhibited in Augustine's work (see Wittgenstein 1953, pp. 1ff.), that forces Augustine to a subjectivist theory of time. According as it is combined with other premisses, the Reference Theory forces its adherents either to platonist, to conceptualist or subjectivist, or to nominalist theories of time. That is, the situation is like that with respect to universals (see further chapter 9). By abandoning the Reference Theory and its associated logic, these old options can be left behind.

Other very different arguments for the subjectivity of time derive from the theory of relativity. A first crude argument is this: Only relativistically invariant notions are objective (cf. Stein 1968, p. 18). Time is not a relativistic invariant. Therefore Time

---

of one is past to part of the other. But Time, insofar as it implies a total chronological ordering of space-time, is not well defined. As Stein puts it (somewhat less satisfactorily)

> ... although *temporal relations* have a meaning in relativistic theory, *time* as a separate concept has no role and no meaning for it (Stein 1968, p. 9).

It is this that leads him subsequently to say, mistakenly, that Time is meaningless, and Minkowski to imply that Time is unreal.

It is a serious question, however, whether the account of time presupposed should be accepted. According to Stein, "the quotient set of space-time under a certain ... chronological ordering ... *is* just what is called 'time'" (p. 9). This is just false. Moreover ordinary set uses of "time" do not require relativistic invariance. Time as an abstract can be perfectly well defined from predicates that are relativistically local.

is not objective. What is not objective is subjective. Hence Time is subjective. The argument is valid, but several of the premisses are incorrect. Objective notions are by no means limited to those that are relativistically invariant (r-invariant). Most of our ordinary objective notions are not r-invariant, e.g. being a table, now being a unicorn, etc. If a special sense of "objective" is introduced – what really happens in expositions of relativity theory – then the assumption that what is not objective is subjective fails seriously (it fails anyway on usual senses). Furthermore Time defined as an abstract is, like most mathematical notions, r-invariant; it is also not a subjective notion in any good sense. A second even cruder argument is this: Such central notions as simultaneity, succession, the present, etc., which are not invariant can only be accommodated as local notions (with the very frugal accommodation of the received theory of relativity). Local notions are always relative to observers, and so are subjective. Hence central temporal notions are subjective, as Augustine contended. The second premiss is false. Observers are dispensible; on the special theory, inertial frames without observers suffice.[1]

3. *The future is not real.* Purely future objects and times do not exist (yet at any rate). Thus the future, which concerns these items, does not exist either, and by the time they come to exist, they will be not future but present. Hence the future does not exist, and in this important regard the future is not real. Of course the future is not bogus or an artifice or a human concoction.

The main arguments for the prevailing philosophical position that the future is real, depend, like those for the reality of time, on the Ontological Assumption, and collapse with its removal. Smart's case for the reality of the future, for example, consists primarily of repeated applications of the OA. Thus the following Parmenidean argument:[2]

> If an object *a* acquired reality, there would have to be a time at which it lacked reality and a time at which it possessed reality. But to lack reality (or to have any other property), it would have to exist, i.e. be real, which is a manifest contradiction (Smart 1979, p. 3).

[1] Godfrey-Smith 1976 has added sustenance to this argument by proposing to define "the present" – and thereby indirectly all tensed discourse – in terms of the class of events *seen now* by an observer. This introduces an apparently subjective element to many central notions. The definition should be rejected on other grounds as well: its chauvinism, its conflict with the ordinary notion of present.

[2] The Parmenidean element emerges even more strikingly subsequently:

> Notice how hard it is even to state this notion of the unreality of the future (p. 9).

Beyond the confines of the Reference Theory however, there is no problem at all in stating and arguing the thesis that the future is unreal.

Hence, what at any time exists, what will exist, is real. Therefore the future, since comprising what will exist and future times, is real. But withdraw the OA, the contradiction disappears, an object can lack reality without existing and object a can consistently acquire reality having lacked it – in precisely the way that an object can come into existence (as explained and defined in §2). For according to Smart, "the notion of 'real' is fundamentally just that of existence" (p. 9).

Other arguments for the reality of the future draw on the theory of relativity and the assumption that "real" is relativistically invariant. Thus, to illustrate with a very simple consideration, since an event which exists now and so is real is in the future for another observer, the future event should be accounted real by invariance. The short reply to this sort of argument, and elaborations of it (some mentioned below) is that "exists", and "real" when it means the same, is not r-invariant. Although this is the obvious and commonsense thing to say given the received theory of relativity, it is regarded as outrageous. Godfrey-Smith goes so far as to say,

> It would be sheer *hysteron proteron* to suggest that existence depends ... upon the dating system representing the local proper time of some observer (1976, p. 245);

though *when* something happens and *the present* can be so specified, "existence is objective" (p. 244), i.e. r-invariant. The force behind the assumption that existence just *must* be r-invariant is that the alternative leads to violations of the Reference Theory: what does not exist (in certain systems) has properties, stands in relations, objective relations no less, etc. Thus, as far as noneism is concerned, it is a force that does not move. The alternative of going along with the Reference Theory, which there is no reason to follow, leads directly to chronological platonism: everything that at some time exists, exists.

Putnam's much-discussed arguments (in 1967b) for the reality of the future (and of everything in the world) is similarly met. The kernel of the argument[1] is,

p1. Whatever exists now is real

[1] Adapted from Stein 1968, p. 13. Stein's formulation, unless taken as very compressed, is insufficient. It is not merely required that "x is real for y" be transitive (see p. 18); also required are principles ensuring observer-simultaneity guarantees reality-for, and that reality-for transfers reality (and then transitivity is otiose).

p2. Consider the relation R such that xRy iff x is simultaneous with y in the observer's coordinate system. Then if z stands in the relation R^{C} to w (where R^{C} is the transitive closure of R) and z is real then w is real.

p4. (Relativistic fact) For any a and b of space-time, there is some c in space-like relation to both (i.e. such that c is comparable with neither a nor b under the chronological ordering).

Now let a exist and b be arbitrary, e.g. any thing or event in the future. By p1 a is real, and p4 supplies an element c, such that for some observer a and c are simultaneous and for some observer b and c are simultaneous. Hence aRc and cRb so $aR^{\mathrm{C}}b$. a is real so, by p2, b is real.

But why accept p2? Putnam's argument supposedly derives it from a heavily and obscurely qualified principle grandly entitled "There are No Privileged Observers". But no defence is made of the grand principle and the route to p2 is in doubt. In short, Putnam really gives *no* reasons for adopting p2. Stein suggests that "as a leading principle, and part of the explication of reality" p2 (in *his own* form) "is very reasonable". But if its consequences in the argument setting are looked at, it appears most *un*reasonable. Most ordinary things counted as real, such as trees and trucks and TV sets, are *not real*, since, like existence-now, they are not suitably invariant under the changes of chronological perspective allowed and required in the reality of the future argument. Putnam has not, then "soundly refuted the doctrine of the unreality of the future" (Smart 1979, p. 8).

Moreover for the conclusion he eventually arrives at that "future things are real" (p. 247), Putnam could have much simplified the messy argument. For given (a) future things are things that will exist (p. 240); and (b) what sometime-exists, or in Putnam's terms "*tenselessly exists*", is real, it follows that future things are real. But why accept (b)? Because if we extend existence (now) to a notion that is r-invariant what we arrive at is sometime-existence, and r-invariance is required for reality. We are back with the basic assumption behind it all: but that assumption has already been rejected.

4. *Alleged relativistic difficulties about the present time and as to tense.* The difficulties alleged – they come down to the point that such notions as *now*, *the present* and those of tensing are not relativistic invariants – are not peculiar to the noneist theory presented. No theory of time that properly speaks some nonrelativese is really immune to these often spurious difficulties.

The main problems for the noneist theory outlined, as for most theories that rely on tensing revolve around the notion of *the present*, or *now*.[1] Thence they spread, since tenses are oriented on the present, to tenses, and through E0 to existence. Several of the problems arise however from an error – because the present has been mistakenly identified, often by relativists, with the class of events (absolutely) simultaneous with some event now (and here), with the instantaneous world state so specified. Certainly the latter account leads to severe difficulties given the rejection, by relativity theory, of (real) instantaneous connections and the repudiation of a relation of absolute simultaneity. However while the Newtonian theory *did* justify the identification, it was *not* part of the commonsense account; and, reflecting that, it was not incorporated in dictionary definitions of "the present" and "now". In fact the attempt to foist an analysis of the present in terms of a class of simultaneous events on the layman or the women-in-the-street (an analysis which has only to be stated to look wild) is largely an imposition backwards from twentieth century relativistic thinking. Neither the class notions nor the universal simultaneity notion had much popular currency before this century, and do not seem to have infiltrated recent temporal notions or assumptions. Ordinary judgments about or involving the present or now are almost always *context bound*, indeed contextual indicators are important in evaluating judgments involving such (egocentric) particulars. It seems clear furthermore that contexts concerned were almost always local. To force an extrapolation beyond a local region on someone who says "The sun is shining now" lacks solid ground. Similarly for most ordinary nonmetaphysical judgments of a tensed type: they are "local" claims. And counterintuitive results ensue if an attempt is made to render them nonlocal, invariant.

The theory of relativity in no way upsets these commonsense context bound notions of time and tense: on the contrary, they are regularly used in expositions and discussions of relativity theory. Stein, for one among many, thinks that the special theory of relativity is incompatible with the ordinary notions of time and tense (1968, pp. 17–8) and, worse for the present theory, with the thesis, varying E0, that all and only things that exist now are real. However the demonstration of the latter incompatibility turns upon assuming again that "real" is r-invariant. Stein has a further argument:

[1] There are parallel problems for any theory that would do away with tensing in obtaining tenseless dated predicates.

> If ... we were to insist ... that *only* things that exist now are real – , we should be led by an argument like Putnam's to conclude that *for any event, it and it alone is real*; and then, ... we should have the interesting result that special relativity implies a peculiarly extreme (but pluralistic!) form of solipsism (p. 18).

Not so – unless we recklessly buy assumptions already rejected. Fortunately there is no need to reconstruct the argument to arrive at these assumptions; for Stein earlier informs us (p. 15):

> in Einstein–Minkowski space-time *an event's present is constituted by itself alone.* In this theory, therefore, the present tense can *never* be applied correctly to "foreign" objects. This is at bottom a consequence ... of our adopting relativistically invariant language ....

If we do, if we take an r-invariance vow. Then existence now is existence here-now, so what is real also (by the hypothesis) is here-now, and, given r-invariance, what is real cannot be extrapolated (for reasons already seen, that different observers would disagree about the extrapolation). But even more ardent exponents of relativity do not restrict themselves to an r-invariant language. Nor could they without *gross* impoverishment of their discourse. And why should they? No solid reasons have been given, none are advanced by Stein, though he admonishes Rietdijk and Putnam for not deploying r-invariant notions (indeed he describes their doing so as a "fallacy"!).[1] Should it be to maintain objectivity, perhaps? That we have dismissed. To avoid parochialness? That we can do, where necessary, by using r-invariant language; after all *we* are not excluded from doing so. Moreover, when the r-invariance requirement is rightly abandoned we can extend our dating systems beyond our inertial frame, in a way that is not arbitrary, and it is a moot point whether such extensions are conventional in a damaging way. There is no decisive reason then for not continuing to speak in the ordinary non-r-invariant ways, and excellent reasons for continuing to so speak.[2]

[1] The clinching objection is sometimes taken to be that what is real cannot depend on choice of a coordinate system. Existence or genuine properties cannot be wiped out by change of a coordinate system. What is real *now* depends on choice of a frame; what is at some time real does not. Nothing is "wiped out" in this sense; a different cross-section is taken of ongoing processes. All ordinary tensed predicates, which yield genuine properties do presuppose a frame. But there is no theoretical problem, given the Lorentz transformations, in reexpressing what is said using the predicates in any other frame (of the special theory).

[2] There are excellent reasons also for viewing with suspicion philosophical pronouncements made on the strength of the theory of relativity. One reason for this is that relativity theory was formulated and its

5. *Time, change and alternative worlds.* According to von Wright, there

> appears to be a double dependence of time on change. Time presupposes change epistemologically, and change presupposes time logically (von Wright 1969, p. 21).

The theme of interdependence of time and change is indeed a common outcome of verificationist thinking. But remove that setting, allow duly for alternative possibilities, and the main arguments for the interconnections fail. Consider, for example, von Wright's argument that time presupposes change epistemologically, that "time is a change propagated flow" (p. 17). The core of the argument is as follows:

> If the world remained in all its features identically the same on two successive occasions, how should we know that they are two occasions? We cannot use the ticking of a clock or the disintegration of an atom or any other event in nature to distinguish them, for those ways of counting time require that change takes place. If the world under consideration really is "the whole world", there is no room for the required changes (p. 15).

The argument is defeated by counterfactual, or alternative world, considerations. It is enough to say that there *could* have been a clock, and if there *had been* a clock then (ideally) no other change would have occurred than its recording of succession, or better, that from an alternative world like it but for a clock, the lack of change could be verified. So what is right is not "that in a world without change the concept of time would have no application",[1] but only a much weaker thesis to the effect that time presupposes the possibility of change. Von Wright's Kantian way of stating his thesis – "If we 'think away' change from the world, we cannot think of the world as persisting in time" (p. 189) – is similarly vulnerable. Some of us can violate the conditional by thinking of other worlds where the changelessness of the world is recorded, much as we can envisage other worlds without change upon which we can impose the local temporal grid.

With the alternative world semantical picture it is not difficult to solve the "well known puzzle to which the assumption that time depends on change gives rise":

philosophical consequences drawn out in a heavily positivistic ethos. Nor has the situation changed much recently: modern philosophical commentary mainly comes from philosophers with marked empiricist bias.

[1] A claim that is also stated in the misleading form "Time cannot exist without change".

> Assume the world came to a complete standstill, that no further change took place in it. ... Shall we also say that *time comes to an end*, that after that time there is no time, indeed no "after"? (p. 17).

Yes, according to von Wright:

> *The appearance to the contrary* is due to the fact that we refuse to take the idea quite seriously or, rather, that we combine it with another idea, inconsistent with it, of someone, a "god", being there to contemplate this dead world *in time* (pp. 17–8; emphasis added at beginning).

There is only an inconsistency if the someone is in the world. There is no inconsistency in having an ideal observer in a different world, perhaps with its own time, contemplating our now static universe. That is perfectly in order, as the relation of contemplation is intensional and induces no interference in the static world. In short, contrary to von Wright, the requisite "machinery" does not have to "go on somewhere *else* in the universe". Only a one-world view would oblige such a conclusion.

The argument that change presupposes time logically is nearer valid, but again not decisive. The argument is that "change without time involves contradiction", attributions of contradictory features to one and the same object. To escape contradiction, however, it is enough to find a difference, a factor, which is "time-like". Thus, for example, a hardline positivistic exponent of Einstein–Minkowski space-time like Stein, who jumps to the conclusion that "although *temporal relations* have a meaning in the theory, *time* as a separate concept has no role and no meaning for it" (1968, p. 9), can nonetheless accommodate change consistently. The differences in factor which resolve contradiction do not, that is, imply a difference in time in the intended sense; only a very stunted notion of "time" which need not meet expected temporal ordering conditions results.

6. *Limitations on statements about the future, especially as to naming objects and making predictions?*. According to Prior (1957, p. 34, 1967, pp. 143–5), and apparently to Ryle (1954, pp. 25–7), one cannot say, for example, "It was the case 100 years ago that A.N. Prior did not exist". But the sentence is surely significant, it can easily be symbolized in a neutral chronological logic, for example thus: $(Pt)(t < 0 \mathbin{\&} t \oplus 100 \text{ years} = 0 \mathbin{\&} {\sim}(\text{A.N. Prior } E, t))$, and it certainly appears to be true. So one can say it, and say it truly. Prior responded that

> if "A.N. Prior" is a logically proper name, one cannot *truly* say this, though one *can* truly say "It is not the case that it was the case 100 years ago that A.N. Prior exists". One can also truly say "It was the case 100 years ago that there was no A.N. Prior".

We have already observed that the implied thesis, that the logical status of a name depends on whether the bearer exists or not – a thesis, forced by the Ontological Assumption, that *might* legitimately be abused for making logic parochial – has severe effects on logical theory. A much more drastic position is however sometimes taken with respect to future objects.

While Prior amended his thesis to hold that there were limitations on naming purely past objects, limitations confined to *logically proper* names, a much more sweeping thesis has been stated with regard to purely future objects, namely,

(α) One cannot name a purely future object.

There is no restriction to logically proper names. A corollary is a generality of predictions thesis,

(β) Predictions about purely future objects are (irreducibly) general, and more generally,

(γ) There are no particular facts about purely future objects.

All these theses are advanced by Godfrey-Smith 1978.[1]

The main source of these theses is again the Reference Theory. This is clear enough in Ryle 1954, where appeal to or reliance on the Ontological Assumption underpins all the arguments, e.g. "When I predict the next eclipse of the moon, I have indeed got the moon to make statements about, but I have not got her next eclipse to make statements about" (p. 27), and more generally (though Ryle is a little devious about this) the "logical" reason why one cannot give individual mention to future objects or make singular predictions about them is that "one has not got them to make statements about" (p. 27). The reliance is also indicated by the "queer logical fix" Ryle finds himself in (p. 25) trying to speak of things that were prevented, averted, and so on, from happening or existing, analogues

[1] These theses may also be found in other sources; a stronger version of (β), with "purely" deleted, was advanced by Ryle. Godfrey-Smith argues, as against Ryle, that we can make predictions about particulars if they have begun to exist and so are nameable, and also that Ryle is mistaken in treating future tensed statements as about events, a point that removes some of Ryle's examples.

of the "negative existential" "problem": we cannot, according to Ryle, claim about any particular event that *it* was prevented! The reason is said to be that

> What does not exist or happen cannot be named, individually indicated or put on a list, and cannot therefore be characterised as having been prevented from existing or happening (pp. 25–5).

Here is an open-ended list: Sherlock Holmes, Kingfrance, Rapseq, Ruritania, 2001, ... Or consider again the cyclone code-named *Thales* which did not eventuate. And so on.

Godfrey-Smith is more straightforward, and his 1978 abounds with referential theses, which are unsupported and presented as if they were obvious. But – at least as seen from outside the referential paradigm – they are false and counterexamples or counter-considerations are not difficult to find. Here are some examples: "nonexistent items clearly cannot be the subjects of singular propositions" (p. 18); "for the prediction to be singular it is necessary (only) that its subject expression refer to an existing individual" (p. 18, brackets added); "we fail to precognise the future ... because there is just nothing 'there' to causally act on the present" (p. 17); "before an individual exists no statements at all can be made about it" (p. 20); "in a world in which Gilbert Ryle did not exist it would be impossible to know that *he* was absent from it" (p. 20); "after, and only after, an individual has come to exist can it be taken as the logical subject of a proposition" (p. 21); "in the case of future 'individuals' there is nothing about which information might be gathered" (p. 23); "the supposition that a future 'individual' might have a different life-'history' to the one that it will have is one that cannot intelligibly be entertained" (p. 24).

There are three sorts of considerations Godfrey-Smith appeals to in defending (α)–(γ), all of a referential cast. Firstly, there is direct appeal to the Ontological Assumption. There are no entities for the facts of (γ) to be about, hence (by the OA) (γ) holds.[1] This confirms (α) since there are no particular facts about such nonentities one could hardly name them, or particular facts would be forthcoming. In any case the Ontological Assumption yields (α) directly since purely future objects do not exist.[2] For if the objects could be named one could talk about what does not exist. Secondly, thesis (α), and so

[1] That there are particular facts about future objects implies, by OA, there exist future objects.

[2] There is a severe tension in Godfrey-Smith's work (1976 and 1978) as to what exists. On the one hand, when confronted with relativity he insists that existence is objective and r-invariant; on the other hand, he contends that purely future objects do not exist ("There are no future individuals", 1978, p. 24). But what is future terrestrially is present or past from other chronological perspectives. R-invariance will

(β) and (γ), has been bolstered by the causal theory of names. For on that theory future objects cannot be named without, what is presumably impossible, inverse causal chains leading from the future back to the present (in the one local frame).[1] The (referential) causal theory has, however, like the Reference Theory, already been criticised and found seriously wanting, and accordingly rejected (see 1.14 [Vol. 1 $_{—Eds.}$]).

Thirdly there are the inevitable arguments from "want of distinct identity", the arguments of Peirce and Edwards against future and possible individuals (arguments refurbished in the latter case by Quine, and discussed in chapter 3).

> The difficulty in treating Ryle and a future individual before he existed is that it could not be intelligibly asked at those times before he existed which future individual or unactualised possible he then was (p. 21).

Why not? "Will Gilbert Ryle be red-headed?", "Will he become a philosopher?" are perfectly significant, and intelligible, questions, which could have been asked before Gilbert Ryle was conceived. The future Gilbert Ryle can be singled out furthermore by descriptions which contingently identify him.[2] The absurdities we are supposed to be led to in treating Gilbert Ryle "as having once been an unactualised possible" (p. 22) are not stated, so they are hard to assess.[3] In trying to make his identity objections stick, Godfrey Smith *has* to fall back firstly on the Ontological Assumption itself ("at present we have no way of knowing *which* individual this will be because there is at present no individual for the prediction to be about" (p. 22); "we could not rule out the possibility of mechanical malfunction or failure in ... the stone cutting machine of incredible precision ..., as a result of which the individual to which we were apparently referring, as it turned out, did not come to exist" (p. 24)); and secondly (as the latter quote indicates) sceptical

accordingly lead to the conclusion that nothing exists, contradicting his claim that present (and past) individuals do exist.

[1] As one of the quotes above indicates Godfrey-Smith thinks there can be no such chains (Godfrey-Smith 1976, p. 17).

[2] Godfrey-Smith's objections to appeals to descriptions (p. 22) can be met by a distinction of identity determinates, in much the way that some of Kripke's objections to use of identifying descriptions can be met (see also 1.14 [Vol. 1 $_{—Eds.}$]).

[3] The "parallel" from Edwards has nothing to do with future objects, and insofar as there is any absurdity at all in Edwards' numerous examples, this derives from questions as to the distinctness of supposed indiscernibles. The "worries" Prior claims are "surely senseless" are not, but are intelligible, and the alternatives considered by the "fortune teller" are "at that time" "distinct", as but a little alternative worlds semantics will reveal.

considerations ("however detailed we make the prediction it will never be sufficient to distinguish (a unique) possible competitor to satisfy it", p. 23; "any description of the statue is still general. There are an indefinite number of surfaces which might have been revealed by the labours of the sculptor", p. 23). In fact there are definite descriptions that could have been used in 1899 to uniquely and unproblematically contingently identify Gilbert Ryle, and it is simply not material that it is logically possible that other individuals might satisfy those descriptions (as they might). Moreover there are soon-to-exist individuals that nothing short of scepticism about the future (backing a resolve to hang on to (α)–(γ)) would stop us naming and ascribing properties to, namely all those code-named or individually labelled objects assembly-line factories are producing.[1] Consider a precision number plate machine, which has just stamped out plate "HVP 731", and is about to produce a plate just like HVP 731 except that it will bear the label or (as the assembly line workers who have all had a training in arithmetization say) name "HVP 732". HVP 732 can be described in very considerable detail, there are many particular facts about it. Its features appear to refute (α)–(γ). Godfrey-Smith would say the machine might break down. Everyone on the line and elsewhere is determined, even if it does break down (contrary let us say to all the evidence since the machine is in perfect working order, has its own power supply and plenty of fuel, etc.), to get it going, by hand power if necessary, just to refute (α)–(γ). Well, one could imagine a surprise nuclear explosion upsetting these plans, cosmic interference, etc., but now the theses depend for their defence on a scepticism about the future (that should be rejected on other grounds).

7. *Fatalism and alternative futures.* Ryle introduces his strong generality of predictions thesis – the thesis that while, or even if, singular statements can be made about the past and present, statements concerning the future are irreducibly general – in the course of and in order to deal with a fatalist argument. But the generality thesis is quite unnecessary in order to deal with the argument Ryle calls "the fatalist argument". Though Ryle accounts the argument "a severe and seemingly rigorous argument" (p. 16), the argument he presents involves an elementary modal fallacy, namely (a variant on) the argument $\sim\Diamond(A \mathbin{\&} \sim B)$; but A. Therefore $\sim\Diamond\sim B$;[2] and fails once the fallacy is exposed.

[1] Other examples are provided by dates. Next year is called "1980". 1980 is an individual year; it is presently purely future. There are many particular facts about it, e.g. necessarily it succeeds 1979, contingently some sales of *this* book (itself a future object) will occur in it, etc.

[2] The modalities need not be logical ones. "It is impossible that" can be treated as determinable in the argument.

Ryle's formulation of "the fatalist argument" considers first the particular statement q_0, namely "Ryle coughed and went to bed on a certain Sunday (viz. 25 January 1953)", and the main trick is pulled before Ryle goes on to generalise. Since q_0 is true it was certainly true on the day before.

> Indeed, it was true a thousand years ago that at certain moments on a certain Sunday a thousand years later I should cough and go to bed [$T(-1000)q_0$]. But if it was true beforehand – forever beforehand – that I was to cough and go to bed at those two moments on Sunday, 25 January 1953, then it was impossible for me not to do so [$T(-1000)q_0 \rightarrow \sim\Diamond\sim q_0$]. There would be a contradiction in the joint assertion that it was true that I would do something at a given time and that I did not do it [$\sim\Diamond(T(-1000)q_0 \;\&\; \sim q_0)$]. This argument is perfectly general. Whatever anyone ever does, whatever happens anywhere to anything, could not *not* be done or happen, if it was true beforehand that it was going to happen [$T(-t)A \rightarrow \sim\Diamond\sim A$]. So everything, including everything we do, has been definitively booked from any earlier date you like to choose (p. 15).

The particular case relies on the fallacious argument: $\sim\Diamond(T(-1000)q_0 \;\&\; \sim q_0)$; so $T(-1000)q_0 \rightarrow \sim\Diamond\sim q_0$. But $T(-1000)q_0$. Therefore $\sim\Diamond\sim q_0$.[1] In the general case Ryle does not bother to state a main premiss, the truth $\sim\Diamond(T(-t)A \;\&\; \sim A)$, but slides immediately to the false $T(-t)A \rightarrow \sim\Diamond\sim A$.

So much for Ryle's argument; but can the damaging assumption $T(-t)A \supset \sim\Diamond\sim A$, or something like it, be independently defended? The answer is Yes, by a no-branching-futures assumption of some form. Less obscurely, the assumption can be enforced by:

(a) a Diodoran account of necessity, or

(b) a no-alternative-worlds thesis.

Consider (a). An apparently damaging assumption, $(t \leq 0)T(-t)A \supset .\ \sim\Diamond\sim A$, can be sustained by defining necessity thus: $\Box A =_{\mathrm{Df}} (t \leq 0)T(-t)A$.[2] But firstly, the definition

[1] The argument, as formalised, also involves the fallacy: $\sim\Diamond(A \;\&\; \sim B)$, so $A \rightarrow B$. But this fallacy is readily avoided by replacing $A \rightarrow B$ by $A \mathrel{-\!\!3} B$ or $A \supset B$. (The argument need not trespass beyond the elementarily consistent, so (γ) [i.e. the rule of detachment for material implication —*Eds.*] is admissible.)

[2] The alternative definition $\Box A =_{\mathrm{Df}} (t < 0)T(-t)A$ is inadequate, because (at least on a discrete time theory) $\Box A$ does not guarantee A. For something may upset A between the last previous moment and now.

– not Diodorus's (likewise faulty) definition which was in terms of all times – does not define necessity at all, but only a variant of Prior's functor H, "It was always the case that". Secondly, even if the definition is accepted, only a "fatalism" without bite would result. For the only sense in which it could be established that something could not not happen would be the relatively trite one that it was always true that that thing did happen. No necessity is thereby imputed to the happening; in no serious sense could it not have been otherwise. Thesis (b) likewise tends to collapse modality, e.g. necessity to truth or some qualified "sort" of truth such as tenseless truth or omnitemporal truth – insofar at least as exponents of (b) are prepared to admit or tolerate modal notions at all. Again only a "fatalism" without sting emerges. If however (b) is combined with a limited account of necessity (certain grades thereof) in terms of logical truth (e.g. truth in all classical models), then once again fatalism only results by commission of a fallacy: for truth at all previous times is no guarantee at all of logical truth.

Fatalism suggests that the future is fixed, already fixed, and now unalterable. This view, and the determinism with which it tends to converge, is encouraged by a set of faulty and misleading pictures such as the already written Book of Destiny (considered by Ryle, pp. 15–6), as opposed to the (also misleading) metaphor of the Moving Finger (of Omar Khayyám), or the rolled out carpet, as distinct from the rolling out carpet. As Prior puts it:

> ... what is the case already has by that very fact passed out of the realm of alternative future possibilities into the realm of what cannot be altered (1962, p. 241).[1]

But in order to explicate the respect in which the future alternatives are closed as what was future becomes present and past, alternative cases or worlds have to be (neutrally) spoken of and quantified over. Such is anathema in referential quarters. Thus for instance, Smart (1979, p. 12):

[1] But it can also be made less platonistically, by replacing "realm" by a neutral term. Prior also puts it thus, in explicating Aristotle's questionable formulation of the point:

> when it passes from the future into the present and so into the past, a thing's chance of being otherwise has disappeared.

Certainly it is logically, also physically, possible that a which is (in fact) f [or was f] is not f [or was not f]. *Future possibilities* are a subclass of physical possibilities.

> It makes no more sense to talk of altering the future than it does to talk of altering the past. ... There are no more alternative futures than alternative pasts.

The second claim conflicts with the first. What Smart means by the somewhat misleading claim that it makes no sense, is what he shortly says, that one "cannot alter the future". For, so the argument runs, the future is what will be and, tautologically, what will be will be. The argument, and its elaborations, e.g.,

> Whatever change I make or do not make in the manuscript is in the future part of that manuscript (p. 12),[1]

deliberately miss the intended point – deliberately, because no-alternative-worlds positions cannot adequately render intelligible (and sometimes true) discourse about altered and alternative futures. Consider such claims as that Bismarck's unification of Germany altered the future course of European history. Assuming the claim is true, it makes good sense to say that what Bismarck did altered the future, that is, to elaborate, altered the future from what it would otherwise have been, i.e. without his doings. Or slightly differently again, but for Bismarck things would have been different; an alternative (and better) future would have eventuated. Naturally, the familiar explanation makes use of counterfactuals (which can in turn be explained through, and help explain, alternative cases), but that's alright. More generally, the weak form of fatalism Smart advances, a form built into the closed futures of Minkowski diagrams, is upset by counterfuturefactual considerations. Things could go differently in the future in a way they could not go differently in the past, the future can be altered as the past cannot, the future is open in a way the past is not.[2] Such differences, incidentally, between past and future along with the alignment of cause and effect disclose – what is often supposed to be a mystery – the direction of time. It is this asymmetry of time, together with the openness of the future, that have encouraged even referentially-inclined philosophers to contrast the reality of the past with the unreality of the future. Neither however are real.

[1] There are no alternative futures because whichever alternative comes to pass is the future. But that in no way rules out alternative possibilities, all it shows is that one alternative among the possibilities will be how things are.

[2] The outlines of the requisite semantical picture are clear enough even if the details are not. The picture is like that of an indefinitely growing tree that, like some eucalypts, drops its branches as it grows: the main unbranched stem is the past, the present is where the branching begins to occur, alternative futures are the branches, one of which is, at least in initial stages, strongly indicated as leading. A little more exactly, in the model there are worlds which are alternative possibilities for the present whose pasts coincide.

Ryle's and Smart's arguments are but two of the many arguments that seem to lead to fatalistic conclusions.[1] One of the reasons why fatalism has managed to maintain its logical grip, and its charm, is that it comes in so many forms, with different arguments. Wherever one is cut down another – so it seems on first acquaintance with the problem – springs up in its stead. However the arguments are of limited variety, and one and all fallacious, the logical versions turning on modal fallacies. Following Cahn (1964, p. 295) fatalistic arguments and the positions they lead to can be roughly divided into: (i) *logical* fatalisms, according to which it is necessary that the future be what it is going to be and impossible that it be any thing other than what it is going to be (in these days of reductions of such modalities to the logical, it is usually taken that the necessity is logical necessity and the impossibility logical impossibility); and (ii) *empirical* fatalisms, according to which our deliberations and efforts are pointless and do not influence the future.

The argument Ryle presents gives a fatalism of the logical sort. An even simpler fallacious argument of this sort is this: Necessarily what will be will be. Therefore, what will be necessarily will be. The fallacy, $\Box(A \rightarrow B)$, so $A \rightarrow \Box B$, is the same as that which proves that whatever is known must be necessarily true, whence knowledge can never be of the merely contingent.[2]

For an example of (ii),

> Consider the proposition that Mr A will become a great philosopher. This proposition is either true or it is false. If it is true, then Mr A will become a great philosopher, no matter what else he does. On the other hand, if it is false, then, no matter what else he does, Mr A will not become a great philosopher. Therefore, it makes no difference whether Mr A studies hard or not, since in either case, his studies will have no bearing on the outcome.

All the argument as far as the "therefore" establishes, given that "no matter what else he does" can be tacked on to each disjunct, is that Mr A will or will not become a great philosopher no matter what else he (in fact) does. But this is analytic (as $A \vee {\sim}A \vee B$).

[1] A rather different argument can be based on the theory of special relativity: see Stein's criticism of Rietdijk's "proof" of determinism (Stein 1968, p. 13).

[2] Aristotle's sea-battle argument can likewise be seen as involving a distributional fallacy, of an intensional functor over a disjunction: roughly, where D represents "It is determinate that" from $D(A \vee {\sim}A)$ to $DA \vee D{\sim}A$.

What follows the "therefore" makes a contingent statement (a contingently false one perhaps). Hence the argument is a nonsequitur, since a necessary statement cannot imply a contingent one. For $A \Rightarrow B \supset . \Box A \supset \Box B$.[α] It may however be contended that precisely what the argument shows is that, contrary to appearance (and to semantical modellings of the conclusion), the conclusion is necessary. But it is not necessary because it is falsifiable: there are possible circumstances where it is false, where it makes a difference, one way or the other, what else Mr A does (and *would* do). Suppose for instance, Mr A commits suicide. This no doubt destroys what small prospects Mr A had of becoming a great philosopher, and so makes a difference. Hence the acclaimed conclusion does not follow. In this case it is false that Mr A will become a great philosopher. Does it not matter what else Mr A does? In one respect No (the "what will be will be" respect). In a more important respect, that of accounting for the falsity assignment, it matters considerably that Mr A did what he did. It is false that Mr A will become a great philosopher, *because of* what he did, *not* no matter what he *did*.

[α] Where "$\Rightarrow$" represents entailment. —*Eds.*

Chapter 3

On what there isn't[1]

Most things do not exist. For every thing that exists, e.g. Professor Smart, there are several things that do not exist, abstractions beginning with the property of being Professor Smart and the set of properties of Professor Smart. And there are a great many abstractions other than those directly generated by things that exist. These truths we hold to be elementary, and if not self-evident they can be argued for. Quine, however, in a very bold stroke, has stolen much of the terminology we ordinarily use to state, and argue, these elementary facts – and as far as most philosophers are concerned, he has got away with it.

The theft is evident from the first lines of Quine's "On what there is" with the discussion of which this chapter is primarily concerned. The ontological problem is there said to be formulable as "What is there?", and answered "Everything". But the ontological problem in question is the problem as to what exists or (perhaps differently) of what has being, which is not just a very different problem from the more easily answered question as to what items there are, but a problem which is not truly answered "Everything" since many things do not exist. The theft is of the English expressions "what", "there is", "is", "thing" and "everything" which are commonly enough used, without existential import, to consider and talk about items that do not exist and have no being. Just consider "what is" questions concerning fictional objects or the objects of false theories, e.g. "What is a hobbit?" "What is phlogiston?", or such questions as "What is an impossible thing?", "What is a merely possible thing as distinct from one that exists?" It is not merely

[1] This chapter is a noneist commentary on "On what there is", the first essay in Quine's FLP. All page citations without further detail are to FLP.

The original version of this chapter was revised. Due to a mistake by the publisher, the editor's name 'Dominic Hyde' was mistakenly appeared as a co-author in all the chapters on Springerlink and third-party websites. The correction to this chapter is available at https://doi.org/10.1007/978-3-030-26309-6_7.

D. Hyde (ed.), *Noneist Explorations I*, Synthese Library 415,
https://doi.org/10.1007/978-3-030-26309-6_2

ironical then that Quine should subsequently (p. 3) magnanimously give away the word "exist", claiming that he still has "is".[1]

Once the stolen goods are restored it is no great feat to resolve many long-standing, but gratuitous, philosophical puzzles, beginning with the Platonic riddle of non-being, that "non-being must in some sense be, otherwise what is it that there is not?" (pp. 1–2). Consider some thing, a say, that does not exist, e.g. a is Meinong's round square. Then *what* does not exist is in this case a, but it in no way follows from "a is non-existent" that "a exists". Such non-entities as a need have no being in any sense. It is basically because whatness and thinghood have been illicitly restricted to what exists or has being that a puzzle seems to have arisen: for certainly we contradict ourselves if we say that what has being does not have being. There is no contradiction however in saying that what is a thing or object, e.g. a, may have no being in any sense; and this dissolves what Quine nicknames *Plato's beard* without using, or blunting, Occam's razor. For Occam's razor to stay sharp requires only that entities not be multiplied beyond necessity; but no multiplication of entities has been made, no bloating of the universe (of what exists) has occurred. Indeed noneism enables a very substantial *reduction* in what is said to exist, so that what is said to exist can coincide with what really does exist, namely only certain particular objects now located in space.[2] But, more to the point, Occam's razor embodies various muddles of the very sort that we are concerned to remove. In particular, Occam's dictum that entities (or differently objects) should not be multiplied beyond necessity supposes that it is in our power to increase or decrease the number of entities (or objects): but of course in *that* sense – as opposed to the destruction or creation of objects by one's activity – it is not. What we can increase or decrease is not what exists but what we *say* exists, what we (choose to) *talk about*, and what our theories *commit* us to in one way or another. So the dictum confuses what exists with what we (choose to) talk about or what we, or our theories, *say* exists – a confusion that runs through into modern criteria for ontological commitment.

Because a has no being there is no cause to try, like Quine's philosopher McX, to assign some kind of being to a, e.g. ideational existence as an idea in men's mind. Pegasus and the Pegasus-idea remain, as they are, distinct items: Pegasus is a horse,

[1] Nor is ontology exactly independent of lexicography, in the way Quine suggests.

[2] It was Meinong's thesis that any existing object has a more or less definite location in space and/or time. It is a corollary that abstract objects do not exist. See further chapter 9 [Vol. 3 —*Eds.*].

the Pegasus-idea is not, since ideas are not (significantly) horses; Pegasus does not exist, but the Pegasus-idea is presumed to; and so on.[1] The false dichotomies, spawned by empiricism, that Quine relies upon in putting the mythical McX on the spot can equally be dispensed with, e.g. that whatever has being exists either spatio-temporally or as an idea in men's minds. Since a does not exist, it does not exist *in* this or that way. Like other alleged problems about a, the problem about how or in what way a exists vanishes given that a does not exist.

The initially commonsense noneist position being advanced is not to be confused either with that of Quine's other non-existent philosopher, Wyman.[2] For it is not maintained that Pegasus, for instance, "has his being as an unactualized possible", (p. 3) or *as* anything else, since he simply has no being. More generally, the transformation of "c is d" (e.g. "what I am thinking of is Pegasus" or "Pegasus is an unactualized possible") to "c

[1] Detailed arguments that ideas are different from the objects they present were given by Meinong. One of these, set out more formally, as in EP, p. xxv, runs as follows:

1. Ideas are, by their very nature, of something (some object).
2. Ideas, when they occur, exist.
3. If ideas were identical with their objects, all their objects would exist whenever someone was having an idea of them.
4. But there are objects which never exist (e.g. the *perpetuum mobile* and Pegasus), yet there are ideas of them.
5. Therefore, ideas are not identical with their objects.

Elsewhere (1959, p. 199) Quine himself makes a similar distinction:

> ... to identify the Parthenon with the Parthenon-idea is simply to confuse one thing with another; and to try to assure there being such a thing as Cerberus by identifying it with the Cerberus-idea is to make a similar confusion.

Yes, to confuse one thing with another. Quine's "essential message" in 1959, §33, repeated over and over, is however that

> Some meaningful words which are proper names from a grammatical point of view, notably "Cerberus", do not name anything (p. 202),

otherwise briefly referred to as "the mistaken view that 'Cerberus' must name something". In fact, but not, of necessity, "Cerberus" does name something, not the Cerberus-idea, but Cerberus. There is no mistake: "Cerberus" names Cerberus, whence, particularising, "Cerberus" names something. Quine's message is a plea to have us restrict quantifiers to existentially-loaded ones; for it is true that "Cerberus" does not name anything existent. There are excellent reasons, as we have seen, for ignoring such pleas, for not so limiting quantificational apparatus. Nor does the removal of existential restrictions involve any of the mistakes Quine imagines he finds in taking nonentities in the domain of quantifiers or as named: there need be no confusion of meaning with naming (though meaning can be explicated through interpretation – which is wider than naming – in worlds); there need be no confusion of meaning with things talked about (but *naming* is a subspecies of *being about*); there need be no appeal to attempts, inspired by the Ontological Assumption, to make nonentities exist somehow (e.g. as shadows of entities, or names) or somewhere (e.g. in the mind, in myth or fiction); and so on.

[2] Philosophical legend has it that Wyman is modelled on Meinong, but the serious discrepancies between Wyman's position and any but a hearsay Meinong position cast some doubt on the legend. Wyman is somewhat more like Russell 1937.

has its being as *d*" (e.g. to "what I am thinking of has its being as Pegasus" or "Pegasus has his being as an unactualised possible") is to be rejected. Nor is it maintained, as by Wyman, that Pegasus is (p. 3). The expression "Pegasus is" is, like "Pegasus likes", deviant in many English idiolects (including mine): its incompleteness is suggested by questions like "is what?". Insofar as "is" functions intransitively, "x is" means, as the OED indicates "x exists". In short, *we* can give Quine the intransitive use of "is", and so (as in the title of this essay) of "isn't". Since the noneist position certainly does not claim that Pegasus exists, or, as Wyman puts it, subsists (i.e. exists, though perhaps in a very low level way), it does not claim, in Wyman's way, that Pegasus is.[1] It follows, using the first account Quine gives in FLP of commitment to an ontology (on p. 8), that noneism is *not* committed to an ontology containing items that do not exist, such as possibilia and abstractions, since, for each such x, it is true that x is not.[2]

As regards impossible items such as the round square cupola on Berkeley College Wyman's position differs drastically from noneist alternatives.[3] Unlike "purple happy number", there is nothing meaningless about "round square cupola", which is what Quine has Wyman say (p. 5). This does not mean that noneists admit "a realm of unactualizable impossibles": "realm" like "universe" carries ontological overtones. Although some

[1] Meinong's position diverges from the stream of noneism here followed: for according to Meinong possibilia such as Pegasus subsist. Insofar as this implies more than that possibilia are possible – in some or other of the commonly confused senses, e.g. that they could exist or that the supposition of their existence leads to no contradiction or that their characterisation leads to no contradiction – it should be rejected as misleading.

[2] Observe that the result *differs* from that obtained by applying Quine's other and better known criterion for ontological commitment, according to which ontological commitment is determined through preparedness to quantify (p. 12). In these terms noneism *is* ontologically committed to what it maintains does not exist, e.g. possibilia, abstractions and so on. A corollary is the inadequacy of the quantificational criterion that "to be is to be value of a bound variable" (p. 15). The inadequacy of Quine's case for the criterion is discussed below.

[3] Including Meinong's position; see, e.g. EP, chapter 2. The same supposed dilemma as in FLP is colourfully presented in Quine 1959, p. 202:

> Having already cluttered the universe with an implausible lot of unactualised possibles, are we to go on and add a realm of unactualised impossibles? The tendency at this point is to choose the other horn of the supposed dilemma, and rule that expressions involving impossibility are meaningless.

That is almost certainly not the main historical tendency; and it is a tendency to which there are serious objections (see Slog). The supposed dilemma is no dilemma, not for the reason Quine offers, that "there need be no mystery about attributing non-existence where there is nothing [existent] to attribute it to" (and we have observed the inadequacies in reductions of talk of nonentities to talk of entities), but because the first option can be restated neutrally. There are, and need be, no additions to the universe (of entities) of new realms: the population, or number of elements, of the entity-universe is unchanged. Talk of "cluttering", "implausible lot", etc., reflects inappropriate referential thinking.

items such as round squares are impossible, and we can accordingly form non-null sets of impossibilia, these sets do not exist, any more than their elements or than other abstract sets.

It is widely thought nonetheless, despite the differences between Wyman's position and those of a more Meinongian turn, that Quine's scholastic objections to Wyman's unactualised possibles do serious, perhaps irreparable, damage to *all* these positions. Thus, for example, Kenny (1968, p. 169): "these objections, I think, make untenable the notion of Meinongian pure objects". This is far from the truth, as we now try to show. Quine's main charge (p. 4) is that possible objects are "disorderly" and "well-nigh incorrigible", and the basis of the charge is to be found in the suggestion that "the concept of identity is simply inapplicable to unactualized possibles". But *the very same notions of identity* – most importantly, extensional identity – *and distinctness that apply to entities apply likewise to nonentities.* Identity is, as always, reflexive, symmetric, transitive, holds given indiscernibility, and warrants qualified replacement. The criterion for identity of nonentities is, as for entities, coincidence in extensional properties. Thus, for instance, Hercules and Heracles are identical, though some people did not and do not know this. The criterion for distinctness is that of positively differing on extensional properties. Thus, for example, Pegasus is distinct from Thunderhead because Pegasus has the (extensional) property of being winged and Thunderhead does not. Hence nonentities *can* "meaningfully be said to be identical with themselves and distinct from one another" (p. 4). Moreover, far from "the concept of identity being simply inapplicable to unactualised possibles", precisely the same criterion as that given, and ordinarily used, is presupposed classically – on one common theory – in such results, for example, as that Pegasus and Chiron are one and the same because they have the same traits (namely none). But, strictly, what is true of nonentities classically depends on the theory of names and descriptions adopted.[1] On Russell's theory of descriptions all nonentities are identical, indeed all statements concerning unactualised possibilia b and c are indeterminate because it is false that $b = c$ and false that $b \neq c$. At this point a latent inconsistency in Quine's stance becomes apparent. For it cannot both be false – as it is on Russell's theory which Quine endorses (pp. 5–6)

[1] For instance, on Quine's ML, which includes a Fregean style theory of descriptions much inferior (in all but technical ease) to Russell's theory, nonentities have the most amazing properties. For example, Pegasus is identical with the null set – so that concept of identity is certainly applicable – and has all the same properties, e.g. Pegasus exists, but has no members, Pegasus is a subset of every set whatsoever, and all the natural numbers are simple logical constructions of Pegasus. The data concerning nonentities may be a little soft – but it is not *that* soft.

– and meaningless, in the sense of concept inapplicability – as Quine suggests it is – that nonentities are identical and not distinct.

The criteria for identity given, whether nonclassical or classical, also serve to undercut Kenny's overstatement (1968, p. 168):

> The most serious – indeed the insurmountable – objection to Meinongian pure objects is that it is impossible to provide any criterion of identity for them.

But, despite the ready availability of criteria, the objection, that nonentities have no (clear) identity conditions, is repeated over and over in the literature. Another recent example, where the objection is used as ground for putting aside Meinong's theory, may be found in Linsky (1977, pp. 35–6, our transposition):

> Meinong comes nearest to capturing ...
> our intuitions about reference in natural language and his theory does not seem to lead to contradiction as it is widely supposed to do. What disturbs us about his ontological population explosion, I believe, is that these objects have no clear identity conditions. Is the present king of France identical with the present king of China? There seem to be no principles which can be used to provide an answer to such questions. One answer is as reasonable as the other and this makes the very notion of an *object* seem misapplied here.

Apart from the very first claim, this requires correcting claim by claim. Although Meinong's theory, if carefully (re-)formulated, does not lead to contradiction, the account Linsky gives (1977, p. 34) does lead to contradiction (given only a very minimal logic):

> But Meinong insists, against Frege, that what phrases of this form [$\ulcorner(\imath x)(\phi x)\urcorner$] denote is always $(\imath x)(\phi x)$ .... The insistance that "$(\imath x)(\phi x)$" denote the right thing leads immediately to a special case of the independence of *Sosein* from *Sein*, for it entails that $\phi(\imath x)(\phi x)$ is always true for any choice of ϕ.

That for every ϕ, $\ulcorner(\imath x)(\phi x)\urcorner$ denotes $(\imath x)(\phi x)$ is distinct from the Independence Principle[α] (as varieties of free logic will show) and does *not* entail $\phi(\imath x)(\phi x)$ (as neutral logic

[α] Discussed at length in Vol. 1, Ch. 1 §3. —*Eds.*

countermodels will show). The unqualified Characterisation Principle $\phi(\imath x)\phi x$, which Linsky (mistakenly) attributes to Meinong (p. 33, as well as p. 34), is an exceedingly damaging principle, and contrary to Linsky (p. 35) does lead to propositions of the form p & $\sim p$. For consider the object $(\imath x)(Rx$ & $\sim Rx)$, a for short. Then, by the assumption principle, Ra & $\sim Ra$.

There is no "ontological population explosion" under Meinong's theory: to suggest so is to misrepresent the theory. The identity *conditions* for objects are clear, even if, as in the case of existing objects, it is not always clear, or determinate, whether certain objects are the same or not. Since these ordinary identity principles provide an answer to questions, such as Linsky's about the kings, there certainly are principles which supply answers to such questions – and similarly theories of descriptions such as Russell's provide answers, even if wrong ones, e.g. the king of France = the king of China. In fact the king of France is Leibnitz-distinct[α] from the king of China, since someone can think of one but not the other; and a theory which gave a different result would hardly be reasonable. And on Linsky's assumptions about Meinong's theory, the two kings are presumably extensionally distinct since one is king of France and the other not (but of China): thus as they differ in extensional respects they are in fact distinct. Accordingly there is no solid evidence, on the basis of identity-conditions, that the notion of *object* is misapplied in Meinong's theory.

As with identity so for likeness and similarity, essentially the same accounts that apply to entities apply to nonentities, two items being alike if they have sufficiently many extensional properties in common. It is on this basis that we say that a dryad and a naiad are alike, and much more alike than a unicorn and a centaur. Thus some possible things are alike, but as in the case of entities, alikeness is in general not sufficient for identity. These points answer all Quine's alleged difficulties except those concerning the various men in the doorway. Briefly, the concepts of identity and distinctness, likeness and difference, are applicable to nonentities, and the criteria for their application are the same as in the case of entities. Hence, too, set-theoretical notions are applicable to nonentities as well as entities, and numerical concepts apply. As Locke and Leibnitz argued against the scholastics and Frege reiterated (1950, p. 31), various classes of possibilia can be counted and numbered. A non-actual man in the doorway belongs to the three element class consisting of Pegasus, Heracles and a non-actual man in the doorway (not as Frege's

[α] I.e. "not fully indiscernible"; cf. Vol. 1, Ch. 1 §11. *—Eds.*

own theory, adopted in Quine's ML, would have it, to a one element class comprising the null set). Similarly, it may make good sense to ask "How many objects of a given type have a given property?" even where some or none of the objects exist.[1] The main problem, not special to nonentities, lies in determining which properties the objects in question have.

Consider now a nearest open doorway, and consider an arbitrary fat man who never has existed, e.g. Mr. Pickwick. Ask whether Mr. Pickwick is in that doorway. The answer is, as a matter of observation, No. In literal contexts the answer is the same in the case of every other merely possible fat man. Hence the answer to the question "How many (merely) possible fat men are in that doorway?" is: Zero. The same answer may be expected on more theoretical grounds. By a familiar, and classically accepted, thesis attributed to Brentano, a merely possible item cannot stand in entire physical relations, such as *being in* or *standing in*, to actual items. Hence merely possible men of any variety, fat or thin, bald or not, cannot stand in actual doorways. Thus the answer to each of Quine's numerical (how many?) questions is: Zero. There are zero possible men in that doorway, zero possible fat ones, and exactly the same number of merely possible thin ones.

The answers given to these last questions are, again, exactly those classical orthodoxy should supply, even if the reasons for the answers are of a somewhat different cast. Since classical orthodoxy has, with its very limited quantificational apparatus, serious difficulties in expressing its answers, let us articulate them for it. Since no nonentities exist, of whatever kind, there are exactly as many nonexistent fat men as thin men, namely none, so none can be standing in any actual doorways.[2] In short, classical ortho-

[1] Quine appears to be suggesting that such questions as "How many possible fat men are there in that doorway?" make no sense, and, presumably, that corresponding assertions of the form "There are n possible fat men in that doorway" are meaningless. For the question is significant if the corresponding indicatives are. And *prima facie* the indicatives are significant (and transformations can be used and arguments constructed which reveal that the sentences are significant): they contain no category or type mistakes. So they cannot be convincingly written off as not well-formed: even if they strike the uninitiated as odd they require logical accommodation. Accordingly such questions, and indicatives, are just as much a problem for classical logical positions as for the nonclassical positions they are directed against.

[2] On a popular alternative account, they are all identical with the null set (or entity), which does not stand in any actual doorways. The result is as before. But other accounts from out of the classical stables differ, e.g. Hilbert's theory supplies no answer, and even Russell's theory strictly applied gives no answer to those questions which include the adjective, "possible". Let it not be taken as an objection to nonclassical approaches, then, that different theories provide different answers to the questions.

doxy can already supply answers to what are reckoned the hardest of Quine's questions – giving the lie to such charges as that identity, difference and numerical properties cannot be meaningfully attributed to nonentities, and so removing the ground for the further charges of disorderliness and incorrigibility.

It is not easy to avoid the impression that one of the many reasons why Quine's questions have been thought to cause especially severe difficulties for Meinongian style theories is because "A possible man is in that doorway" has been confused with "A man is possibly in that doorway", that *de re* modalities have been confused with *de dicto* modalities (in *one* sense of that vexed distinction).[1] As the second claim, like the pure *de dicto* claim "It is logically possible that a man should be in that doorway", is usually true for open unoccupied doorways (and let us suppose *for* the selected one such), *if* the conflation were correct the answer to the question "How many possible men are in that doorway?" – now the question "How many men are possibly in that doorway?" – would seem to be "At least some" since Smart is possibly in that doorway, and the determination of the exact number would – on the doubtful assumption that it is determinate at all – become a knotty, though hardly "insoluble", problem. But the conflation is not correct: the sentences have different semantical analyses, the first stating a relation of *being in* in the actual situation whereas the second states such a relation in one possible world.

Part of the interest of Quine's question comes from the number of different answers it has been given by different philosophers, different answers induced partly by different theories of nonexistent objects and partly by different questions, not only the *de re* question we have so far been concerned with, namely:

(1) How many merely possible (nonactual) men are in that actual doorway,

but also the questions:

(2) How many men are possibly in that doorway?

(3) How many men can possibly be in (be crowded into) that doorway?, or, more precisely, what is the largest number of men that can possibly be in that doorway?

Question (2) splits into different questions according as the collective quantifier precedes or follows the modal operator, namely:

[1] According to Lewy (1976, pp. 32–6) Quine does confuse *de re* modal statements with *de dicto* modal statements. Unfortunately on the point of issue it is perhaps Lewy who is confused.

(2a) Of what number n of men is it true that for those n men it is possible that they are in that doorway (together)?

(2b) Of what number of men is it true that it is possible that those men are in that doorway?

The answer to question (3) sets a bound on the answer to question (2). No doubt the answer to (3) is at least, Many, but a more exact answer depends on how small humans can be and what shape and how large the doorway is (and also on the type of modality). The answer to (2b) is then: any number between zero and the bound. For let k be any such number: then the statement "it is possible that k men are in the doorway" is true. Accordingly the best answer to (2) is: it is indeterminate,[1] though the indeterminacy is bounded.

Without doubt some conflation of modalities is encouraged by ordinary discourse. Consider, for example: "Some cloud and a possible thunderstorm are forecast for Canberra this afternoon", where the apparent *de re* modality has an intended *de dicto* expansion. A confusion of modalities also seems, at first sight, to occur in Parsons' answer to Quine (1974, p. 572):

> ... when Quine asks about "the [merely] possible fat man in the doorway", he uses a definite description which, on this account [because the uniqueness clause is not satisfied], fails to refer – for there are *many* possible fat men in the doorway.

It is logically possible that many fat men are in the doorway, but it would be quite invalid to infer that many possible-fat-men *are* in that actual doorway – nor does Parsons' analysis support such a claim.[2] For properties such as those of possibility and actuality figuring in the interpretation of "possible fat men" and "actual doorway" are (in the intended sense) extranuclear, and so are not characterising. But what has been said is not what Parsons means. There is an ambiguity in "possible fat man", depending on whether "possible" goes into the description, as supposed above, or not, i.e. is regarded as consequential. On the latter construal, there are on Parsons' theory infinitely many possible

[1] This is the answer to Quine's question arrived at in the original version of *Exploring Meinong's Jungle*, Routley 1967. It now appears to be an answer to a different, if easily confused, question, namely (2b).

[2] Parsons' other answers to Quine's "embarrassing" questions (Parsons 1974, note 21) are just fine (and happily in agreement with mine).

fat men in that doorway (indeed the cardinality is presumably nondenumerable), one for each consistent set of nuclear (i.e. roughly, characterising) properties which includes at least the properties of being a man, being fat, and being in the doorway.[1] Infinitely many, irrespective of the size of the fat men, and that some of them will be giants who fill the doorway or more! The result is implausible (deeper reasons for dissatisfaction with, and ultimate rejection of, Parsons-style accounts will be presented in the final chapter [i.e. chapter 12, Vol. 3 —*Eds.*]).

It is bound to be objected that any Meinongian-style theory will generate many, very many, possibilia standing in that doorway. Consider the n possible fat men standing in that doorway, for an arbitrarily selected number n. Then surely, by the Characterisation Postulate, some n possible fat men are standing in that doorway? Emphatically, No. The postulate has only a carefully restrained role on any theory that can claim coherence: it does not (at least in the noneist theory advanced) warrant ontological arguments or the like, or the establishing of new entire relations between what exists and what is merely possible. Nothing, of course, stops the design of (generally less satisfactory) theories with more sweeping Characterisation Postulates. Some of these theories would, like Parsons' theory, give different answers to some of Quine's questions. This is unsurprising, rather as it is unsurprising that different classical-style theories identify Pegasus with different objects, or admit different universes of sets.

A rather similar reply can be made to the related objection that we can tell stories which describe fat men in the doorway. A tells one story with 3 fictional fat men fitted into the doorway. B out-does A and tells a tale in which 10 fat men are squeezed into the doorway. C tells an even taller tale ... . How many fat men are in the doorway? On A's story 3, B's 10, and on C's tale, 98 say. But in reality, as before, 0. For the

[1] And as one may admit for *Parsons doorways*, Parsons doorways being the doorways of Parsons' theory and having properties supplied by that theory. That is, a Parsons doorway is like Holmes' London – only the source book for such an object is Parsons' theory, not the requisite group of Sherlock Holmes stories. There are many interesting questions one can ask, and answer, about Parsons doorways, e.g. Do any exist? The answer is No; for if one did a merely possible object could stand in entire physical relations to an entity, contravening the Brentano thesis. Could an existing object, e.g. Parsons, stand in a Parsons doorway? On Parsons' *theory* the answer is Yes, given that Parsons is fat enough (in fact a Parsons doorway will, on the theory, contain as many existing fat men as are standing in the given doorway). But, in fact (i.e. on the theory here elaborated), the answer is, again, No; for Parsons would stand in the doorway along with, and next to, various merely possible objects, again contravening Brentano. However Parsons' Parsons, i.e. the set with exactly Parsons' nuclear properties could be in a Parsons doorway (e.g. in the world of Parsons' theory); but now the Brentano thesis is not violated, for, in particular, the set correlated with Parsons does not exist.

world of A's story is not the real world. No assumption postulate applies: we cannot directly determine such characteristics of the real world as what stands in actual places by storytelling.[1] For these reasons too Rescher's answer (in 1968) to Quine's how many questions, namely, "As many as are described", is inadequate (Rescher's answer is a much better answer to question (2) above, which may be the question Rescher intended to answer). Each of A, B and C describe different numbers and not all can be right, since 3 men differ from 98. That is, different, and inconsistent, descriptions may be given, and different, and inconsistent, riders may be added at a later time. Moreover descriptions are not the only way of specifying possibilia; they may, for example, be inferred from a theory. That the storytelling-assumption line cannot be right emerges more clearly if the storytellers use actual fat men in their accounts, e.g. A's story is about Herman Kahn and two other modern Falstaffs. A story about actual figures does not make that story true, except in appropriately indicated fictional contexts. As it is with entities, so also it is with nonentities.

In fact some of the differences between entities and nonentities have been much exaggerated, especially by the enemies of the nonexistent, empiricists and idealists alike.[2] For nonentities are not as chaotic, as indeterminate, or as lacking in independence as they have been represented as being; while at the same time entities are not as totally independent, as free of indeterminacy and vagueness as has been made out. The following cloud parody, which can be reworked for a great many other natural entities, is intended to draw out these points:

> ... The slum of entities is a breeding ground for disorderly elements. Take, for instance, the cloud in the sky above; and, again, the adjacent cloud in the sky. Are they the same cloud or two clouds? How are we to decide? How many clouds are there in the sky? Are there more cumulus than nimbus? How many of them are alike? Or

[1] This is part only of a larger story concerning the truth of fiction and storytelling. Of course it is true that Mr. Pickwick wore gaiters, that Sherlock Holmes lived in London, and that phlogiston is a heat substance; and it is true that James Bond stood in, or at least passed through, various doorways; and this need not conflict with the fact that no such objects exist or ever existed. The reconciliation is, as chapter 7 tries to explain, in the first place, by way of contextual differences – fictional statements are contextually intensional – in the second and associated place, by way of duplicate subjects, e.g. Bond stood in the doorway depicted in the film (which is true iff, in the world of the film, Bond stood in the given doorway), and, in third and most important place, through reduced relations.

[2] Of course, when it comes to attempts to *discredit* nonentities, the reverse happens: features of entities, such as unqualified reliability, are inappropriately transferred intact to nonentities.

> would their being alike make them one? ... is the concept of identity simply inapplicable to clouds? But what sense can be found in talking of entities which cannot meaningfully be said to be identical with themselves and distinct from one another? These elements are well-nigh incorrigible ... I feel we'd do better to clear the slum of entities and be done with it.

And so to parody Kenny also: these objections make untenable the notion of an entity. However, what should be removed is not the slum of entities and nonentities, but the classical logical economy which has reduced these solid dwellings to slums.

There are several corollaries and further points. Firstly, many of the problems that are taken to be insuperable in the case of nonentities arise equally in the case of entities, especially natural objects such as clouds and storms and waves, mountains and waterfalls and forests. But the problems are not usually seen as – and should not be seen as – discrediting entities. Thus a double standard is being applied. Questions which are realised not to present insuperable problems for entities are taken to do so in the case of nonentities, which are required to be determinate, distinct, and so on, in a way that entities are frequently not. But recall all the decision questions for entities that Wittgenstein and Wisdom introduced us to (see especially Wisdom's neglected 1953), and add some more, e.g. How wide is Mt. Egmont? Where do its slopes end? How long is a leech? How long is Plato's beard? Is this a new wave? How many mountain peaks are in the range? Questions as to precise boundaries, in particular, are very common with natural entities: these are *sometimes* settled by decision or convention, and sometimes not. Sometimes they call only for cheerful indecision.

An upshot is that common philosophers' paradigms of entities and their resulting pictures of the universe (of entities) need adaptation or, better, replacement. The paradigms of entities have too often been artifacts such as furniture and office equipment which, because human artifacts, do have sharp boundaries and determinate numerical properties, in contrast to natural objects, which frequently, in advance of specific decisions, do not. The paradigms have encouraged dictums, such as Quine's "No entity without identity" designed (unsuccessfully) to rule out such things as attributes, which seriously applied exclude many natural objects as entities. It is the dictums, not the entities, that have to go. Also to be rejected as decidedly misleading is the familiar philosophers' picture of entities as the "furniture of the universe".[1]

Why has the Identity Problem been thought to be so severe for nonentities, far more problematic than for entities. There are a number of different sources for identity anxiety, and in order to see where the sources of anxiety lie it is important to separate out these different sources for the alleged Problem. Several different aspects of the noneist theory are relevant to deal with the different sources. Thus some anxieties are resolved by use of indeterminacy, some through the theory of extensionality and identity in intensional frames, and some by making use of features which come from the Characterisation Postulate. There are at least these cases:

1. Anxiety arising from indeterminacy of identity. Some identity claims concerning nonentities are indeterminate, e.g. which of the various Faustus's of the literature are the same. From this point of view identity is simply on a par with other features of nonentities. It is felt however that this reveals an arbitrariness and perhaps chaoticness about nonentities because the property in question, namely identity, is a logical one. It is felt that the fact that some identities concerning nonentities are indeterminate makes nonentities unsuitable objects for logical treatment. This is not so, any more than it is so in the case of entities. It is simply that a satisfactory logical treatment will have to allow appropriately for indeterminacy. Further, this particular sort of worry should be resolved once indeterminacy and the way it is treated are grasped; and in fact it should be seen as a superior feature of a theory that it can take up and explain the data on which the anxiety is based, rather than simply using it as a reason for rejecting nonentities as outside the scope of a logical theory.

2. Several worries derive from the issue of criteria of identity for nonentities. The first worry arises because no distinction is made between contingent and necessary identity; it is assumed that identity relations between nonentities must be necessary identities (e.g. identity of concepts), giving rise to the mistaken charge that nonentities are nothing but concepts, and thereby making them unsuitable for intensional analysis and sacrificing much of the very substantial point of having nonentities for intensional analysis (on this see chapter 8 [Vol. 3 $_{—Eds.}$]). That the assumption is mistaken should be evident from elementary contingent identities, such as "Pegasus is what I am thinking about".

[1] For an elaborate recent sketching in of this misleading picture, see Findlay 1963, pp. 328–9, on "the universe's undeniable furniture", much of which does not exist. See too Bunge 1977.

Necessary identity is rightly perceived as generating serious problems, but the options are not perceived. The difficulties are resolved by the theory of extensional identity (as explained in Slog, §3.5, pp. 160ff., and in chapter 1 [Vol. 1 $_{—Eds.}$]), which applies to nonentities just as to entities. Then "no therapy of concepts" is required for the rehabilitation of nonentities (p. 4); nor is such a reduction at all desirable.

A worry remains. It is thought that one cannot have contingent identities between nonentities because this is identity of reference and in the case of nonentities there is no reference to be identical. This problem is resolved (in the theory of identity) by distinguishing identity of reference from extensional identity, that is identity over extensional properties. Referential identity, which can only apply truly to existing items, is defined in terms of coincidence *of* entities *in* extensional respects: it is extensional identity of entities. Thus if a and b are referentially identical "a" and "b" have interchangeable referential occurrences.[1] Since expressions about nonentities have no referential occurrences in true statements, nonentities cannot have identity of reference. But they can still be extensionally (or contingently) identical, since they have extensional properties, and extensional identity of nonentities is coincidence of extensional properties.

3. Perplexity arising from failure to see that nonentities can have extensional properties, with the result that it is thought that any two of them must be the same. The worry is resolved through assumption, in particular through the Characterisation Postulate which assigns extensional features to nonentities on the basis of their characterisation.[2]

4. Anxiety arising from the failure of nonentities to have distinctive identity criteria, different from those for entities. For example, Lambert (in 1976, p. 252) seems to think that each sort of item should have its own distinctive identity criteria. This need not be so. Different sorts of items (e.g. possibilia and impossibilia, or properties and intensional sets) may have the same identity criteria and yet be distinguished by other features, e.g. the assumption of existence leads to inconsistency in the case of impossibilia but not of possibilia, and sets differ categorially from properties in such matters as being able to have members.

[1] For a referential occurrence of a subject both existential commitment and referential transparency are required. And then truth can be assessed entirely in terms of the reference. According to the Reference Theory all *genuine* subjects occur referentially.

[2] Or through equivalents on other theories e.g. through the fundamental (assumption) postulate of Parsons' theory which assigns to each nonentity all nuclear properties of its characterising set.

Quine suspects "that the main motive for including nonentities in the domain of discourse is to escape the riddle of non-being" (p. 4; also 1959, p. 202); but since that riddle can be satisfactorily disposed of, so he thinks, by way of Russell's theory of descriptions without appeal to nonentities (p. 8; also 1959, p. 202), there is no need or ground for such expansion of the discourse domain. According to noneism he is wrong on both counts. Firstly, the Meinongian and noneist solution to the riddle is an incidental, and pleasing, by-product of a theory designed primarily for, and from, the analysis of intensional discourse and discourse about what does not exist (see chapters 1 [Vol. 1 —*Eds.*] and 2 in particular). Secondly, Russell's theory of descriptions is inadequate for such a task; for it sometimes delivers the intuitively wrong truth-value assignments. For example, it is true that Meinong thought that the round square is square, but whatever scope it is given on Russell's theory of description it nevertheless gets wrongly assigned value false. A somewhat different counterexample to the theory is the following truth: If the winged horse Pegasus does not exist I can nevertheless think of him and be aware that he is winged.

Noneists have no taste for grossly impoverished discourse – which is what Quine's taste for desert landscapes (p. 4) comes to – yet find no convincing case for populating the domain of reality with a profusion of abstractions such as sets in their transfinite multiplicity – after the fashion of Quine. (Indeed one has the feeling from Quine's work that there is *no case* for admitting that such objects as sets exist, *except* that the immensely important enterprise of scientifically essential mathematics could not get along without their existence. But, somewhat re-written in neutral form, it can.)

Quine's discussion of the ontological problem of universals in FLP is much less detailed in argument and less conclusive than his discussion of the problem for particulars. The noneist critique of Quine which follows will be correspondingly more doctrinaire and less detailed in arguments for the claims advanced. The noneist thesis is, in direct contrast to McX, that there are no such entities as attributes, relations, classes, numbers, functions, propositions and the like: none of these exist, in any sense. Even so, some items are attributes, others are numbers, and so on; and these nonexistent items play an important, and sometimes essential, role in discourse and can have a major explanatory role. Such a position, anathema to most empiricists, Quine tries, in effect, to rule out as not even an option. In this he is less than successful, as we shall try to show.

Quine's main move is to try to foist upon us a criterion of ontological commitment in terms of use of bound variables. The argument is hardly compelling. While it is true that we can easily involve ourselves in ontological commitments, i.e. commitments to the *existence* of certain things, by maintaining (saying is not enough) that *there exist* such and such things, this is not the only way in which ontological commitments can arise, nor need non-existential quantification commit us ontologically. Quine contends, however, in an inference which looks remarkably like an A-conversion, that "use of bound variables" ... "is, essentially, the *only* way we can involve ourselves in ontological commitments" (p. 12). On the face of it this contention is just false: someone who maintains that such and such thing exists commits himself just as much as someone who maintains that there exist such and such things. Quine's further argument (pp. 12–13) is that the eliminability of names and descriptions shows that names, and descriptions, are "altogether immaterial to the ontological issue". Even were names and descriptions eliminable, the argument is invalid: the support is irrelevant. For "exists" is a referentially transparent predicate, and the paraphrasing of "Pegasus", for example, as "the thing which pegasizes" does nothing to eliminate the commitment but simply rephrases it. That Pegasus exists (or does not) remains true because it is true that the x which pegasizes exists (or not). In symbols, $p = \imath xp(x) \supset . \imath xp(x)E \supset pE$. It is similar with the eliminations of descriptions. Because $xE \equiv (\exists y)(x = y)$ and $xE! \equiv (\exists! y)(x = y)$ we can quite evidently involve ourselves in ontological commitments by way of names and descriptions if we can so involve ourselves through use of bound existential variables. And all the fact that languages shunning names can be designed shows is that in such languages we would lack primitive expressive means of stating ontological commitments through names. In itself this shows nothing about the statements of such commitments in languages which are not so lacking in expressive power. The conclusion is accordingly that Quine's claim as to bound variables being the only way we can involve ourselves in ontological commitments is false.

Nor does the use of quantifiers and bound variables always involve us in ontological commitments: the use of nonexistential or existentially-neutral quantifiers does not (as SE and Slog argue). Use, for example, of the neutral quantifier "something" (which expands to "for some object x...") as in the claim "Something does not exist" in no way commits the claimant to the existence of anything ("anything" can also be used neutrally here). The appealing equation "to be assumed as an entity is, purely and simply, to be reckoned as the value of a variable" (p. 13) is as false as it is simple.

Unfortunately the false criterion pervades much of what Quine has to say about ontology and ontological problems, and renders it unacceptable. This applies in particular to what he has to say about the ontological commitments of conceptual schemes and about the problem of universals. The result in the case of the universals problem is that the noneist positions, according to which we can talk quantificationally about universals though none such exist, is entirely excluded (for more details of such a position, see Routley[2] 1975 and chapter 8 [Vol. 3 $_{\text{—}Eds.}$]). And the separation of neutral quantification from existence, as in noneism, removes what basis such assertions as the following may have had:

> One's ontology is basic to the conceptual scheme by which he interprets all experiences, even the most commonplace ones. Judged within some particular conceptual scheme ... an ontological statement goes without saying, standing in need of no separate justification at all (p. 10).

A noneist conceptual scheme, or theory, may include notions such as those of time and number, which items are definitely not assigned existence, and others where the question of existence is unknown or left open. (And even on Quine's view the latter can happen so long as quantification is eschewed.) Ontology is not so basic after all. For similar reasons fixing upon an overall conceptual scheme does not (contrary to Quine's claim on pp. 16–17) determine an ontology.

Quine attempts to use the relativity of conceptual schemes, and of what he takes to be the automatically associated ontology, to dispose of positions on universals like McX's (p. 10). But the rival scheme Quine sketches is hardly very compelling, and the serious weakness of some of his points becomes apparent if the working example is changed from *redness* to, for instance, *brittleness* or *solubility*. Brittle things have nothing in common "except as a popular and misleading manner of speaking"? The ground for assessments of brittleness extends no further than actual things that are brittle? Properties such as brittleness have no "real explanatory power"? Even more surprisingly, predicates such as "is red" and "is brittle", though meaningful, have no meaning! In "refusing to admit meanings" Quine has thereby deprived himself even of the usual semantics for applied quantificational logic which interprets predicates through universals, either attribute- or set-theoretically. McX, presumably, was not impressed by Quine's attempt to cool down the hot spot he put himself in with his vaunted rejection of meanings, and nor are we.

Meaning does *not* reduce, as Quine hopes we'll allow, to sameness of meaning unless, what is at issue, attribute abstraction is also allowed; but given abstraction, through which meaning can be recovered from sameness of meaning, redness can be retrieved from things being red, and so on. Quine no doubt hopes we'll allow too that "what is called *giving* the meaning of an utterance is simply the uttering of a synonym"; but this (pre-Wittgensteinian suggestion) is a total travesty of the range of things that would count as giving the meaning of an expression, some of which would consist in pointing to an appropriate universal.[1]

When it comes to the universals of mathematics and physical science, as distinct from those of commonsense, Quine's attitude suddenly changes. The "higher myth" of numbers and classes "is a good and useful one" (p. 18). *Truth has vanished*: in trying on one or other conceptual schemes or associated myths, we are only selecting what is simple, economical, useful and serves our various interests or purposes. Important issues among the problems of universals have been lost sight of, such as, what is true in classical mathematics, and which, if any, of the claims made as to the existence of universals, how much of classical mathematics is myth (and how precisely is this analogy cashed out) and how much can be redeemed. Don't anticipate clear answers on such issues in FLP.

Quine supposes that the intermediate and upper ramifications of his conceptual scheme will enable him to communicate successfully, e.g. with McX, on such topics as politics, the weather, and language (p. 16). Given Quine's assumptions as to what can be significantly said and the severe limitations of his conceptual scheme, this should strike one as extremely doubtful, and certainly as nowhere substantiated. Weather forecasts, for example, are frequently decidedly intensional and exhibit remote grades of modality: the intensional assessment of such forecasts, as discussion of the weather may involve, strikes the less credulous among us as even further beyond the pale of legitimate Quinean discourse and admissible myth.

Note: We find much else, of less immediate noneist relevance, to disagree with in Quine's essay, especially in the last pages. We don't agree, for example, with the unsupported claim (p. 19) that the phenomenalistic conceptual scheme "claims epistemological prior-

[1] Findlay makes a similar point (1963, pp. 325–6): "That we often discuss the sense of expressions by equating them with other expressions is of course undoubted: it remains true that the expressions with which we equate them must be *understood*, and understood in that peculiar restrictive fashion which amounts to isolating their sense".

ity". (The reasons for not agreeing include those Austin has given in 1962 and those introduced in the analysis of phenomenalism in the later part of chapter 8 [Vol. 3; cf. esp. §10 —*Eds.*].) We certainly don't agree that "we adopt, at least insofar as we are reasonable, the simplest conceptual scheme into which the disordered fragments of raw experience can be fitted and arranged" (p. 16). That is a slick, and on reflection obnoxious, pragmatico-empiricist reslanting of what is accounted reasonable. Raw experience is not all that has to be accounted for, and truly accounted for. Much depends too on whether or not "fitting" is forcing.

Chapter 4

Further objections to the theory of items disarmed

Theories of objects and items have been – and no doubt will continue to be, so long as they clash with philosophical orthodoxy – subject to a barrage of criticism and objections, often hostile. Some of these are alleged to be fatal or very serious and to show that any theory of objects which do not exist, and certainly of impossible objects, is incoherent in one way or another, or even completely unviable, or involves severe conceptual confusion. Others of these are alleged to show, what is different, that any such theory is otiose or at best uninteresting or not really worth considering. Many of these objections have been made in connection with Meinong's theory of objects. At this stage some of these objections may seem naive; but they are so frequently encountered that they require exposure.

It should come as no surprise that many of the allegedly "commonsense" objections, which are supposed to demonstrate the complete unviability of theories of items, are based, in one way or another, on the Reference Theory, or its components. In such cases it is enough to bring out this fact; for the Reference Theory and its components have already been criticised at length [cf. Vol. 1 $_{—Eds.}$]. In the long run the choice between the Reference Theory and other theories must rest on which theory provides the best explanation of the data. But in the meantime, it is methodologically objectionable to discount rivals of the Reference Theory on the ground that they fail to conform to the Reference Theory, especially when the Reference Theory has shown itself so singularly unsuccessful in accounting for the data, and so capable of generating gratuitous philosophical problems.

There is also another striking feature of many of the objections: either nonentities are expected to behave like, what they are not, entities, or else nonexistent objects and

---

The original version of this chapter was revised. Due to a mistake by the publisher, the editor's name 'Dominic Hyde' was mistakenly appeared as a co-author in all the chapters on Springerlink and third-party websites. The correction to this chapter is available at https://doi.org/10.1007/978-3-030-26309-6_7.

D. Hyde (ed.), *Noneist Explorations I*, Synthese Library 415,
https://doi.org/10.1007/978-3-030-26309-6_3

theories of them are expected to measure up to standards that are rarely or never met in the case of existent objects or theories of them.

Russell's famous objections to theories of objects, that certain objects are apt to infringe the law of noncontradiction and that the theories engender ontological arguments, have already been neutralised (by use of predicate negation and a division of properties); and Quine's well-known problem questions concerning nonexistent objects have been answered. But much remains to be done, for many other objections have been lodged. In what follows all the usual objections to theories of objects to be found in the literature (that have not been assessed in earlier chapters) and other objections as well – indeed all that have so far been encountered that have any plausibility – are considered and, hopefully, disarmed.

§1. The theory of objects is inconsistent, absurd; Carnap's objections, and Hinton's case against Meinongianism.

It is commonly objected that any general theory of nonexistent objects is absurd, because contradictory. Husserl, for example, attempted to prove to Meinong that the notion of a nonexistent object is absurd: all that he really proved, however, according to Chisholm (1968, p. 374), is that it is absurd to suppose that there *are* nonexistent objects. In fact, Husserl did not even prove this much, unless "are" is existentially loaded (i.e. is "are^{E}"), but only that it is logically impossible that nonexistent objects exist. A simple neutral quantificational model shows that Meinong's famous statement (in TO, p. 83, but without the loading shown) "There are objects of which it is true that there are^{E} no such objects", more succintly "Some objects are such that they do not exist" is consistent and leads to no absurdity; and a modal elaboration of the modelling will then show that what Meinong said is logically possible.

The assumptions of the Reference Theory, that all quantifiers are existentially loaded, that all discourse is at bottom *Sein* discourse, not only lie behind Husserl's alleged demonstration of the absurdity of any theory of nonexistent objects, but are at the base of Russell's contention (Russell 1905, p. 482ff.)[1] that *any* theory that admits inconsistent objects is inconsistent, a contention given qualified endorsement by Carnap (MN, p. 65):

[1] Russell's commitment to this contention is further discussed in the next chapter.

> Russell is certainly right in the following respect: Within the logical framework of our ordinary language, we cannot consistently apply the conception of impossible things or even that of possible nonactual things.

This is entirely mistaken: in natural language we can, and do, talk about nonexistent items (as we have seen by way of many examples), and elementary fragments of this discourse – which admit of consistent extension – can be shown to be consistent. What is Carnap's basis for conceding ordinary language, in effect, to the Reference Theory: empirical investigation? No, else he would have begun to find out that the claim is mistaken. Carnap has, as his arguments indicate, made the mistaken assumption that all ordinary discourse is referential, in Meinong's terms consists entirely of *Sein* statements. Carnap argues that any language accommodating nonentities

> must be different from the ordinary one [as] is shown by the following example: In the ordinary language we say: "There are no white ravens and no round squares". In the new language we would have to say, instead: "There are white ravens; however they are not actual, but only possible. And there are round squares; however, they are neither actual nor possible, but impossible" (MN, p. 65).

Several things are wrong with this argument. To begin with it does not establish its point, since the sentence, "There are no white ravens and round squares", can be perfectly well expressed in the formalism of the theory of objects using existential quantifiers (and moreover its equivalence to "No white ravens exist and no round squares exist" – something classical theory cannot express directly – can be shown). Nothing prevents the use of existential quantifiers, as well as neutral quantifiers, in the theory of objects, just as nothing prevents the theory's catering for *Sein* as well as *Sosein* statements. Indeed in these respects the theory simply reflects ordinary language which likewise includes both sorts of quantifiers and caters for both *Sein* and *Sosein* statements. It may be (and certainly *is* in the case of extensional discourse) that existentially loaded discourse such as "There are^{E} no white ravens", can be paraphrased – in terms of wider quantifiers (though *not* in Carnap's way, since the claim says nothing about white ravens being possible objects) – in the example as "It is not the case that some items are white ravens and exist". But that in no way shows that we *have* to use the paraphrase: most English can be translated into German but we don't have to speak German.[1]

Elementary modellings of the sort that vindicate Meinong against Husserl's charges (e.g. the models of SE) at the same time dispose of the naive inconsistency objection that any theory of objects is inconsistent because it implies that there are things that do not exist, which, it is said involves a contradiction: "to say that *something* does not *exist*, or that there *is* something which is *not*, is clearly a contradiction in terms" (Quine ML, p. 50). A contradiction only results however upon reading the neutral quantifier "some" or "there are" existentially, upon conflating "Some things do not exist" with "There exist things that do not exist".

An alternative argument, which depends not on the mistaken identification of the particular quantifier "there is" with the existential quantifier "there exists" but on direct application of the principle of existential generalisation, goes this way: Since "Pegasus does not exist" is genuinely about Pegasus according to the theory of items, Pegasus must be a genuine logical subject open to quantification; but then it follows, by existential generalisation, that there exists something that does not exist! Unfortunately for the argument the particular quantifier "P", "for some", of the theory of items is logically distinct from the existential quantifier "$\exists$", "there exists"; and though the principle $(\exists x)A \to (Px)A$, i.e. there exists an x such that A implies that for some x, A holds, its converse is certainly not valid. Thus too, though the principle (of PG), $A(b) \to (Px)A(x)$

[1] Carnap has an auxiliary argument which, if correct, would show that *Sosein* statements (and nonreferential claims) rarely, or never, occur. According to Carnap, both Meinong and Lewis mistakenly apply their distinctions

> to objects and thereby violate the rule of ordinary language which takes the addition of "actual" to a general noun as redundant. For example, the ordinary language takes phrases like "actual horses", "real horses", "existing horses", etc., ... as meaning the same as "horses", differing from this only in emphasis; and, likewise, "actual axioms" is taken as meaning the same as "axioms" ... (MN, p. 67).

There is no such ordinary language rule; the adjectives "real", "actual", and "existing" are often not redundant, as examples of nonreferential occurrence of subjects in particular show. Thus "Pegasus is a horse" is not (materially) equivalent to "Pegasus is an existing horse", "Meinong believed the round square is round" is not equivalent to "Meinong believed the actual round square is round", "Unicorns are one-horned" is not equivalent to "Real unicorns are one-horned". In any case, "real", "existing" and "actual" have different roles even in referential uses. "Real", for example, unlike "existent", contrasts with "artificial", in "Are these real diamonds?" – another nonredundant use. On the real complexity of "real", see (but on one's guard) Austin 1962, pp. 64–76.

In addition, the principle which in certain cases (referential contexts) admits the addition of "existing" without upset, is not a syntactical one but a contextual principle.

is valid for every subject "b", this in no way underwrites the principle of existential generalisation (EG), $A(b) \rightarrow (\exists x)A(x)$, which is not generally valid, and to which, as we have seen, counterexamples abound. It is not just that nonreferential quantifiers like "P" can be consistently introduced as part of a coherent theory; contrary to Reference Theory propaganda, nonreferential quantification is a frequent, and important, phenomenon in ordinary language, especially in quantification into intensional contexts. In fact quantifiers, like "P", which carry no commitment to existence, are a rather inevitable outcome of the admission of nonreferential occurrences of subjects. For then the variables of quantification will hold places for nonreferential as well as referential occurrences. Since the value of a variable on which a generalisation step is based may in fact be a nonreferential occurrence, only generalisation with quantifier "P" is *always* admissible; generalisation to an existential quantifier can only be adopted where what is generalised upon, the constant or value of the variable, occurs referentially.

In fact wide deployment of the EG principle is yet another way of insisting on the Ontological Assumption. Existential Generalisation is used to argue to the Ontological Assumption as follows: If, for example, it can truly be asserted that a round square is round, then by EG (given "a round square" *is* a logical subject) there exists something that is round. But *what* is this something that *exists*? It is a round square. Therefore a round square exists. Generally by the same argument, if x has a property then for some predicate "f", xf is true; and so $(\exists y)yf$. *What* is this y which *exists*? It is x. Hence x exists. The argument is circular. For one is only entitled to use EG *if* the subject does exist. One can't *first* use EG, and *then* conclude, since one has, that the subject exists. What is correct is not EG but only PG and the free logic principle: $xf \;\&\; xE \rightarrow (\exists y)yf$. The only logical moves that are generally applicable without qualifying premisses, such as that an item exists, are those that are common to both referential and nonreferential occurrences. These form the basis of a neutral logic, with neutral quantifiers such as "P", and "$\sim P$", "for no item". Within this neutral quantification logic it is a routine matter to show up equivocations and slides between referential and non-referential uses of such quantifiers as "something" and "nothing" – examples abound. Consider one conspicuous recent example.

In a very brash stroke, Hinton writes both the Ontological Assumption and Existential Generalization into his discussion of what he calls "Meinongianism": for example (Hinton 1972, p. 99)

> We can independently say that whether the Meinongian likes to think so or not, he is in fact claiming that there exists in Russell's quantificational sense a golden mountain, since he is claiming that at least one object of reference has a property of golden-mountainhood.

As regards genuine Meinongians, what Hinton claims is seriously mistaken, on both historical and factual counts. Hinton's attempt to rubbish Meinongian systems begins with, and depends crucially upon, an error as to the intended meaning of Russell's existential quantifier "$\exists$". Hinton objects to Chisholm's argument that

> Russell's theory of descriptions presupposes the wrongness of Meinong's theory ... for Russell's theory of descriptions requires us to interpret "the golden mountain is golden" as meaning among other things that there exists a golden mountain.

on the (astonishing) ground that Russell's theory does *not* require the latter existence claim. Hinton's case is that:

(a) Russell does encourage us to think of "$\exists$" as saying "there exists", that "is his philosophical view", *but*

(b) Russell's quantifier "$\exists$" does not mean "there exists", but reads neutrally "there is at least one" or "for some", and so captures exactly Meinong's nonexistential quantification (*es gibt*).

Unfortunately for Hinton's case Russell's post-1905 work contains no such separation of quantifiers, "there exists" and "there are", and accordingly does not include a neutral quantifier in terms of which Meinong's claims can be captured. Consider, what is perhaps definitive of the meaning Russell intends to assign to "$\exists$", the statement of PM, *9, p. 127:

> We shall denote "ϕx sometimes" by the notation $(\exists x) . \phi x$. Here "$\exists$" stands for "there exists", and the whole symbol may be read "there exists an x such that ϕx".

In a similar vein, it is said "the symbol '$(\exists x) . \phi x$' may be read 'there exists an x for which ϕx is true'" (PM, p. 15), and "An asserted proposition of the form '$(\exists x) . \phi x$' expresses an 'existence-theorem', namely 'there exists an x for which ϕx is true'" (PM, p. 20). Since Russell's theory of descriptions is formulated in terms of "$\exists$", Chisholm is right and Hinton wrong as to Russell's existential construal of "the golden mountain is golden".

What is happening in Hinton's attempt to foist the Ontological Assumption and Existential Generalisation on the Meinongian should now be clearer. Consider Hinton's summary (p. 101):

> in a nutshell: the Meinongian holds that $(\exists x)(x$ is a golden mountain), where the quantifier has its standard meaning; ...

On a straight factual count this is false. The standard meaning of the existential quantifier "$\exists$" is that given by Russell, "there exists" (see virtually any of the standard textbooks), but the Meinongian does not hold that there exists a golden mountain. Of course, with a neutral quantifier "P", corresponding to that of Hinton's case, part (b), *some* Meinongians would hold $(Px)(x$ is a golden mountain); but that in no way warrants the slide to unacceptable $(\exists x)(x$ is a golden mountain), where the quantifier has its standard, Russellian meaning. A corollary is the collapse of Hinton's case for his thesis (p. 102) that:

> the task of constructing a new logic, in which Meinongian and Anselmian arguments can be formalised to the extent necessary for their appraisal, is imaginary. We already have such a logic –

dear old ordinary Russellian logic. All that is true in Hinton's thesis is that classical quantification logic can be satisfactorily *reinterpreted* neutrally: but further important parts of Russell's logic, such as the theory of descriptions, cannot be.

But let us concede Hinton his nutshell claim for the *non*-standard quantifier "P", in order to see how his arguments against Meinongianism fare.[1] The main arguments when stripped to basics are rather pathetic; for they amount to little more than the following

[1] Hinton, although he frequently gives the impression that he is rejecting all (uncompromising) Meinongianism, e.g. in his parting suggestion that any system that is Meinongian ought to be rejected as such – directs his arguments against what he is pleased to call "resolute Meinongianism". Resolute Meinongianism is intended to capture, the last pages of Hinton's article indicate, Meinongianism which does not disappear under analysis or paraphrase. But Hinton's "resolute Meinongianism" is neither necessary nor sufficient to the task. An exponent of the contextual theory of fictions explained in chapter 7 would be a resolute Meinongian, having statements like "Something is a flying horse" true in a world of myth, without being committed to central Meinongian theses. Conversely, such Meinongians as Parsons and myself, are not "resolute" in Hinton's sense because not prepared to accept his suggestions *One* and *Two*, to the effect some Meinongian statements are "true only in the realm of myth". Both "only" and "in the realm of myth" are objectionable. According to some Meinongians both "Some thing is a flying horse" and "Some object is round and square" are true simpliciter; and neither true in the realm of myth nor true only there. The position here adopted is that truth just is truth simpliciter, that expressions of the form "true in this or that world or realm" are misleading and best avoided. Hinton's account of when a system is Meinongian is similarly defective, since defined in terms of resolute Meinongianism.

shift in onus of proof: Meinongian claims "are unwarranted, because they need a good ground and have none" (p. 105; repeated, with a little embroidery, p. 107). But there are excellent grounds for such claims. Take (from §1.4 [Vol. 1 $_{\text{—}Eds.}$]) any counterexample af to the Ontological Assumption, e.g. "The winged horse is believed by Parsons to be winged", "Pegasus is self-identical according to free logic", "Zeus does not exist"; then particularise, to get $(Px)xf$, e.g. "There is at least one object such that – –". Given his logic Hinton can hardly object to particularisation (his EG), the usual point of resistance. Thus, as there is good independent evidence for the truth of the claim of "Zeus does not exist", there is good evidence for the Meinongian claim that there is at least one object that does not exist.

Hinton does outline one supplementary argument for his rejections of Meinongianism (and simultaneously exhibits his real predilection for the standard existential quantifier, and his prejudice in favour of the actual) in his "curt" criticism of the perennial philosophical idea from which Meinongian claims are, he thinks, *often* derived, namely "the idea that a meaningful description is the same thing as a description to which something answers" (p. 107).[1] The "grim consequences" of this idea are, according to Hinton, firstly, that we are always wrong in seriously saying "Nothing whatever answers to that (perfectly meaningful) description", and, secondly, that "'talking about things that do not exist' could not after all consist in *using, in the grammatical subject position, a description to which nothing answers*" (p. 107). Neither of these consequences ensues, however, given the distinction already observed – that Hinton has been trying to plaster over – between existential and particular quantifiers. Simply represent the critical quantifier "nothing" in each consequence by $\exists$, i.e. *nothing* here amounts to *nothing existent.* Then the perennial idea implies neither that "we are always wrong" – because something nonexistent may answer to the description and this is perfectly compatible with nothing existent answering

In fact none of these defects is material to the main issue, because Hinton's arguments against Meinongianism do not depend essentially on his notion of "resolute Meinongianism". Accordingly the main text is not burdened with the latter bit of technical jargon.

[1] Versions of this thesis are rather easily proved in Meinongian logic, and so in simple adaptations of Hinton's "Russellian logic" (the versions depending on how expressions like "answers to" and "meaningful" are spelt out). One argument is as follows: If $(\imath x)xf$ is significant then some properties, e.g. analytic ones, are true of it. Moreover every property it has it has, so $(\imath x)xf = (\imath x)xf$, whence, particularising, $(Py)(y = (\imath x)xf)$, i.e. something answers to the description. Conversely, if $(Py)(y = (\imath x)xf)$ then, by quantification logic with identity, $(\imath x)xf = (\imath x)xf$ (for let y, be such a y; then $y = (\imath x)xf$ & $(\imath x)xf = y$), so $(\imath x)xf$ has a property, and so is significant. If Hinton were as philosophically neutral as he pretends in trying to have the Meinongian committed to existential claims, he would be much less ready "to take for granted" the wrongness of the perennial philosophical idea.

– nor that talk about nonentities could not be accounted talk about nothing – because of course it is, talk about nothing existent. The supplementary argument accordingly fails.

§2. The attack on nonexistent *objects*, and alleged puzzles about what such objects could be.

What often strikes one (though the impression is something of an illusion) as the most basic objection to the theory of items attacks the notion of a nonexistent item or object. "Nonexistent objects are not *objects*" (Linsky 1977).[1] But Linsky's main ground of objection, a variation of Quine's "No entities without identity", has already been met and identity criteria for objects supplied. More threatening is Linsky's charge:

> The whole idea of an item which does not exist is unintelligible; to say that something does not exist is to say that there is *no* such item, to point to the absence of an item, not to say that some item has the property of not existing. The absence of an item cannot be regarded as yielding any further item, any more than a hole can be regarded as a sort of object. And how can the absence of an item have properties – there is nothing to have them. *What* has the property?

The objection makes the usual assumptions of the Reference Theory, that an item is a pure reference having only referential properties, and that all quantifiers ("no", "nothing", "what") are referential; and if these were true the objection would be correct. But an item is not a pure reference: objecthood is not characterised (only) referentially. To say that something does not exist is to say that there is no *reference*, and to point to the absence of a *reference*, but this does not guarantee that there is no item, and would do so only if an item were a reference and all properties were referential. Since an item can be viewed as the sum of *all* its properties, referential and nonreferential, the absence of reference and referential properties still leaves something to have the properties, and many properties to have, namely those held nonreferentially. Because of this *a nonentity is not a mere nothing.*

An object, even though it does not exist, is a thing, with features, which distinguish it from other things, and also relate it to other things; commonly it is a thing or unit

[1] One might almost as well say "Microphysical objects are not objects". There are not dissimilar puzzles about the identity and difference of some of these.

which can be thought of or apprehended. A nonexistent object may thus be an object both in the sense of the medieval Latin *objectum*, meaning "thing presented to the mind", from which the modern term derived, and in the modern sense of "thing thought of or apprehended as correlative to the thinking mind or subject" (OED). Similarly nonexistent particulars really are particulars and nonexistent individuals are individuals.

Such details hardly provide a "deep" account of object; but then a deep account is not needed to meet the objection. A fuller picture of objects does of course emerge from the discussion of the properties and sorts of properties objects have and can have and in the classifications of the varieties of objects. But even this does not offer a deep account, it does not furnish the quiddity or whatness of objects, or give the essence of objecthood.[1] Nor should a deep account really be expected. For deep accounts (e.g. in terms of material substance, individual essence, primary existence) invariably turn out to be reductionistic in orientation and to unduly narrow the class of objects, and commonly turn out to incorporate the Reference Theory. An example will bring out the general point. One deep metaphysical account of *object*, which runs way back, has it that an object is a material substance, a corpuscular body, usually with certain primary properties. Since everything is an object, materialism is the consequence: all there is, or that exists (since "material" quickly conflates many distinctions) is, ultimately, material, stuff, and "whatever is not stuff is nonsense".[2] As the modern philosopher Hobbes put it, back in the seventeenth century (1909, pp. XLVI, IV):

> ... the universe, that is the whole mass of all things that are ... is corporeal, that is to say body, and hath ... dimensions [and parts

[1] A certain Aristotelian turn is easily introduced into the theory, but the connections do not go deep or cast much light. Firstly, every object has some properties, at least logical ones. It is true that whatever conditions, including logical ones, are imposed on objects in general one can think of objects that *purport* to break them, e.g. objects described as breaking basic logical principles. But such objects do not, despite their specifications, break all conditions. (It would be bleak for a logical theory of objects if they did.) Further, each object will have its own properties, as distinct from those that hold of its kind and for all objects. Therefore, for every object x, x is thus and so, "thus and so" indicating its properties. But thusness is the property of being thus and so (i.e. $\lambda x(x$ is thus and so$)$). Hence every object has the character of thusness. Thusness is, in a sense, the essence of objecthood; thusness gives the whatness.

The emphasis on properties and character of objects does not imply that objects are, in some fashion, bundles or complexes of properties (without the glue of substance). All that is true, as later explanations try to make plain, is that objects may be *represented* as sets of properties (and since a set is a thing, a property adhesive is not really missing).

[2] Flew's phrase; 1971, p. 45.

> which are corporeal] . . . . And because the universe is all, that which is no part of it is nothing; and consequently nowhere.

All objects which do not succumb to materialist reduction – there are *many*, beginning, perhaps more conspicuously, with the abstract objects of mathematics and the impossible objects of logic – are left out on this narrow account (see further chapter 9, §13 [Vol. 3 $_{—Eds.}$]). Objects are the most general items of designation; they include whatever something can be said about.[1] An object is thus whatever can be quantified over; it is something, and it always makes sense to ask *what* it is, and to answer that it is *that*. It is such features, together with the assumption that all designation and quantification (including that of quantifiers like "for what x", "for that x") is existential, that have characteristically led to the objections to theories of objects that they are platonistic and involve damaging subsistence theories.

§3. The accusation of platonism; being, types of existence, and the condition on existence.

A widespread objection to theories of objects such as Meinong's is that they involve unrestrained platonism. Very often this objection is based on a misunderstanding or

[1] Contrary to Findlay (1963, p. 342), it is no "conclusive proof" that it is "unsatisfactory to hold that golden mountains or round squares are *objects*", that there are difficulties in Meinong's theory which he met by his doctrine of the modal moment. It would have to be shown, what Findlay does nothing towards, that the difficulties are inherent in a theory of objects and are damaging.

The rest of Findlay's case (1963, p. 340ff.) against accounting Meinong's objects *objects* will bear no more weight. Objects are not all mere objects of thought and cannot all be assimilated to such objects or to ideas; objects can be the independent subjects of assertings; objects do not have being (Findlay's attempt to attribute being and extra-ontological status to Meinong's objects, p. 343, is in fact inconsistent with his earlier account of the theory of objects). Meinong did not misunderstand the grammar of such terms as "objects".

Findlay even accuses Meinong of falling

> victim to the same prejudice [as] Moore and Russell, the same determination to "analyse" intensional situations into a relation among existents (p. 345, re-arranged);

and to this he attributes "the many absurdities of the theory of objects". But to find "a vicious existential inspiration" in Meinong's theory is completely misguided. The theory of objects need not involve, and Meinong's theory does not involve, any such attempt to analyse intensionality away. The main reason why Findlay arrives at this astounding accusation is that he imports a referential account of relations, which he attributes to Meinong, and combines it with Meinong's perfectly legitimate treatment of intensionality as a relation (though not a "relation" in Findlay's narrow sense). According to Findlay (p. 343) "relations obtain only when all their terms exist, and for such obtaining it does not matter under what description their terms are conceived." Findlay's relations are, that is, entirely referential relations, a small subclass of relations in the customary logical sense. Intensional relations are not, in general, Findlay relations, but they are quite legitimate, and use of them has none of the evil consequences Findlay alleges.

misrepresentation of Meinong's theory, and on taking the theory as claiming that objects are entities, that they all exist in some way, somewhere, etc. For example, it is said that Meinong "countenances impossible things" (MN, p. 65) and that on his theory unreal objects such as the golden mountain and the round square "have some kind of logical being" (MP, p. 169) and in fact "the Universe is populated by a variety of entities with most surprising properties" (Passmore 1957, p. 186). Furthermore there are many oft-reported witticisms, which are the source of much undergraduate amusement, about overlush landscapes, bloated universes, ontological slums, fantastic hierarchies of non-actual entities, population explosions, and so forth. Of course a theory of objects can easily be made to appear ridiculous if it is misrepresented in these referential terms, so that objects are presented as "queer entities", which if not found in the actual Universe are at least furniture of some similar counterpart of the actual world, and if not here are around or out there. Insofar as these sorts of objections are simply based on a misunderstanding of Meinong's position, or what a theory of objects is about, they can be met by pointing out that Meinong did not assume, and often and explicitly denied that, his nonexistent objects exist or have anything approximating to existence: they are neither entities nor queer entities; and have no kind of being or reality. Because very many objects are beyond the actual they are not part of the Universe at all or found anywhere in the world (cf. Meinong's remarks in TO). The richness and diversity of objects far outruns that of entities of the Universe, and reflects rather the richness of the universe of discourse, of what we talk and reason and think about: once this is grasped there is nothing excessive or ludicrous about the richness and variety of objects.

Normally however this sort of misrepresentation is not based merely on misunderstanding, and cannot be cleared up just by restating the theory of objects: the insistence on describing positions like Meinong's in *these* referential terms can only be explained as a further manifestation of the Reference Theory, the assumption that intelligible discourse *must* operate referentially. The assumption, now evident in causal theories of reference, is already conspicuous in Russell's criticism of Meinong (MP, p. 169):

> It is argued, e.g. by Meinong, that we can speak about "the golden mountain", "the round square" and so on; we can make true propositions of which these are the subjects; hence they must have some kind of logical being, since otherwise the propositions in which they occur would be meaningless.

The assumption has of course persisted, and is woven into the bottom of recent causal and historical explanation theories of reference; for example, it is for this reason that Donnellan is led (1974, p. 12) to reject what he calls (at first) "the natural view of many uses of ordinary singular terms":

> If I say, "Socrates is snub-nosed", the proposition I express is represented as containing Socrates. If I say instead "Jacob Horn does not exist", the "natural" view seems to lead to the unwanted conclusion that even if what I say is true, Jacob Horn, though non-existent, must have some reality. Else what proposition am I expressing?[1] The "natural" view thus seems to land us with the Meinongian population explosion.

But the assumption is, as we have seen (in chapter 1 [Vol. 1 —*Eds.*]), mistaken: neither the significance, nor the truth, of assertions of the form af require that a exists (in some way) or has some reality.

As should be well known, Russell confused – in a way that persists,[2] and that has in fact done much damage to the credibility of theories like Meinong's – Meinong's theory, according to which many objects have no being of any kind, with his own early theory, outlined in the *Principles of Mathematics* 1937, according to which being belongs to all objects.

> Being belongs to whatever can be counted. If A be any term that can be counted as one, it is plain that A is something, and therefore A is (1937, p. 449).

[1] The answer is disarmingly simple: the proposition that Jacob Horn has the property of not existing. Jacob Horn does not require some reality to have properties.

Donnellan appears to think (see also, pp. 26–7) that it is a merit of a theory that it maintains an appropriate distance from Meinong's theory of objects (Meinong lives, so to speak, on the wrong side of the tracks), and that if a connection of a theory with Meinongian objects can be turned up, then the theory stands condemned, and that is the end of it (cf. §14).

[2] Thus,

> Some philosophers of a later date, in particular Alexius Meinong and Bertrand Russell at one period of his life, have said that terms like "chimaera" and "golden mountain" stand for objects which possess being of a kind, though not existence (Kneale[2] 1962, p. 262).

William of Shyreswood, whose theory of supposition appeared to allow for terms about what at no time exists and so may have included rudiments of a theory of objects, then gets a warning from the Kneales: "he might find himself committed to some such extravagance"!

But the argument relies on a version of the Ontological Assumption, namely, A is something therefore A is – as if the "something" could be simply detached. If "A is" means, as it ordinarily does where "is" is intransitive, "A exists", then the assumption is open to all the earlier objections to the OA.[1] If "A is" means something less than "A exists", the question is what is meant? For Russell's argument to succeed in assigning being to all (countable) objects, it has to mean more than "A is an object"; for this does not guarantee being (without illegitimate detachment of "an object"). At least it does not guarantee being in a sense with any stuffing, for the being guaranteed is simply being an object. For Russell's purpose "A is" has to ensure "A has being" – whatever this means – so the assumption implies that if A is something A has being, a watered-down version of the Ontological Assumption, and objectionable on this score. The assumption is, moreover, inconsistent with Meinong's thesis (M2) [cf. Vol. 1, p. 3 $_{-Eds.}$], that some objects have no (kind of) being whatsoever. Let A be such an object: then it is true that A is something, namely some object, and also true that it does not have being, countering Russell's assumption. Russell argues however as follows:

> "A is not" must always be either false or meaningless. For if A were nothing, it could not be said not to be; "A is not" implies that there is a term A whose being is denied, and hence that A is. Thus unless "A is not" is an empty sound, it must be false – whatever A may be, it certainly is.

But the argument is a *petitio*, and twice assumes the watered-down version of the Ontological Assumption, that if A has features, such as not having being, A is, A has being. The fact is simply that "A is not an object", if well-formed and significant, always yields a false statement, a statement about A whether or not A has being; for A, though having no being, is nevertheless not no-thing.

A more theoretical argument, designed to convict Meinong of platonism, proceeds by arguing, on the basis of criteria for what counts as platonism derived from the Reference Theory, that whatever Meinong actually said, whatever words he actually used for "exist", his position is tantamount to platonism. This argument proceeds by taking a platonist as a person committed to the existence of items which do not or only doubtfully exist, and then characterising commitment to existence in such a way that Meinong can

[1] The evidence from elsewhere, e.g. the account of the world of being in Russell 1912, p. 165, where *being* is equated with *subsistence*, is that Russell thought of being as a sort of timeless existence.

be represented as committed to the existence, or being, of all his objects. The criterion for commitment to existence adopted may be that a person is committed to the existence of item a if he is prepared to talk about a ("countenance" a), "recognise" a, attribute properties to a, quantify over a or employ it as the value of a bound variable, or admit "a" as the proper subject of a true statement. It is certainly true that Meinong was prepared to do all these things.[1] But it is quite unsatisfactory to argue on this basis that Meinong's position was a platonistic one; for clearly such criteria for commitment to existence depend upon acceptance of the Reference Theory. They not only unacceptably assume the correctness of the Ontological Assumption and other components of the Reference Theory, but involve the further fallacy of assuming that everyone accepts them. For even if the Ontological Assumption were true it would not follow that the position, i.e. philosophical opinion, that nonentities have properties is the same as the position that nonentities exist. For even if p is (materially or strictly) equivalent to q, that someone believes that p, asserts or claims or is committed to the proposition that p, does not imply that he believes, asserts or claims or is committed to the proposition that q. A similar *invalid* principle is involved in the familiar characterisation of the platonistic position as one of being prepared to *talk* (in some way, e.g. quantificationally) about nonentities.

When platonism is adequately characterised, it is clear that Meinong's position in the theory of objects is quite distinct from a platonistic one, both in its motivation and its effects. The motivation for platonism, like that for the other standard positions on universals, is provided by acceptance of the Ontological Assumption, and the problem to which it is directed is the gap which appears between what actually exists and what one can apparently make true statements about.[2] The platonist attempts to bridge this gap by widening his conception of what exists or enlarging his claims about what exists. The nominalist tries to bridge the gap from the opposite direction by fashioning his truth claims to fit what does exist, that is by forbidding any enlargement of the class of objects he takes to exist and dispensing with discourse not reducible to statements about these objects. Conceptualist and idealist positions typically try to bridge the gap by a reanalysis providing new abstract or ideational subjects in place of the disputed subjects; the new

[1] Chisholm, in his valuable defences of Meinong, seems prepared to do only *some* of these things (roughly, those that can be represented in free logic). For example, he carefully avoids quantification over nonexistent objects (see especially 1972) and concedes to Husserl the claim that "it is absurd to suppose that there *are* nonexistent objects", (1968, p. 374).

[2] A fuller discussion of universals will be found in chapter 8 [Vol. 3 —*Eds.*].

subjects are said to exist, though perhaps in a different sense. What all these standard positions have as a common assumption is the Ontological Assumption; and without the Ontological Assumption the standard moves in the Universals game are pointless, as Reid long ago realised (cf. Reid 1895, p. 368ff.; and see also the beginning of chapter 1 [Vol. 1 $_{—Eds.}$]). Meinong's position, like that of Reid, differs markedly from the standard positions, and succeeds in evading many of the difficulties in these positions. For it is able to maintain, like platonism, much of discourse; but, like nominalism, it can maintain a scrupulous account of what exists.[1] Because what exists and what can be the proper subject of a true statement are no longer tied together once the Ontological Assumption is rejected, it is no longer necessary to give the platonist's extravagant account of what exists in order to have or explain true statements in which nonentities appear as proper subjects.

The rejection of the Ontological Assumption does not give Meinong's position a merely terminological advantage or claim to novelty, nor does the position differ merely terminologically – or by an elementary translation – from platonism (as the official positions are inclined, quite mistakenly, to claim). The distinction is not merely terminological because the question of what exists is not, as for instance Quine's criterion for ontological commitment might suggest, completely uncontrolled by conditions: one cannot say what one likes about what exists. For to exist is to be in the actual world,[2] and the logical properties of entities are controlled by those of the actual world. Hence these conditions which are sometimes taken to derive from the logical features of the actual world: *first*, what exists is consistent; *second*, what exists is suitably determinate; and *third*, what exists is unqualifiedly defensible, e.g. if an [the] x which fs exists then an [the] x which fs does f. In short, in the case of an entity we do not require further guarantees about the suitability of its description: the guarantee is provided by its existence.

[1] Quine claims (in FLP) that any attempt to resolve the universals issue along these lines is facile, presumably because it breaks the referential rules of the game – which reference theorists feel are *the* rules of the game. But, to paraphrase Wittgenstein 1953, the deep reason why the common phenomenon of true statements about nonentities such as universals is felt to be such a problem (and a problem to which the standard theories provide no satisfactory solution) is because of the hold of a mistaken theory of meaning; so that a resolution which challenges such a deeply-entrenched theory does not automatically count as facile.

[2] Strictly, it is the actual world considered as the totality of entities in their interrelations, the one true world of the empiricist faith. Because the actual world appealed to is this world there is some (unsurprising) circularity in obtaining the (independently uncoverable) conditions for existence in this way. But for the "derivation" other features of the actual world, drawn from other characterisations, are required.

The platonist then must transfer these conditions to all the objects he claims to exist, a restraint in the treatment of such items which does not apply in the case of the theory of objects. Since Meinong abandons these necessary conditions on what exists in the case of nonentities his position is not platonistic in any good sense; for his nonentities cannot be treated as if they existed. This is especially conspicuous in the case of impossibilia, since they are inconsistent; even the most ardent platonists are reluctant to say that the round square exists. For how can what is *im*possible exist? Similarly the provision of qualifications on the Characterisation Postulate in the case of nonentities shows that such objects are not being treated as existing – because such qualifications are not needed for entities. For existence in the actual world, i.e. possession of a reference, is a sufficient guarantee that a bottom order object is consistent and defensible, that a description which applies correctly to it cannot lead to logical trouble (because the actual world is, logically at least, it is supposed, trouble-free at base).[1]

If the platonist abandons these conditions which control existence, then, even if he continues to use the word "exist", his position is platonistic in a terminological sense only. If, on the other hand, he retains these conditions on his notion of object then, even if he abandons the word "exist", his position is platonistic in that he treats his items as if they existed; he takes over and applies to nonentities an inappropriate logical structure, and one which is *dangerous* when applied to nonentities because essential safeguards are lacking; for example in the case of the Characterisation Postulate, reference is no longer available to provide a guarantee of assumptibility, yet no other safeguards on the descriptions are provided. The real perniciousness of platonism is that, by his existence assumptions, the platonist is enabled to transfer unchanged to nonentities a logical structure suitable only to entities and evolved in that case. Thus the platonist, instead of constructing a new logic for the new class of objects, merely tries to transfer the old one which is entirely unsuitable. The transfer is unwarranted since not even consistency is usually established, or even *can be* established; and it is pernicious, since it may lead to trouble through any of the three necessary conditions for existence. Firstly, the transfer of the determinacy requirement entitles one to expect answers to a great many questions to which there are no determinate answers; hence too the mysteriousness of platonic entities (cf. again Reid). Secondly, the transfer of consistency requirements

[1] The point does not extend to higher order objects which may, in a paraconsistent setting, be paradoxical.

entitles one to expect consistency where none has been established, and mostly where it cannot be established because of inconsistency or because of classical limitative theorems such as Gödel's result. Thirdly, the transfer of assumptibility entitles one to expect that controls of various sorts on admissible descriptions are not needed; and so it leads both to ontological arguments and to unqualified existence axioms like the abstraction axiom in naive set theory.[1] It is not fanciful to see in this transfer to nonentities of the inappropriate logical structure developed for entities, with its absence of adequate conditions on descriptions and guarantees of trouble-freedom, the source of many grave problems which beset modern logic and the foundations of mathematics, particularly the logical paradoxes.

The necessary conditions on existence are neglected in the ever-popular kinds-of-existence doctrines, under which nonexistent objects are alleged to have existence of some other kind. Thus objects drawn from fiction exist in fiction, mental objects exist in the mind, theoretical objects exist as part of their theory, etc. The doctrine admits of ready send-up, for kinds are easily multiplied up; e.g. paradise and holes have geographical existence, God may not exist but he has religious existence, the superego and all sorts of factors and effects exist in psychology, absolute values exist in objective morals, canned peaches exist as grocery supplies, and perfect spuds and indigo-flowered plumwoods have agricultural, or is it horticultural, existence. It is here too that the subterfuge of saying that mathematical and theoretical items have *mathematical existence* is likely to be introduced, in order to escape the argument (of §4) that very many objects of mathematics and theoretical sciences have properties though they in no way exist (i.e. they do not exist). Thus, it is said, aether and phlogiston do not exist, but since they could no doubt take their places in consistent mathematical theories they have mathematical existence (cf. Hilbert). But if they *don't exist* what recommends the terminology "mathematical existence"? Nothing much: the terminology is otiose. The relevant distinction can be more adequately and less misleadingly marked using familiar terms such as "consistent" and "items of a consistent theory" – and, more generally, "has a place in the subject matter of such and such a discipline or theory". The much favoured mathematical existence terminology is, like subsistence terminology, misleading, because items such as phlogiston and point masses do not exist (unless again implausible platonism is admitted).

[1] Bernays argues in 1964, p. 277, that unrestricted platonism just does lead to an unrestricted domain and thence to naive abstraction axioms, and accordingly is refuted by logical paradoxes.

In fact the terminology is unnecessary in a stronger sense. For what motivates the introduction of talk of "k-ly existence" and "existence in k"? The idea is that whatever is talked about must somehow exist. It is the Ontological Assumption once again that motivates the kinds-of-existence doctrine; and with its rejection the need for the doctrine – as a way of guaranteeing existence of a sort while trying to sneak around the problems of platonism – disappears.

Nor is it just that the doctrine is misleading and unnecessary: it engenders its own (gratuitous) problems. Firstly, it leads to violations of the necessary conditions on existence. Consider, for example, the consistency qualification commonly imposed on mathematical existence. The qualification is wrong, because impossible objects can be, and sometimes are, the objects of mathematical investigation and theorising (in theories that are, or may be, inconsistent). But how can impossible objects have, as they should have, mathematical existence? The tenuous linkage of mathematical existence with existence is also challenged in the second problem, that "exists mathematically" unlike "exists", is not transparent, and so does not really express an existence notion at all. Consider the true factual identity, 12 = the number of Apostles. Though 12 exists mathematically, the number of Apostles does not. But how can something exist in a given way if things identical to it do not? Thirdly, there is a much larger problem, the problem of the interrelation of objects with different sorts of existence.[1] How can object a which has l-existence (extensionally) relate to object b which exists k-ly, without l-existence transferring to b and existence k-ly to a, i.e. without the kinds collapsing? And, in any case, how do objects of the different kinds relate and interact?

§4. Subsistence objections.

Many of the same points that go to show that a theory of items is not platonistic also serve to show that the theory need not be, in any damaging sense, a subsistence theory. Yet Meinong is said to have had a subsistence theory – and this has been taken to be very damaging – almost as often as he has been accused of platonism. As regards Meinong's mature theory, the charges are unjustified; even so they have been written

[1] The problem can be restated as a *many-worlds* problem; for each kind of existence there is a domain of objects of that kind, or a "world" in Wittgenstein's sense of a totality of things. The problem, which is further considered in chapter 12 [Vol. 3 —*Eds.*], is examined in some interesting (if perplexing) detail in Passmore 1970, chapter 3, "The Two-Worlds Argument".

into the received history of modern thought. It is true that Meinong's very early theory, with its attribution of *Quasisein* to every object, was a subsistence theory in much the same way as Russell's early theory. But Meinong subsequently rejected the notion of *Quasisein*, and it was his settled belief that many objects have no (sort of) being or subsistence (cf. Findlay 1963, p. 47). To the extent, then, that the subsistence objection to Meinong depends upon the claim that Meinong used, as he is popularly supposed to have, the term "subsist" ("bestehen") to cover all those objects which did not exist, it is textually mistaken. For Meinong only said of certain objects that they subsist; the term was only applied to bottom order objects that at some time existed, and in the case of higher-order objects the subsistence distinction was used to mark out only certain classes of objects; for example, objectives were said to subsist when and only when the corresponding proposition was true, so "false" objectives did not subsist and subsistence could in the case of objectives be said to amount merely to being the case. In fact, by such shifts, the subsistence commitments of Meinong's theory – which is a *selective* subsistence theory – could (as suggested earlier, pp. 5–6 [Vol. 1 —*Eds.*]) be largely eliminated. The result would presumably be a theory, like that of the present work, in which the notion of subsistence did not figure essentially. How, with such theories, does the subsistence objection gain any momentum?

The subsistence objection is the charge that the items of a theory of objects must all be taken to subsist, to exist in some inferior second-class way, if not to exist. To assess the objection it is essential, first of all, to get some agreement over what is meant by "subsist", otherwise it is quite unclear that it is wicked to have a subsistence theory. Subsistence normally means an *inferior, second class* or lowly mode of existence ("subsistence farming", "subsistence level"), and this surely is what is philosophically objectionable about supposing items which do not exist to subsist. They are supposed to be nebulous items in some second limbo. What is really wrong with having a subsistence theory is that it attempts to solve the difficulties engendered by the Reference Theory and the Ontological Assumption by a retreat to a second-class or lower-grade Reference Theory with a lower-grade Ontological Assumption. But what has to be explained and accounted for is nonreferential occurrence, not having a reference at all, not the possession of some lower-grade reference. A subsistence theory would (try to) treat its subsisting items just like existing ones, but as of lower grade. In this sense of "subsistence" a subsistence theory is indeed a bad thing, but in this sense the *Automatic Subsistence Objection*, the view that any theory which treats items which do not exist is thereby *automatically committed* to their subsistence, fails. For of course one can treat them without treating them as second-class existents, without treating lack of reference as second-class, lower-grade reference. To attribute a property to a non-existent item is not thereby to assume it to subsist if it is the *appropriate* sort of property for a non-existent item to have – we can, for example, say that we are thinking about a round square, the number 7, Queen Hatshepsut, or simply something which does not exist, without in any way supposing these items to exist in a lowly way, or to subsist. To suppose otherwise is to make the Lower Grade Ontological Assumption, that whatever has a property must subsist. The motivation for this, as for the whole Automatic Subsistence Objection, is the supposed need to supply a reference – if not a first-grade obvious reference, then a second-grade one. The items must be taken to subsist, it is thought, because it is impossible to conceive of the business operating any other way than referentially – so if there is not obvious, explicit, immediate first-grade reference there *must* be non-obvious, or mysterious, remote, disguised or second-grade reference – subsistence. In this form the subsistence assumption is motivated by and draws its strength from the Reference Theory. There is no mystery, for example, unless we try to account for nonentities in referential terms.

There are two ways the automatic objection may be developed. On the one hand, "subsistence" may be used in the sense of lower-grade existence (and a criterion specified, perhaps, or some contrasts provided, so it is clear what fails to have lower grade existence); but in this case the items of the theory do not automatically subsist, and some of them may not. It is similar when subsistence is taken, as by some philosophers, as involving a dishonest existence assumption; though subsistence is then objectionable, the theory of items is not guilty of it, certainly not automatically. If, on the other hand, "subsistence" is taken to mean merely the attributing of properties to some items which do not exist, or the treatment in the theory of such items, it is quite unclear that one is involved in treating such items as if they existed or as having second-class existence. Hence, although on this low redefinition of "subsists", a theory of items *would* be a subsistence theory, what is now wrong with being a subsistence theory other than its preparedness to violate the Ontological Assumption? The answer is Nothing: not much can be made of the use of the word "subsist" or "being" (even if it did occur in the present theory) to cover those objects which do not exist, unless it can be independently argued that there is something unsuitable about the treatment accorded "subsisting objects". But there are things wrong with the adoption of such a low redefinition of "subsists" which breaks most of the usual conditions on, and connections of, subsistence. Most important, even a second-grade or low-level variety of existence, such as subsistence, has to meet the general conditions set out on existence in order to count as a variety of existence. But this the redefined notion fails to do, e.g. "subsistents" such as impossibilia are not consistent, so the redefinition is inadequate.

Although the Automatic Subsistence Objection fails (ultimately because it is based on the Reference Theory), other more specific subsistence objections purporting to show that various theories involve subsistence are not thereby invalidated. Such subsistence objections will try to show that a supposed nonexistent is treated as if it did exist, or too much like an existent item, particularly by the attribution to it of properties which only an existent item could have. This sort of *Damaging Subsistence Objection* can have real force. For if some properties are, as it is reasonable to suppose, held in virtue of an item's existence or if they are referential properties, then nonexistent items cannot have them. To attribute such properties to these items, then, would be to treat them incorrectly as if they did exist; and a theory which included such properties might well be labelled a subsistence theory, and criticised on the ground of including such properties. Notice how such a subsistence objection is backed up, in terms of appropriateness of treatment of objects, and inappropriate assimilation of nonentities to entities. Meeting such subsistence objections is a more painstaking task; for what they really call for is further elaboration of the theory of items, especially consideration of what sorts of properties are appropriate to nonexistent items and what inappropriate, and which properties are determinate of which nonentities and which are not. As to the second, a well-attested fact, the theory of items must take account of, is the indeterminacy or incompleteness of nonentities. Once the revolutionary step of allowing nonexistent items to have genuine properties has been taken, an immediate problem is to prevent them having, determinately, too many properties, and the wrong sort of properties. (It is primarily a problem of the conditions on assumptibility, and of the coherence of theories with stronger Characterisation Postulates.) To solve this problem is to meet damaging subsistence objections.

To meet these objections is to expose a theory to other objections. If nonentities do not have too many properties then they certainly have too few, at least for logical and scientific purposes. The apparent dilemma is of course met by a theory in which nonentities have just the right properties, neither too few nor too many, but just those they in fact have. Such a theory may be hard to come by, but it is perhaps not so difficult to approximate to it.

A closely related group of objections attempts, then, to exploit the features by which nonentities *differ* from entities in order to discredit them, and to show that they are unsuitable for logical or scientific treatment. The mistaken referential assumption behind this is that the logic and behaviour of nonentities must parallel that for entities, and that because nonentities do not satisfy the relevant conditions they are unsuitable for treatment by any logical theory. But that nonexistent objects are their own sort of thing, and that they certainly differ from entities in important respects, does not render them unintelligible, illogical, disorderly, illbehaved, incorrigible ... arbitrary, capricious, and at best a source for jokes. Two characteristic features of nonentities are, however, regularly employed in arguments designed to establish their defectiveness, *firstly* that they cannot stand in physical relations to entities, and *secondly* their incompleteness or indeterminacy.

§5. The defects of nonentities; the problem of relations, and indeterminacy.

From the first "defect" of nonentities spring the jokes, e.g. "Watch out for the nonactual automobiles as well as the actual ones when you cross the road!", "You won't starve: you can eat a possible breakfast!", "I can't afford an actual swimming pool this year so I'm putting in a nonactual one instead; it's cheaper and less environmentally damaging!" – and puzzle questions, e.g. "How many angels can stand on the point of a needle?", "How many possible fat men are standing in that (actual) doorway?". Both the jokes and the puzzles purport to show that nonentities are not to be taken seriously, and that from a theoretical standpoint nonentities are incorrigible (FLP, p. 4). But the fact, emphasized first in modern times by Brentano, that nonentities cannot collide with, stand in, or be eaten by, entities both helps to explain why it is a joke to say "Watch out for the nonexistent cars at the next intersection!" and why it is silly or sick to worry about collisions with nonactual automobiles, and enables questions like Quine's to be answered (as in chapter 3). For if merely possible objects cannot be physically related (in an entire way) to entities then, as argued, the number of merely possible fat men standing in an actual doorway is zero. It would however be quite invalid to conclude from the fact that nonentities cannot have properties such as these, namely entire physical relations with entities,[1] that they cannot have *any* properties, or that *all* attempted attributions of properties to nonentities must be as strong as these.

The second "defective" feature, indeterminacy, is highlighted by a range of questions which do not seem to have (non-arbitrary) answers, e.g. "What colour was Meinong's round square?", "Who was the mother of Hecuba?", "How many days before he died did Polonius have a haircut?" (Peirce), "How long are the sides of the Triangle?". On the basis of such "awkward" questions it is variously charged that theories like Meinong's admit any number of "insoluble" problems, and that nonentities are capricious, chaotic, incoherent, disorderly, well-nigh senseless, and inappropriate for treatment by any serious theory (see especially Findlay 1963, pp. 56–8, 340ff.; also Quine FLP). According to this class of objections nonentities are defective objects, quite unsuitable for a place in logic or serious theory. Findlay, for instance, considers it a fatal weakness of Meinong's theory of objects that it admits any number of "insoluble" problems – problems which arise because some items are not determinate in all respects. The incompleteness of nonentities, which is the basis of the "unanswerable" questions and "insoluble" problems, is taken not merely as a difference, but as a deficiency. The objections are based indirectly on the Reference Theory, because they assume that all respectable objects, including of course mathematical objects, must satisfy referential conditions, in particular the completeness and determinacy conditions that entities are supposed to satisfy. It is the failure of nonentities to measure up to these referential requirements which is alleged to make them defective. Thus Findlay assumes that because nonentities break some of the rules for referential behaviour, there are *no* rules which govern them, thus presupposing a false choice between referential rules and logic, or none.[2] Similarly Quine (WO and FLP) assumes that nonentities must be completely determinate, individuable and countable, to

[1] It is assumed (as in chapter 3) that Quine's questions, like Brentano's points, concern "ordinary" entire relations, which are not totally beyond the range of their referential theories. However through reduced relations nonentities can be "physically related" to entities, though they do not stand in entire physical relations to entities. The point, which is important, is explained in 1.21 [Vol. 1 $_{—Eds.}$]and further developed in chapter 7.

[2] It would, in any case, be very surprising if no rules or logic governed the behaviour of nonentities; for then we should scarcely be able to converse about them as satisfactorily as we do in natural language. Findlay's objections not only offer a false all-or-none choice, but further involve an elementary confusion between the behaviour and properties of the items a theory treats and the behaviour and properties of the theory itself. A theory of chaos need not be chaotic, a theory of inconsistency inconsistent, a theory of vagueness vague, a theory of approximations approximate, and so on. The theory of items can treat of the indeterminacy of nonentities and the insolubility of questions about them without *itself* raising insoluble questions or itself being radically indeterminate. And such a theory could be "scientific" in a worthwhile sense, that of attempting to account for admitted facts.

be other than logically disorderly elements, that is in order that logical operations such as quantification are applicable to them. But because nonentities do not satisfy referential requirements it does not follow that they satisfy no logic. For logic is not referentially restricted any more than it is extensionally restricted.[1]

The theory of objects can simply admit – without incurring *any* damage – the indeterminacy of very many statements about nonentities, and that very many questions about nonentities, though by no means all, have no determinate answers: it is part of the data on which the theory is built. The problems are resolved within the theory by explicit allowance for and treatment of indeterminacy, e.g. through the theory of negation. Nonentities are not determinate in every respect, e.g. the present king of France is not determinate as to baldness or wisdom, though the present bald king of France is; they are entirely physically disconnected from entities; they are not always determinately distinct, and consequently aggregates of them are not always determinately countable. But they are not thereby discredited or excluded from logical investigation, any more than transfinite numbers are discredited by not having all the features of finite numbers, e.g. the division properties of finite numbers.

What the examples reveal are features of nonentities that a satisfactory theory of them should reflect or bring out, much as the so-called "paradoxes" (of Bolzano) of the infinite assembled features of infinity that were subsequently brought out and explained in a coherent fashion by Cantor's theory. In a somewhat similar way indeterminacy and other features of nonentities might be called "paradoxes" of the nonexistent. "Paradoxes" of the infinite are but a case of these. A satisfactory theory of items should be able turn these "paradoxical" features to advantage.

The fact that some statements about nonentities are indeterminate, while others are not, is then part of the data that any adequate theory in the area would try to take account of, and it is a virtue of Meinong's phenomenological theory that it goes some way towards organising and explaining this data. For what can be said truly or falsely

[1] The widely disseminated false contrast between referential (or extensional) logic on the one side and no (formal) logic on the other, has been encouraged both by referential logicians and by anti-logic philosophers (e.g. in their differing styles, the later Wittgenstein 1953, Marcuse 1964, and Strawson ILT). The result has been considerable harm to logic, especially formal logic, which has been quite erroneously seen either (often by conservatives) as useless for philosophical investigation or, very differently (by self-styled radicals), as reactionary and a tool of capitalism or of the establishment.

Part of the case for this false contrast has been examined and found wanting in Slog, 4.5. This work is intended, among other things, to elaborate the argument in Slog.

about a nonentity and what is indeterminate of it is not an arbitrary or random matter, Findlay, despite his accusations of randomness and chaos, has no hesitation in picking out certain statements about nonentities as indeterminate and others as definitely true or definitely false. Indeed, as regards indeterminacy phenomena, Meinong's theory has very substantial advantages over its standard rivals such as Russellian theory. For the latter, for example, makes none of the requisite pre-analytic discriminations; it assigns not only indeterminate statements but all statements about nonentities, many of which are true, to the same limbo of chaos and falsehood.

If the assumptions on which the defectiveness and incompleteness objections are based were, *per impossibile*, satisfied, nonentities would be just *like* entities, and so would have to exist or subsist. Then of course the objections would be telling, since nonentities do not exist. The objections do tell against any theory which tries to treat nonentities just like, or even very like, entities, and they tell against any logic which does not sufficiently differentiate nonentities from entities. To put the point another way, a theory of nonentities may very easily be overlavish in what it permits one to assert truly or sensibly about nonentities. In favour of the austerity of Russell's theory and classical logic, it can at least be said that they do not allow the wild ascription of all sorts of (or *any*) properties to nonentities. But in fact the classical theory, embodying as it does the Reference Theory, is quite unequipped to treat nonentities, even after reinterpretation of the logic. For there is, for example, no ready-made way of marking out genuine indeterminacy and thus handling the indeterminacy features of nonentities.[1] Once it is conceded however that nonentities are not so like entities, the objections begin to crumble. Not just that, but providing some account of nonentities and their properties helps *explain* features of the more detailed charges, the jokes, the puzzles, and so on.

§6. Nonentities are mere shadows, facades, verbal simulacra; appeal to the formal mode.

Though recognising the indeterminacy of nonentities resolves several problems, it soon leads to new objections. For, it is claimed, the incompleteness of nonentities reveals that they are merely shadows of their descriptions or projections of language, since their prop-

[1] Nor are means provided for saying that the statements about nonentities that Quine, Goodman and others want to say are nonsense are indeed nonsense – a point which leads Quine into difficulties (see Brady and Routley 1973).

erties end just where the properties of their descriptions do. This leads to a dilemma: nonentities must either be treated as complete "proper" things (in the material mode), in which case all the difficulties about indeterminacy arise; or, if incomplete, they are really disguised descriptions, and should really be treated in the formal mode. Hence too Ryle's accusation that Meinong's objects are "only the verbalised simulacra of genuine entities" (1972) and that Meinong's theory was a result of referentialising semantics (1971, p. 225ff.), Findlay's criticism that Meinong's objects are merely incomplete facades (1963, p. 341), and Carnap's contention (MN, pp. 65–8) that all Meinong's material mode distinctions should have been made and would better be made in the formal mode. These all amount to attempts to referentialise, in yet another way – namely, as really statements about symbolism – discourse about nonentities.

The material mode, in the usual sense, comprises statements about items, whereas the formal mode comprises statements about descriptions or symbols. The distinction is like that between using versus mentioning a description. But clearly the notion of material mode is open to the same ambiguity as the notion of reference itself, that is as between statements about items or statements referring to items. And since the material mode/formal mode distinction is intended to be exhaustive, the material mode cannot be identified with the referring mode unless one is prepared to adopt the view that all nonquotational use of subject expressions is use to refer, which is just the Reference Theory. Unless one smuggles in the Reference Theory, the full material mode must consist of all statements in which subjects occur nonquotationally, that is which are about items which are not words. Yet all *the* above *objections* to Meinong's project *identify the material mode with the referring mode.*[1] If there is, as we have argued and as Meinong believed, an irreducible nonreferring material mode way of using subject expressions, the choice between referring to an item or referring to its description, which the objections presuppose, is a false dichotomy, and one furthermore that is based squarely on the Reference Theory.[2]

All this batch of objections assumes that truths of the sort Meinong was concerned with could, and should, be handled in the formal mode. Even Carnap, although

[1] Hence also Ryle's objection, that Meinong was the arch reference-theorist because his theory allowed true material mode statements about nonentities, is misplaced, because it relies on just these assumptions – assumptions which are not only mistaken but which would certainly have been rejected by Meinong.

[2] So also was much of Carnap's (misguided) enterprise (in LSL), which aimed at either dismissal of or formal mode reanalysis of much "material mode" discourse.

strictly obliged by his principle of tolerance to tolerate the nonreferring material mode of Meinong's theory, thought that it was preferable and less misleading to carry out such an enterprise in the formal mode. That it is preferable presupposes, what is false, that it is possible. It is impossible first, because not all material mode statements about nonentities have formal mode counterparts. For not all the properties of an object correspond to properties of its descriptions. Neither intensional statements about nonentities nor quantified statements about nonentities always have formal mode counterparts. To believe that a ghost is a disembodied spirit is not to believe anything about any word for ghost in any language, nor is it obvious that there is any other statement in the formal mode which holds whenever such a material mode statement holds. (For telling objections to formal mode analyses of intensional statements, see Church 1950; also Pap 1958 and EI). Similarly quantified statements in the material mode, such as "Whatever is a nonentity is indeterminate in some respects", and "Some nonentities have properties" have no direct formal mode equivalents.[1] Secondly, even where nonreferential statements in the material mode *do* have formal mode counterparts, these counterparts do not have the same function as, the same sense or properties as, and cannot replace the material statements to which they correspond. Although it is true that a ghost is a disembodied spirit iff it is true that "ghost" is somehow co-extensive with "disembodied spirit", these statements do not have the same function and we cannot replace the first statement by the second inside an intensional context. As Pap pointed out (1958, p. 201), formal mode statements do not preserve the logical properties of their material mode "equivalents", e.g. "roses are things" is inconsistent with "But roses are not things", whereas "'Rose' is a thing-word" is not. The statement that a ghost is a disembodied spirit is necessary, whereas its formal mode analogue is not. Because a major aim of such a theory is to present an analysis of intensional properties, modal and intensional counter examples to replacements cannot simply be dismissed.

Because formal mode statements cannot always replace material mode nonreferential statements, Meinong's enterprise is distinct from Carnapian semantics, and the distinctions and theses of the theory of objects cannot, despite Carnap's claim (MN,

[1] The reason for this is that variables in the material mode become constants in the formal mode, on the usual referential theory of quotation which accompanies the account of the formal mode. This point could be avoided by liberalising the formal mode through introduction of quotation functions: but such a move would trangress the Reference Theory, because quotation functions are not transparent, and because values of the functions are not restricted to entities (see Tarski's objections to quotation functions, discussed in Goddard and Routley 1966).

pp. 66–7),[1] "be taken care of" satisfactorily in the formal mode. Carnap's procedure, at this point, does however dispose of Findlay's reckless assertion that

> Modern "semantics" is nothing if not liberal, and only ignorance of Meinong's detailed views prevents it from giving them hospitality. Meinong's round square could be stitched, with complete seamlessness, into the fabric of Carnap's *Meaning and Necessity* (Findlay 1963, p. 327).

Carnap's semantics, which is far from liberal, being basically referential, would have to be extensively modified to include generally either impossible states-of-affairs or impossible objects.

Similarly, objects are not just shadows of descriptions or names (pace Quine 1959, p. 198, p. 202), because one cannot project discourse about objects onto discourse about their descriptions or names. Not all the properties of nonentities derive from, or even correspond to, properties of their descriptions; and the identity conditions for objects are quite different from the identity conditions for descriptions. Nonentities are not shadows of descriptions or of any other entities (linguistic or not). For consider the metaphor more closely. Shadows stand in a physical relation to that of which they are shadows; so shadows of entities are entities, and shadowy entities are entities. Nonentities therefore are neither shadows of entities nor shadowy entities. Similarly objects are not half-entities, twilight entities (*pace* Kripke), or any other sort of underworld entities. The temptation to expect some sort of queer entity corresponding to objects or designated by material mode uses of subjects, to see statements about nonentities as referring to nonactual entities, is simply a hangover from the Reference Theory. The feeling that there is something very mysterious in talking in the material mode about objects which do not exist, that such objects must be taken to exist in some *strange* way, likewise derives from the grip of the Reference Theory. The demand that we explain *away* objects, either as descriptions or as references, or as attempted combinations such as things-as-described (a reference with a label on it), is simply the demand that we reduce objects to something more "familiar" – references – when they are irreducible, and simultaneously the demand that we reduce nonreferring uses of subjects to referring uses.

[1] Carnap relies upon, what is quite inadequate given the generality of his claim, the transformation of a few specific examples.

§7. Tooley's objection that the claim that there are nonexistent objects answering to objects of thought leads to contradictions.

Due recognition of the indeterminacy of nonentities also leads, so it is claimed, to serious objections to Meinongian objects and Meinongian semantics, i.e. semantics which make essential use (including quantification over) nonexistent objects. Objections of this sort-relying on quite similar principles for assigning properties to nonentities and as regards disjunction have been independently advanced by Tooley 1978 and Williams 1969. Their objections, which are important, will be considered in turn.

> The claim that there are *indeterminate objects*, in the sense required by a Meinongian semantics, seems to lead to contradictions (Tooley 1978, p. 5).

The claim does not, in its own, lead to contradiction. For indeterminacy can be defined and consistently handled in Russell's theory of descriptions, and a Meinongian-type theory can be modelled in simply modified versions of the Russellian theory (enlarged by neutral quantifiers).

Tooley's argument to contradiction turns on the following further principles and assumptions (not all of which are really required):

(I) Where a is a consistent nonexistent object, Pa is true iff Pa is entailed by the characterisation of a.

(II) Where a is a consistent nonexistent object, Pa is false iff $\sim Pa$ is entailed by the characterisation of a.

(III) Where a is a consistent nonexistent object, Pa is indeterminate iff neither Pa nor $\sim Pa$ is entailed by the characterisation of a.

The defence of principles (I)–(III) offered is that some principles are required, and that these "seem very natural". The framework is extended by a connective *or*, satisfying the condition

(*) $(P \text{ or } Q)a$ is true iff Pa is true or Qa is true (stipulative truth condition for *or*).

Although Tooley asserts that "given principles (I), (II) and (III), the claim that there are nonexistent objects answering to objects of thought, etc., leads to contradictions" the argument takes for granted, as well as (*), which is substantive, these matters:

(0) The following object c is a consistent nonexistent one: the golden mountain which is red *or* green.[1]

(0′) Being red or green is part of the characterisation of c.

(0″) Being red is not part of the characterisation of c, and being green is not part of the characterisation of c. (*Part of* here means *included in* or *entailed by*).

The argument to inconsistency is then as follows:

(1) "The golden mountain which is red *or* green is red *or* green" is true, by (I), (0) and (0′). Further, by (I), (III), (0) and (0″),

(2) "The golden mountain which is red *or* green is red" is not true, but indeterminate; and

(3) "The golden mountain which is red *or* green is green" is not true, but indeterminate.

"But then (2) and (3), together with (*), which simply specifies how '*or*' is being used here, entails", what contradicts (1), namely

(4) "The golden mountain which is red *or* green is red *or* green" is not true.

Observe that principle (II) is not used, and that principle (III) is not really required; But (0)–(0″) *are* required. What is established this is:

Proposition. (I), (*), (0), (0′), (0″) *are inconsistent.*

Proof. (after Tooley). By (I) and (0),

(I′) Pc is true iff Pc is part of the characterisation of c.

Thence using (0), that c is red *or* green is true, and, using (0″), that c is red is not true and that c is green is not true, so by (*) that c is red *or* green is not true. Contradiction.

Corollary. (I), (*), (0) *and* (0′) *(relevantly) entails* ~(0″).

Thus the argument by no means establishes Tooley's unqualified thesis, that "an approach which treats so-called non-existent objects as genuine objects which have properties leads to contradictions" (p. 2). For the argument for the thesis also *uses* (*),

[1] This assumption is far from obvious; e.g. if the mountain is golden *in colour*, red *or* green seems excluded: but the example could easily be changed, e.g. delete "golden" from the description.

(0)–(0′), and by the corollary these assumptions undermine (0″). Accordingly the thesis is *not* established by the argument.[1] Only an approach which makes *further*, and unwarranted, assumptions leads to contradictions. And in this respect there is nothing terribly special about nonentities: an over-determined theory of entities will also result in inconsistency (as various semantical paradoxes reveal). The idea that existence of objects guarantees consistency is something of a myth: it relies on such questionable assumptions as that the actual world (considered as everything that is the case) is consistent.

Not only is (0″) false in the context of the other assumptions: also other principles are false, in particular (I) *is false*. For consider (I′), given (0). By (0), that c is consistent is true, and similarly that c is is nonexistent. But consistency and nonexistence are not part of the characterisation of c; hence (I′) is false, and similarly (I). So much for the "naturalness" of (I), which (so far as I know) has never been adopted in a theory of nonexistent objects: it is open to immediate counterexamples.

(I) can, of course, be repaired. But satisfactory repairs destroy Tooley's argument to inconsistency. There are two (connected) directions repairs can take: either the class of predicates admitted in (I) is restricted, or the biconditional of (I) is weakened. Unless the class of predicates is severely restricted, the biconditional of (I) should be weakened to a conditional. The reason is very simply that a nonexistent object may have a wide range of features that are not just part of its characterisation, but are in part consequential upon it, e.g. logical features, intensional features, etc. The argument to contradiction breaks down with the weakening: (2) and (3) no longer follow. To approximate the (faulty) argument more closely a biconditional form is required: it will have the following shape, the exact form depending on the type of nonentities introduced:

(IP) where P is a suitable (characterising, nuclear) predicate and a is a nonexistent (consistent) object, then Pa is true iff Pa is part of the characterisation of a (P is one of the predicates of the set correlated with a: Parsons 1975, p. 75; etc.).

Then the problems about nonexistence and consistency vanish, because the corresponding predicates are not suitable. But at the same time the argument to inconsistency vanishes:

[1] Elementary semantical modellings reveal that the Tooley thesis is false. For example, the model of chapter 5 proves consistent a theory of nonentities, which assigns elementary properties to the nonentities, and which treats nonentities as genuine objects – in the sense that they have unrestricted subject roles and are quantifiable over.

given that everything else is properly teed-up Tooley's argument now shows that "is red *or* green" is not a suitable predicate. Specifically:

New Corollary. (IP), (*), (0), (0), (0′) *and* (0″) *entail "is red* or *green" is not suitable.*

That is not quite the end of the matter. There are problems which can affect almost any theory, about throwing in assumptions like (*) which may not conservatively extend the theory.[1] Even if a new postulate such as (*) is presented as merely giving the sense of some new operator, such as *or*, it may add new theses to the original theory, and in really bad cases it may result in triviality. An example of the latter was the addition of an unscoped indefinite descriptor to Russell's theory (see 1.22 [Vol. 1 $_{—Eds.}$]). The point (or part of this difficult point) comes out a little more sharply with negation. Define *strict negative predicates*, by analogy with Tooley's strict disjunctive predicates as given in (*), thus:

(†) (*not* P)a is true iff Pa is not true.

Now consider the object which is R and *not* R, d say. Then, by half of (I) (and a conjunction clause), Rd and (*not* R)d, whence, by (†), Rd is not true, and contradiction.

The correct conclusion seems to be that in a *consistent* theory of objects, strict negative predicates are decidedly unsuitable; not characteristic (or nuclear). Nothing of course stops someone from introducing (or trying to introduce) such predicates; but because of their collapsing role (among other things) they cannot function in corrected versions of (I). Furthermore – and this may be harder to take because it calls into question a popular critical method in philosophy – an exponent of a theory may refuse to admit, and be justified in refusing, new predicates or operators; for example, an exponent of relevant logic is justified in refusing admission of connectives which destroy relevance, even if they only conservatively extend the theory (for some indication of reasons why see Belnap and Dunn 1981).

§8. Williams' argument that fatal difficulties beset Meinongian pure objects.

In place of the (golden) mountain which is red *or* green, Williams considers (in 1969) the Polygon which has either an even or an odd number of sides. In place of (I), Willi-

[1] See further the discussion of the limits of postulation in RLR, chapter 15.

ams applies the following principle, extracted from Kenny to determine truth-values of statements like the above:

(P) "A pure object of type F possesses just those properties which are possessed necessarily by any F" (69, p. 55).

Williams contends that "fatal difficulties beset the account of Meinongian 'pure objects'" (p. 55; my rearrangement). But the supporting arguments are simply invalid: the main argument effectively distributes a universal quantifier across a disjunction. Williams considers these propositions (about the ideal or pure object, the Polygon):

(2) The Polygon has either an even or an odd number of sides;

(3) The Polygon has an even number of sides;

(4) The Polygon has an odd number of sides;

(6) The Polygon is not a polygon.

For the purpose of his argument Williams specified "F" as "polygon" and "a pure object of type polygon" as "The Polygon"; and he applies the principle derived from (P):

(Q) For any property G, it is true that the Polygon (more generally: the individual n) has G iff every polygon (every individual n) necessarily has G.[1]

Since it is true that

(2′) Necessarily any polygon has either an even or an odd number of sides,

(2) is true; but since

(3′) Necessarily any polygon has an even number of sides, and

(4′) Necessarily any polygon has an odd number of sides,

are both false, neither of (3) and (4) is true. (In fact, by (P) and Williams' specifications they are false, in view of the sense of "just".) Williams draws these conclusions:

> If we can go directly from saying that (3) and (4) are not true to saying that they are false . . . , then we have a straight contradiction. Even if we cannot, we have a breakdown in inference; for one would

[1] The switch from "any" to "every" is made in order to resolve correctly scoping problems in complex sentences.

> suppose that from (2) it followed that one or other of (3) and (4) must be *true* .... Further ... from the fact that neither (3) nor (4) is true, together with (3′), there follows ... (6) (p. 56, my rearrangement).

But none of these damaging conclusions follow. Firstly, the conjunction ∼(3) & ∼(4) & (2) is not a contradiction any more than its mate ∼(3′) & ∼(4′) & (2′) is a contradiction. A contradiction would result only given the invalid move from (2) to

(7) Either the Polygon has an even number of sides or the Polygon has an odd number of sides.

The move is invalid because (2) is true, but (7) is false, applying (Q), since

(7′) Necessarily either every (any arbitrarily selected) polygon has an even number of sides or every polygon has an odd number of sides,

is false. Indeed given (Q) the move from (2) to (7) is tantamount to the invalid distribution move from (2′) to (7′). Thus, too, one would not expect that (7) followed from (2), nor therefore that the truth of one or other of (3) or (4) followed from (2). Likewise given (Q), (6) does not follow from the cited premisses; for

(6′) It is not necessary that every polygon is a polygon

does not follow from the conjunction (2′) & ∼(3′) & ∼(4′). Since every polygon is necessarily a polygon, (6′) is false; but the premisses from which (6′) is supposed to follow are true. To establish non-deducibility formally a quantificational model suffices, a model with a two individual domain $\{a, b\}$ and with necessity interpreted as an identity functor, where, in obvious notation, *pa* & *pb*, *ea* & ∼*oa* and ∼*eb* & *ob*. Then (2′), ∼(3′) and ∼(4′) are all validated but (6′) is not.

What goes wrong with the intuitive argument for (6) can be seen by expanding the argument as follows: Suppose, on the contrary, that the Polygon is a polygon. It follows from (2′) – given some assumptions –

(2″) For every object η, if η is a polygon then η has either an even or an odd number of sides,

whence

(7″) For every object η, if η is a polygon then either η has an even number of sides or η has an odd number of sides.

Hence, by instantiating "η" in (7″) by "the Polygon" and detaching, (7) follows, contradicting (3) and (4). Thus the supposition that the Polygon is a polygon is false, and (6) follows. However the argument is flawed at two points. First, the inference from (2″) to (7″) simply generalizes the fallacious inference from (2) to (7), as instantiations of (2″) and (7″) with "the Polygon" reveal. In short the argument for (6), with its distribution move from (2″) to (7″), involves the same sort of fallacy as that in getting from (2) to (7). Secondly, within the given framework of assumptions, the step from (2′) to (2″) is fallacious. Principles (P) and (Q) are supposed to provide elimination schemes for statements about pure objects such as universals in favour of statements about unpeculiar well-behaved objects, individual entities, say. But if so (2′) is at best about every individual entity and not about every object, and only supports the the inference to

(2‴) For every individual entity x, if x is a polygon then x has either an even or an odd number of sides,

and not to (2″). But the Polygon, being a pure object, is not an individual entity; so the needed instantiation of "x" in (2″) with "the Polygon" is inadmissible, and the argument for (6) again breaks down.

Attempts to salvage Williams' argument will almost certainly appeal to the principle

(R) For every object η, if η has F or G then either η has F or η has G.

But while this principle holds for such disjunctively complete objects as individual entities, it does not hold for incomplete objects, such as universals. For example, though it is true of the object, the Triangle, that it is either isosceles or scalene, it is not true, as Meinong explained in detail (Mög, p. 170ff.), that either the Triangle is isosceles or the Triangle is scalene: as regards the equality of its sides the Triangle is indeterminate.

Williams' arguments fail, then, to establish what he (erroneously) thinks is wrong with the whole [Meinongian] enterprise: that it is utterly misguided to try to represent as singular and categorical and about peculiar objects, propositions which are indeed about what they seemed to be when we started: general and hypothetical, and not about peculiar objects (Williams, p. 56). The arguments, since invalid, do nothing to show that enterprises like Meinong's are misguided. Moreover ideal geometrical objects, far from being peculiar, are quite familiar to geometers and many of their properties are now known; their only "deficiency" is that, unlike some commonplace entities, they are incomplete in various respects. And unfortunately for Williams' final claim the assumptions (P) and (Q) from which we started are not generally true: they hold only for a restricted class of properties (cf. the discussion of (I) in §7 above). For instance, the proposition that the Wheel has a long history is not truth-functionally equivalent to the proposition that necessarily any (every) wheel has a long history, that the Fox is four-footed is not equivalent to, what is false, that (necessarily) any fox is four-footed, that the Lion is tawny is not equivalent to the proposition that every lion is tawny, and similarly for such propositions as that the Bull is fierce, that Pythagoras was thinking about the properties of the Triangle, and (*pace* Ryle 1971, pp. 49–51) that Punctuality is virtuous, since, even if it is, punctual murders and atrocities are not virtuous.

Meinong's theory of incomplete objects is a going enterprise; but the standard opposition theories ought to represent a degenerating research programme. For it has yet to be shown, after all these years, if it can be shown at all, that statements about universals can always be eliminated, preserving relevant properties, in favour of hypothetical statements about "ordinary" objects. All we have been offered by those who would eliminate "systematically misleading" statements about universals (e.g. Ryle 1971) are sample eliminations, which however exemplify schemes such as (Q), which break down if applied generally. And it would appear to be a corollary of Ackermann's demonstration of the unsolvability of the elimination problem for second-order predicate logic (see e.g. Church 1956, p. 304) that a general elimination of polyadic universals in the proposed style is impossible.

§9. Further objections based on quantification and on features of truth-definitions.

It is a commonplace, but entirely mistaken, objection to theories of objects that since quantification over nonentities is essential to them – otherwise it would be impossible to state many principles holding of nonentities – the theories are committed, if not to the existence of their objects at least to their subsistence. For, so it is claimed, quantification requires at least countability, differentiation and determinacy of items. In order to quantify over objects the objects would have to have these properties, and would accordingly, it is said, exist, or at least subsist.

The assumption that quantification involves determinacy or other referential features is completely false. The logical relations of "every" and "some", connectives and variables, captured in the formalism of quantification logic are not restricted in their applications to entities. Quantification logic applies – as well as to items like hills and clouds, species and genera, holes and sounds which are not always nicely distinct and countable – to nonentities which violate all referential demands. Moreover quantification logic can be consistently enlarged, as has been observed, by explicit assumptions about nonexistence and indeterminacy, e.g. by theses to the effects that certain items do not exist, do not subsist, and are indeterminate. That quantification logic is not limited by Reference Theory demands is also revealed, not only by the objectual semantics for Q (given in 1.16 [Vol. 1 $_{-Eds.}$]), but by newer semantics for quantificational logic, for example by domainless and contractional semantics (on these see, e.g. Slog, chapter 7). The standard referential semantics for quantificational logic can be conveniently divided into two parts, (1) a referential account of the truth of atomic wff, e.g. of "$(a_1, \ldots, a_n)f$", in terms of the entities referred to (in the model) by the subjects "a_1",…"a_n" standing in the relation specified by "f", and (2) a recursive specification of the truth values of complex wff in terms of the truth values of less complex wff, and in the case of quantified wff in terms also of the references of subject terms. Now there is nothing in this account that prevents the replacement of reference relations by other contraction relations such as aboutness relations. In place of the correspondence picture in terms of which "The morning star is bright today" is true iff what "the morning star" refers to, Venus, has the property of brightness today referred to by "is bright today", the contractional semantics says that "Pegasus is red", for example, is true iff what "Pegasus" is about, namely Pegasus, has the property of being red which is the conversion of the predicate "is red". Domainless semantics goes further (see especially Routley 1971). Noting that except for the recursion clause for quantifiers in (2) an analysis of the truth of atomic wff is unnecessary, it offers no analysis, referential or otherwise, of the truth of atomic wff, but simply assigns truth values *en bloc* to atomic wff, and it changes the recipe for truth value assignments for quantified wff, essentially to the following: $(Px)B(x)$ holds in T iff $B(x)$ holds in T for some x. The rule for the universal quantifier "U" is analogous: $(Ux)B(x)$ holds in T iff $B(x)$ holds in T for every x. In domainless semantics all the distinctive features of referential semantics are eliminated: none are necessary.

It is all very well, it will be objected, to escape the problem of a truth definition for quantificational logic by assigning truth values *en bloc* to atomic wff. But the question of a truth definition arises nonetheless for the theory of items: no logical theory is adequate unless underwritten by a satisfactory truth definition (thus Tarski, Quine, Davidson, and many others). However a satisfactory truth definition does not have to be – and for richer languages will not be – a referential one; otherwise the demand is simply yet another underhand way of trying to enforce the Reference Theory. Moreover, a detailed account has already been offered of how initial wff can be assessed in the case of expressions where subjects do not occur referentially. Finally, the theory of items is far better placed to provide a full truth definition than any referential account; for it, unlike the Reference Theory, can provide a truth definition which accounts for intensionality and inexistence.

The truth problem, the problem of seeing how it is that statements whose subjects occur nonreferentially can be assessed for truth (a problem already answered) is at the bottom of a good many objections to the theory of items. For the idea – a theme of the Reference Theory – that only what exists can really have things true of it, is extremely persistent.

§10. Findlay's objection that nonentities are lawless, chaotic, unscientific.

Like the Hades of Virgil's description, the realm of the nonexistent is "a vague, vexed region". It is indeterminate in many ways, and what is indeterminate is lawless and chaotic. But what is lawless and chaotic is no fit object for a science. Hence, firstly, there can, contrary to Meinong, be no scientific theory of objects, and, secondly, theoretical science can hardly be concerned with the investigation of such objects. But the first premiss involves a *some* to *most* fallacy: that nonentities are indeterminate in some respects does not show that they are indeterminate in most or all respects, which is what the accusation of lawlessness and chaos seems to require.

It is the erroneous assumption that nonentities are lawless and chaotic which also lies behind the objection that there is no role for science in the investigation of nonentities and that nonentities are, at least for this reason, uninteresting and unimportant. Thus, for example, Findlay (1963, pp. 56–58): "*Aussersein* remains as a whole, too chaotic to be studied scientifically",[1] "We can hardly hope to find in [nonexistent objects] a fruitful field for scientific investigation", "Such ... objects cannot interest a science which is always striving to discover law and system in its material". Scientific investigation by no means exhausts intellectual investigation, and even if investigation of nonexistent objects did prove (contrary to the evidence of chapters 10 and 11 [Vol. 3 $_{—Eds.}$]) of little interest to science it could still be of much literary and philosophical interest – as it is.

The case for the interest and importance of the investigation of objects, for the theory of objects, does not rest on the humanities alone. Much of theoretical science, and all of pure mathematics, is an investigation of the systematic properties of nonentities (specially chosen nonentities to be sure, but nonentities all the same); and these are scientific areas of no mean importance. The *prima facie* case[2] for this undoubtedly controversial thesis has already been alluded to, in §4. Any judiciously-chosen theoretical physics textbook will supply many examples of nonexistent objects which are the prime objects of study. Particle mechanics, for example, whether classical or relativistic, studies primarily mass particles, which have no dimensions and so do not exist; the theory of media investigates rigid bodies, perfectly elastic bodies and so on which have no actual counterparts; the theory of gases studies ideal gases; fluid dynamics ideal fluids; and so on. True, the behaviour of actual fluids, gases, and solids approximates, to greater or lesser degree, to various of the ideal objects. Nonetheless, it is the ideal objects that are the main objects of theoretical analysis. On the face of it then, Findlay's assumption that science does not investigate nonentities is seriously mistaken.[3]

The fact that much theoretical investigation concerns what does not exist is, furthermore, no merely accidental matter. That theoretical science characteristically studies nonentities (or *scientific fictions* as they are sometimes misleadingly called), is, as earlier writers like Bentham insisted, a conceptual necessity. Theoretical science could not pro-

[1] Findlay's (very mixed) objections to Meinong's theory of objects are not altogether consistent. For instance *Aussersein* is presented both as "incapable of scientific treatment because of its excessive richness" and, a few lines later, as "a strange sort of desert in which no mental progress is possible".

[2] The case stands up against criticism, as we shall begin to see in later chapters; see especially chapters 10 and 11 [Vol. 3 $_{—Eds.}$].

[3] These remarks also serve to undercut Mach's objections to the rationalists in geometry, that the same arguments which show that geometry is not about actual circles, triangles, etc. but rather ideal ones, would prove, what is absurd, that physics cannot be about real physical objects. The same objection is used by Grossmann (MNG, p. 162) against Meinong. The reply (which Meinong did not make) is that theoretical physics is not directly about real physical objects; it is about ideal objects to which real objects may, to some degree, approximate. The rationalists were right about geometry, and a similar thesis is correct as regards theoretical physics.

ceed without talk of such objects as species, classes, properties, attributes, idealisations, and so forth. Part of the reason is that science investigates and tries to locate relations between universals, whereas all that exists is particular. Another main part of the reason is that, far from being chaotic and lawless, certain sorts of nonentities are far more well behaved, less chaotic and more law-abiding than entities, which are never so prettily regular.

§11. Grossmann's case against Meinong's theory of objects.

Grossmann has suggested that Meinong's theory of objects is not a going enterprise, that "against the most powerful arguments" even Meinong came "very close to abandoning the most distinctive thesis of his doctrine of the *Aussersein* of the pure object" (1974, p. 81).[1] What will quickly emerge, however, is that these arguments are not powerful at all, but for the most part familiar and tried theses drawn from the Reference Theory, and, further, that they give no occasion to abandon the distinctive theses of the theory of objects.

Grossmann begins his case (in 1974, pp. 69–73) against Meinong's theory of objects (against what he, Grossmann, calls "Meinong's doctrine of the *Aussersein* of the pure object") by supposing that there is a serious dilemma, avoided by Russell's theory of descriptions, one horn of which Meinong "boldly embraced". The alleged dilemma is a version of the "problem" of negative existentials:

> Consider the fact that the golden mountain does not exist. Now, either the constituents of this state of affairs have being or they do not. If they do, then the golden mountain has being. But this conclusion flies in the face of common sense. If they do not have being, then the relation of being a constituent of a state of affairs must be of a peculiar sort: it must admit of terms which have no being. But this conclusion, too, may *appear* to be unacceptable (Grossmann 1974, p. 68, my italics).

Grossmann manufactures the "dilemma" by trying to make out that there is something rather peculiar about relations which relate entities to nonentities. Grossmann finds

[1] Grossmann's main arguments against Meinong's theory of objects are concentrated, not in his book on Meinong (Grossmann MNG), but in his article Grossmann 1974.

such relations peculiar because they violate the Ontological Assumption, because they relate nonentities to entities, and so assign properties to nonentities, that is, to put it differently again, admit nonexistent objects as constituents of states of affairs. But there is nothing extraordinary about the relations; to suggest there is equally "flies in the face of common sense" (as Reid explained: cf. 12.1 [Vol. 3 $_{-Eds.}$]). Given common-sense there is no dilemma.

This example – with bogus difficulties generated by an insistence upon the Ontological Assumption, which are then supposed to be solved by Russell's theory of descriptions – illustrates well the main method of Grossmann's critique of Meinong (in 74). At each stage in his critique Grossmann sets up conventional difficulties by using facets of the Ontological Assumption Meinong would (or should) have rejected, and then falls back on Russell's theory as a viable alternative account, which solves the problems with which Meinong is supposed to be grappling. But as we have seen, the Ontological Assumption is mistaken, and Russell's theory is neither viable nor a satisfactory substitute for a theory of objects. An adequate analysis of declarative discourse – which has to take intensional and inexistential statements seriously – cannot fall back on Russell's theory.

Rejecting Russell's theory of descriptions, however, removes the lynch pin in Grossmann's case against Meinong. Consider Grossmann's argument against what he calls Meinong's "second thesis" – namely nonexistent objects are constituents of certain states of affairs. The argument runs in essence as follows (p. 69): Meinong's second thesis "clashes head-on with Russell's theory" (this is true, since Russell's theory excludes nondenoting descriptions as genuine subjects); "Russell's general view is the correct one"; therefore Meinong's thesis is not true. Given that Russell's theory fails, detachment is excluded.

The remainder of Grossmann's case against Meinong's second thesis is that "Meinong's argument against Russell's rejection of the second thesis is not sound"! Yet otherwise Grossmann admits that he "know[s] of no decisive argument against Meinong's second thesis". But somehow he manages to conclude that he "must reject Meinong's second thesis" (p. 72).[1] One reason is that he thoroughly mistakes what is required in order to reject the second thesis:

[1] A good criticism of Grossmann's attempt to show that the second thesis is false and best avoided (at least at one "level of analysis") through a "Russellian theory" is given in Griffin 1978, pp. 2–5. Compare the *prima facie* case for the Independence Principle, p. 32ff. above: for it is this case, as regards bottom objects, that Grossmann fails to appreciate.

> Only if this state of affairs (*Ghosts do not exist*) is of the "subject-predicate form" does Meinong's second thesis stand up (Grossmann 1974, p. 72).

Firstly, the thesis is a particular claim, and so not refuted by way of one example – especially when very different sorts of cases, such as "Zeus does not exist" and "Sherlock Holmes was a detective", are outstanding. Secondly, even where paraphrases which preserve main logical features are available – as with "ghosts do not exist" which can be rendered, preserving leading modal features, "It is not the case that there exist things which are ghosts" – the paraphrase does not demolish the second thesis, for which it is enough that the sentence may be properly represented in subject-predicate form (*not* that it has to be).

It follows, since the second thesis is tantamount to the core of what Grossmann calls the "third thesis", that "nonexistent objects have some quite ordinary properties" – the core being obtained by deleting the not unproblematic qualifying phrase "quite ordinary" – that Grossmann has no decisive objection to the thesis, inconsistent with Russellian theories of descriptions,

(3*) nonexistent objects have some properties.

The equivalence of the second thesis and (3*) follows, for instance, thus: If, in accordance with the third thesis, the golden mountain *is* golden, i.e. the golden mountain has the property of being golden, then the golden mountain is a constituent of the state of affairs *The golden mountain is golden.*[1] More generally, if object a has some property ψ then a is a constituent of the state of affairs that a has ψ. Conversely, if a is a constituent of the state of affairs that a has ψ, e.g. the golden mountain of the fact that the golden mountain does not exist, then a has property ψ (and also the property of being a constituent of the state of affairs that a has ψ, etc.).

Grossmann wants, however, to use "property" in a much narrower way, in conformity with the Ontological Assumption; and his main thesis is that nonexistent objects, such

When it comes to higher order objects, such as states of affairs, Grossmann changes sides and sides with Meinong against Russell: "Meinong's position turns out to be the correct one in the long run" (p. 73). For some states of affairs which do not exist are constituents of certain states of affairs, e.g. p ("A philosopher's stone exists") is a constituent of "Someone wished that p". Thus, according to Grossmann, the second thesis is correct after all – only the objects involved are higher order ones. In a similar way the third thesis, which Grossmann also rejects, is correct.

[1] Grossmann's propositional representation of states of affairs is simply followed at this point.

as merely imagined objects, have no properties (i.e. the OA), but "are merely imagined to have properties" (Grossmann 1974, pp. 74–5; also MNG, p. 164). In particular, the golden mountain (the working example in 1974) is not golden, but only imagined to be golden or thought or conceived of as golden. Grossmann bypasses Meinong's argument that nonentities must have certain characteristic properties,

> that the object whose existence [is denied] must have certain properties and indeed certain characteristic properties. Otherwise, the judgement that the object does not exist would have neither sense nor justification (UA, p. 79; Grossmann's translation) –

an argument that is more satisfactory when generalized to show that unless nonentities did have extensional features the commonplace attribution of intensional features would be without focus (see 6.4). Instead Grossmann plunges almost immediately into psychological issues, what is before our minds when the golden mountain is conceived, its existence denied, etc. – as if it was in these reaches that Meinong's grounds for the Characterisation Postulate are to be located. But the main grounds are to be found elsewhere, as the argument cited shows, and a slip Grossmann makes helps reveal:

> It seems to me to be true that it is the golden mountain rather than *some other entity*, for example, the round square, which is before our minds when we deny the existence of the golden mountain (Grossmann 1974, p. 74, my italics).

For if the golden mountain is some other object than the round square (as it certainly is), then they must have different properties; yet according to Grossmann "nonexistent objects do not have properties"; so in particular, the golden mountain and the round square do not have the properties that make them distinct objects.

Grossmann's argument that all that can be truly said is that "the golden mountain, *as it is before someone's mind*, is golden", that (what says nothing more according to Grossmann 1974, p. 74) the golden mountain "is *thought* of as being golden (that it is conceived of as being golden; that it is imagined to be golden, etc.)", not that it is golden, rests on a false analogy. The comparison is with "the earth, *as it is before someone's mind*, is flat"; but the case is dissimilar since "the earth is flat" is not an instance of the Characterisation Postulate. Were the analogy repaired by replacing "the earth" by "the flat earth", Grossmann's argument would fail; for then more could truly be said, e.g. that the flat earth is flat. Of course it does not follow from the fact that someone thinks or believes something is so, e.g. that the golden mountain is golden or flat on top, that it is. But the question at issue is what more can truly be said. Grossmann's appeal to what follows from his psychologistic premisses is not to the point; for what does not follow may yet be true, e.g. for other reasons. However, Grossmann proceeds – he has to proceed if his argument is to carry weight, though he subsequently adduces another of Meinong's grounds for the third thesis – as if Meinong's ground, and his *only* ground for claiming that the golden mountain is golden, is that Meinong imagines it as golden. Indeed Grossmann contends that it is clear that his diagnosis is correct, on the basis of the following considerations:

> How does [Meinong] distinguish between the properties which a non-entity has and those that it does not have? Meinong admits that a desk which he makes up in his imagination has only those properties which he himself imagines it to have. ... this is the most revealing admission Meinong could possibly make (MNG, p. 164; the points are reiterated in Grossmann 1974, pp. 74–5).

For, Grossmann concludes, it is not that the desk has properties "but rather that it is imagined to have" the properties. But Grossmann's argument is seriously flawed. Firstly (what is not uncommon in MNG), he has not reliably reported what Meinong admitted, but operated with his own erroneous elaboration of what Meinong did say; secondly, he has jumped to a conclusion, which no doubt suits his critical purposes, which the actual evidence from Meinong in no way warrants. In this case Grossmann's construction from Meinong's text, presented in Grossmann 1974, pp. 74–5, is easily exposed. According to Meinong,

> I can make up a desk in my imagination which has the most outstanding features and which does not exist anywhere in the world. If I do not include in my thought any cost of its production, then there is no justification for attributing this cost to it, while in a real case nothing would depend on whether or not I had thought of the cost, since it would not be missing in any case (Stell, p. 46).

Upon which Grossmann comments (1974, p. 75):

> The imagined desk, then, is said to have all those properties and only those properties which Meinong includes in his imagining of it (and we may assume, all those further properties which it must have if it has the former).

The comment is not coherent. If the desk has all the properties Meinong imagined it to have *and* further properties which follow or flow from these, then it does not have *only* the properties Meinong imagines it to have – contradicting the "most revealing admission" Meinong is alleged to have made, and blocking Grossmann's immediate inference

> from this distinction between the properties which the desk is imagined to have and the properties which it is not imagined to have, however, is not that the desk *has* the former and *does not have* the latter, but that it is *imagined to have* the former but is not imagined to have the latter (Grossmann 1974, p. 75; the same move is made in MNG, p. 164).

The basic flaw in this inference is in the shift from "the desk (which is imagined) has such and such properties" to "the desk (which is imagined) is imagined to have such and such properties", a shift Meinong's statement does not legitimate, and which is obviously invalid; for its general form is from A to "It is imagined that A". Consider the desk Meinong imagines, or a mountain I visualize: the mountain is green, forested, tropical and basaltic. That is, I visualize a mountain which is green, forested, tropical and basaltic; I do not visualize a mountain which is imagined to be green, imagined to be forested, etc., as Grossmann would have it. The same goes for Meinong's desk. For example, the desk is such that it does not exist anywhere in the world; it is not that it is merely imagined not to exist anywhere in the world. Grossmann is entirely mistaken then in his claim that

> my objection to Meinong's third thesis is as strong [our objection to Meinong's principle of the independence of so-being from being is just as vehement] as my conviction that there is a distinction between what a thing really is and what it is merely imagined to be (Grossmann 1974, p. 74; also MNG, p. 164).

While the distinction is sound, the objection is not; so in objective strength (as distinct from subjective vehemence) there is much separating the cases. An important point that emerges is that Grossmann is insinuating – what he only later states outright,[1] and what is rather evidently false – that Meinong has confused "(nonentity)a has property ψ" with "(nonentity)a is imagined to have property ψ". The theory of objects need not make, nor need it depend upon, such an elementary confusion.

Grossmann makes an alternative attempt to explain why Meinong held that nonentities have properties – as if Meinong's straightforward explanation was not to be credited, and a different explanation which could be explained away must be found. This time Meinong is accused (1974, p. 75) of confusing, not *an imagined thing is golden* with *a thing is imagined to be golden*, but what is no more probable, "The property of being golden is part of the complex nature of being a golden mountain" (which is true) with "the golden mountain is golden" (which Grossmann says is false): Meinong

> confused the relationship between a complex property and one of its parts with the exemplification nexus between an individual and one of its properties (p. 74–5),

that is, Meinong confused inclusion with instantiation. (The accusation is repeated in Dyche 1976, who follows Grossmann on this matter).[2] Grossmann's grounds for the claim that Meinong makes such a confusion (which is turned into a confusion of a *complex of property instances* with a *complex property*) is however largely speculative (though in MNG with an historical element drawn from Meinong's very early work); and the speculation involves various connections which Meinong would rightly have rejected, e.g. of the identification of the ideal object, Triangle, with a certain complex property, or nature (p. 75).

Grossmann tries to use the distinction between inclusion and instantiation to rebut Meinong's argument from truths of geometry for the third thesis. Very simply, the (Ideal) Triangle is three sided, and has many other geometrical properties, but it does not exist. If ideals somehow reduce to complex properties, as Grossmann and Dyche would like to think, then the argument can be defused: apparent instantiation and attribution of properties becomes inclusion (and, most important, the Ontological Assumption is upheld). Grossmann's approach is (rather characteristically) oblique: he does not directly

[1] "... Meinong eradicates ... the important distinction between what a thing is and what it is thought to be" (MNG, p. 166).

[2] Dyche's thesis that Meinong's nonexistent objects are never particulars but always natures rests upon decidedly tenuous evidence: it is briefly examined and rejected in 12.2 [Vol. 3 $_{—Eds.}$].

criticise Meinong's, or Descartes', argument that the Triangle "has properties irrespective of whether it exists in reality" (MNG, p. 161), but instead sketches a contrasting position, which he defends against Meinong. But the contrasting position (set out in MNG, p. 160; Grossmann 1974, p. 75) involves replacing statements about individual triangles and the Triangle by statements about the complex property of being a triangle (i.e., Triangularity) something that cannot always be done, and something Grossmann makes no attempt to show can be satisfactorily done. Thus he does not meet Meinong's argument at all. What we are given instead is an iteration of referential principles, of the Ontological Assumption in particular (MNG, p. 160, p. 161, p. 163), and a comparison of the cases of the Characterisation Principle (which Grossmann does not separate from the Independence Thesis) with inclusion principles, e.g. the "falsehood" that the round square is round gets its plausibility from the truth that the property of being round and square includes the property of being round. Meinong has arrived, it is suggested, at his mistaken Independence Principle by confusing complex properties for which an inclusion analogue of the principle does hold with individuals for which no such principle holds (MNG, p. 161). Meinong's assignment of properties to nonentities is even said to "indicate that he conceives them in analogy to complex properties" (p. 161) – which is to deploy the contrasting theory in a really self-guaranteeing way. This is a travesty of Meinong's complex position which does little justice to it.[1] There are, as Meinong recognised, many nonentities distinct from properties, and the assignment of properties to them does not depend upon conflation of them with, what are mostly quite distinct, properties.

Grossmann gives no argument against the fourth thesis he considers false, that being (existence) is never part of an individual; for he is quickly diverted to consider two fundamental assumptions, which in "combination ... present Meinong with a cluster of insurmountable difficulties" (1974, p. 78). These assumptions[2] – which are inconsistent with the fourth thesis – are:

[1] So also is Grossmann's repeated attempt to foist on Meinong a reduction of nonentities to complexes of properties, e.g. p. 167. The very limited satisfactoriness of this representation of objects is considered in chapter 12 [Vol. 3 $_{—Eds.}$].

[2] Both (F1) and (F2S) are also ascribed to Meinong in Grossmann's MNG, e.g., p. 221. But it should have been evident from a passage that Grossmann quotes from Meinong, that Meinong only accepted the combination of (F1) and (F2S) in the case of genuine or ordinary determinations of so-being:

> In regard to every genuine or, so to speak, ordinary determination of so-being, it is in my power, according to the principle of unlimited freedom of assumption, to pick out – by means of adequate intention – an entity which in fact has the determination of so-being (Mög, p. 282).

The passage also illustrates however Meinong's tendency to confuse (F1) with (F2S) and (F2).

(F1) An object "has all the features with which it appears before a mind" (an unqualified CP); and

(F2S) "One can think of the existing golden mountain ... just as readily as one can think of the golden mountain" (p. 78).[1]

Whence by (F2S) and (F1), the existing golden mountain is existent, that is, exists; but it does not exist. Any theory which includes both assumptions (F1) and (F2S) is inconsistent: so Grossmann's variant of Russell's argument goes. To which Meinong replied, correctly, that the existing golden mountain "exists just as little as" the high golden mountain (Stell, p. 17). He would similarly have replied, correctly, in response to Lambert's simple variation on Russell's objection (Lambert 1974, p. 308) that the round square which exists does not exist:[2] see Meinong's statement that it is false that the golden mountain which exists exists (Mög, p. 178).

Such replies show two important things. Firstly, Russell's famous second argument against Meinong, as commonly stated, is invalid. The argument is: "If the round square is really round, as Meinong claims, then the existing round square must also exist" (MNG, p. 158). But on Meinong's truth value assignments (as on models of Parsons and Routley) the antecedent is true, but the consequent false. Secondly, (F1) is not only false, but not maintained by Meinong.

Grossmann's case against Meinong depends essentially however on (F1): so what is his evidence that Meinong adhered to (F1)? In fact Grossmann presents no textual evidence that Meinong did maintain (F1), but alludes to evidence (Mög, pp. 287–8) that Meinong questioned (F1) and more satisfactorily formulated variants of (F1), such as that an object has all the features it presents itself as having, is presented as having, thought of as having, and so on. There is, moreover, as Griffin remarks, a sizeable body of evidence that Meinong did not hold (F1) or its variants.[3]

[1] (F2S) is a special case of the principle of unlimited freedom of assumption for objects, namely

(F2) One can "pick out – by means of adequate intention – any object" at all (cf. Meinong, Mög, p. 282);

more exactly, for any significant object $\tau x A$ (described using some descriptor τ) one can think of or conceive $\tau x A$ and assume it to have arbitrary features represented in A. Principle (F2) and other associated principles are studied, and adopted, in 6.4.

[2] Where the objection concerns "'to exist' in the ordinary sense of 'being there' (*Dasein*)" (Stell, p. 17), what Meinong calls "factual existence" or "actual existence". Meinong presents, in Mög, p. 181 ff., a detailed discussion of what he describes as the "even more cumbersome 'the A that exists, exists'".

[3] This also meets Lambert (1976, p. 253):

> To consider just two examples: Meinong gives a careful discussion of the possibility of erroneous judgements and perceptual illusion (e.g. in Erfgl., pp. 30ff.). If it were the case that objects have all the properties they are thought of as having, erroneous judgements about which properties they have could hardly occur, in which case Meinong's discussion of such judgements would be pointless. Of even greater importance to Meinong's philosophy is his distinction (see GA, vol. 2) between the (mental) content and the (non-mental) object of a presentation. He argues for this distinction by noting properties of objects which are not properties of contents and vice-versa. If (F1) were true it would seem that every feature of a content would be a feature of the corresponding object, and thus the distinction between content and object (or at the very least Meinong's arguments for it) would be lost (Griffin 1978, p. 31).

There is also other more direct evidence: firstly, that adduced above from Meinong's reply to Russell's objection, secondly from Meinong's doctrine of the modal moment, thirdly from Meinong's deployment of the distinction between nuclear and extranuclear properties (both these matters are taken up at the end of 5.1), and fourthly (but closely connected with the preceding points), from Meinong's sayso:

> I have formerly appealed to the exceptional position of "existential predication" in relation to the "existing round square" and "the existent golden mountain" (naturally by "existing" is meant "(f)actually existing" ...) .... It would be appropriate, therefore, to consider whether this analytical judgement [of the form "the ψA is ψ"], whose apparent natural territory is the *Soseinobjective*, is also permitted meaningful application to the *Seinobjective*. The above examples show, in advance of any theory, that the answer must come out in the

---

> ... concerning the matter whether or not "the full assumption postulate" [i.e. an unrestricted CP] is an integral part of Meinong's theory, I disagree with Routley. He cites no textual evidence that it is not, and, on the contrary, Russell thought, and I think, there is abundant evidence to the contrary.

I have cited some of the evidence (some of which was alluded to in the article Lambert is criticizing), and doubt that there is abundant evidence to the contrary. Let Lambert cite his evidence, and his evidence that Russell (not altogether a reliable authority, especially since his commentaries on Meinong ceased before much of Meinong's important work was published) thought there is abundant evidence to the contrary.

> negative. There is a quite understandable sense to talk of a golden mountain which exists; however it is false to say, on the strength that, of this object that (in an ordinary manner of speaking) it exists (Mög, p. 278; cf. also UA, pp. 70–1).

(F1) is thus rejected by Meinong. Therewith Grossmann's case appears to fail, and many other cases against Meinong collapse.

Grossmann does however present an important argument that Meinong has to adopt (F1) in an unqualified form – an argument which, if it worked, would seem to show that *any* theory of pure objects is committed to (F1). The argument is that (F1) cannot be restricted to a subclass of properties such as

> "ordinary properties": an [object] does not only have all the *properties* with which it appears before a mind, it has all the *features, characteristics*, etc., *of whatever kind* with which it appears before a mind; otherwise it would not be *this particular* [object] that appears before the mind. As they appear before a mind, the golden mountain, the existing golden mountain, and the nonexisting golden mountain are three distinct entities. Meinong is therefore forced to admit that the existing golden mountain is indeed existing, just as it is indeed golden .... (Grossmann 1974, p. 78; cf. also MNG, p. 160).

Grossmann uses essentially the same argument (in MNG, pp. 159–160) to show why Meinong could not really adopt the obvious move (which he did consider) against Russell's second objection to the theory of objects, of simply saying that existence is not a (genuine, ordinary, ...) property, namely that he would not have been able to explain why thinking "of an existing round square is not the same as to think just of a round square" (p. 160).

In neither form is the argument sound. That the golden mountain (the "gm" for short), the existing golden mountain the "egm"), and the nonexisting golden mountain (the "negm") are *intentionally* distinct, does not show that Meinong is forced to admit the egm is existing. For other features – intensional features naturally – than existing may serve to distinguish the objects adequately. In fact what Grossmann thinks Meinong is forced into would *not* serve to distinguish the gm from the negm, since both are nonexisting. Similarly, the response to Russell that existence is not an ordinary property, and so is not assumptible, is open to Meinong. That thinking of an existing round square differs from thinking of a round square may be explained, not by the existing of one and the nonexisting of the other, but by the intensional differences of the objects. For example, the existing round square presents itself as existing, or as Meinong sometimes puts it, has suppositious existence, whereas the round square does not, hence they are Leibnitz-different. And Leibnitz-difference is enough for thoughts of them to be different: extensional difference is not required to separate objects in highly intensional settings.[1]

It is worth noting that rival classical logic theories are in substantially worse position as regards making the requisite distinctions, including Russellian-style theories which, so Grossmann supposes, avoid Meinong's difficulties. Consider the statements (a) G thinks of a [the] gm, and (b) G thinks of a [the] egm, which may well differ in truth value. To apply Russell's theory of descriptions these have first to be paraphrased, for instance, to take the shift Russellians commonly adopt, to (a′) G thinks that a [the] gm exists, and similarly for (b′), and thence, given the only plausible scoping, to (a″) G thinks that there exists a [the] gm, and similarly for (b″). But on Russell's perceptions (b″) always has the same truth value as (a″) for both apply the functor "G thinks that" to $(\exists x)(xg \;\&\; xm)$. The requisite discrimination is lost. Once again other theories of descriptions, including stronger free description theories, fare *considerably* worse.

Despite the evident superiority, then, of a theory of Meinong's sort, Grossmann thinks Meinong's position is weak (p. 79). The egm presents itself as existing ("appears before the mind as containing the property of existing" in Grossmann's terms), but does not exist. Why not – and this is supposed to bring out the weakness – make a similar distinction in the case of "ordinary properties" such as golden: the gm presents itself as golden, but is not really golden (as a simple extension of a classical theory might well allow)? The answer to this objection, which is not a simple answer, takes us back over ground already traversed: in particular, why it is that it can, and must, be conceded that non-existing objects have some properties (1.4, 1.5 [Vol. 1 —*Eds.*]), and what distinguishes characterising properties from others such as existence (1.17 [*ibid.*]). But it also includes

[1] Of course exactly what is said depends on how the theory of objects goes. If a theory excludes assumptibility altogether where the properties are not "ordinary", as does Parsons' 1974, the egm and the gm are extensionally distinct anyway since the gm is golden but the egm is not. A problem does not arise at all in the form Grossmann has assumed: for Grossmann has supposed that one maximized on assumptibility as it were, so that the egm is golden and a mountain. There are, needless to say, other problems for the Parsons-style approach, e.g. that like many marketed theories of descriptions there is no difference between the egm and the existing round square. That problem is solved however by use of intensional differences or by, what Parsons does adopt in 1978, use of Meinong's "watered-down" predicates.

ground yet to be traversed in detail; especially why the characterisation of an object can determine some of its features, but not others, such as its ontological status (this issue is taken up again in 6.4 ff.).

Grossmann does not try to press this objection, but passes at once (1974, p. 79) to a recasting of the second objection in terms of objectives, or more exactly of states of affairs. Consider the object, the egm, the state of affairs that the state of affairs that the gm exists obtains. By (F2) one can suppose this and by (F1) then, since the state of affairs obtains, the gm exists. There can, Grossmann contends, be no escape, as there was in the case of "pure objects", by way of "pure objectives", which do not contain their mode of being in the way that there was with "pure objects"; there is no presented (or watered-down) obtaining, for "objectives themselves must contain their mode of being", on pain of a vicious regress otherwise (p. 79). Grossmann's argument depends once again upon assuming that Meinong adopted (F1). Since he did not, the argument fails. Quite apart from this, a theory of objects is by no means obliged to say that objectives or propositions must contain their mode of being. Whether a contingent proposition such as "the gm exists" is true or not depends – not on higher order propositions – but on how the real world is; and its truth conditions can be given, without vicious regress, by a neutral theory of truth. The objective in question, *the existence of the gm*, does not contain its own mode of being: it does not, and strictly cannot, determine whether it itself obtains.

To avoid the difficulties (F1) and (F2) engender in combination Meinong abandons, so Grossmann contends (1974, p. 80), "the wrong assumption", (F2) and not (F1). Not so; for Meinong does not adopt (F1).[1] However, Meinong does – it appears, though the evidence is not unequivocal in view of what Meinong goes on to say – qualify (F2): "this principle of freedom of assumption now requires in fact, if I see it correctly, a limitation in regard to the modal moment" (Mög, pp. 283–284). This certainly seems to be a mistake (though not quite the extravagant mistake Grossmann suggests, since it only affects objects and objectives into whose attempted characterisation the modal moment is built). For nothing appears to prevent one thinking of, and making judgements about, the objects with respect to which one's freedom is supposed to be limited, such as the

[1] The same point undercuts Grossmann's argument (1974, bottom half of p. 80) against Meinong's suggestion that one can think in a roundabout way of the existence of the golden mountain – where existence carries the modal moment. For the argument uses (F1).

golden mountain that exists (cf. MNG, p. 222). The reason that Meinong thought the limitation on (F2) is required is, it seems, that he did not always clearly separate (F2) and (F1). (Of course if (F1) and (F2) are not separated then (F1) is similarly qualified.)

While it is true that "many of Meinong's difficulties [in Mög, p. 278ff.] disappear if the first fundamental assumption ... is rejected" (Grossmann 1974, p. 81; the omitted words are "and, hence, the third thesis"), it is false that rejection of (F1) as false, as not holding for *all* features, involves Meinong in "abandoning the most distinctive thesis of his doctrine, the *Aussersein* of the pure object", the third thesis (p. 81). Grossmann is entirely mistaken in thinking that the third thesis entails (F1) (as he evidently does; see also the quotation above with "and hence, the third thesis" inserted). An object may well have *some* properties (indeed all ordinary or characterising properties) without having *all* the properties it is conceived of as having (or with which it appears before the mind). Grossmann has relied here, and in drawing out the lesson of his discussion, upon an invalid inference from "some" to "all": it is not obvious that objects need not have any of the features with which they appear before the mind, only obvious that they cannot have all the features they are presented as having.

Finally, Grossmann's triumphal conclusion in 1974 (p. 81) that Meinong himself came (in Mög, pp. 287–88) "very close to abandoning the most distinctive thesis of his doctrine of the *Aussersein* of the pure object", that "while it is true that the property of being round is a part of the round square, Meinong explains, it is not true that the round square is round", is but a travesty of what Meinong has to say. But this depends on misconstruing Meinong's complex discussions. Whether Meinong did come to say anything approaching what Grossmann attributes to him – a matter taken up again in chapter 12 [Vol. 3 $_{-Eds.}$] – it is evident that Grossmann has adduced no substantial reasons for abandoning a theory of objects.

§12. Mish'alani's criticism of Meinongian theories.

Mish'alani's criticism of theories of a Meinongian cast, in his 1962 paper "Thought and object", has sometimes been taken to do irreparable damage to any theory of objects. It is not difficult to show that it does not, and that what criticism touches a theory of objects is readily avoided. Mish'alani begins by taking for granted existentially-loaded quantifiers: thus his initial assumption, "when someone thinks of the golden mountain, it is not true that there is something of which he thinks" (p. 185). Wrong: particularisation as in A(the golden mountain) $\supset (Px)A(x)$ – is not restricted; it does not exclude functors such as "Someone thinks of"; so it is true of some object, when someone thinks of the golden mountain, that he is thinking of it. Subsequently Mish'alani backtracks, and admits the introduction of wide quantifiers which may range over nonexistent objects (p. 187); but he does not observe that this may be used to undermine his initial claims. Instead he proceeds (p. 188) to present a larger theory, main theses of which are ascribed to Meinong. The first four theses of the theory presented are among the central theses of the theory of items, set out on p. 3 [of Vol. 1 $_{—Eds.}$]. These theses pass uncriticised:

> Not that I think that the first four are correct; only that I can think of no good reason at the moment for maintaining that they are incorrect (p. 189)!

It is the following tenet of the larger theory that Mish'alani considers mistaken:

> there is no difference between ...
>
> (1) S thinks of *the* G if, and only if, (i) S thinks of something, and (ii) what S thinks of is a G, (iii) ... ,
>
> and
>
> (3) S thinks of *the* G if and only if, $(Px)(S$ thinks of x, and x is a $G, \ldots)$ (rearranged from p. 189 and p. 186).[1]

That is, by transitivity,

(M) S thinks of something and what S thinks of is a $G \ldots$ if, and only if, $(Px)(S$ thinks of x and x is a $G \ldots)$;

and it is (M) that equates (1) and (3). Mish'alani wants to assert (1), but to reject (3) on the basis of counterexamples like the following: M thinks, let us suppose, of the man who murdered FDR (i.e. Roosevelt). By (3), for some x, of whom M thinks, x murdered FDR; hence, by considering the converse of the last relation, FDR was murdered by someone, and so was murdered. But FDR was not murdered. Thus (3) is false (p. 190). The argument supposes firstly that a principle of converse holds for the murdering relation, that if xRy (x murders y) then y is R-ed by x (y is murdered by x), and secondly that the relation is a Brentano one, that if yRx and y exists then x must exist. Though these

[1] It is assumed that the dots in (1) and in (3) are filled out in equivalent ways.

assumptions (primarily the first) have been questioned, e.g. by Chisholm and Parsons, and do fail for reduced and hyphenated predicates, a theory of objects need not dispute them.[1] For (3) is not a tenet of the theory of items given, or of any theory of objects that has a suitably restricted Characterisation Principle.

Consider the crucial part of (3) for the counterexample, namely

(3C) If S thinks of *the* G, then $(Px)(x \text{ is a } G)$,

a consequence of (3). According to this half of (3), a definitely thought of object has the properties ascribed to it; an object presented in thought has all the properties it presents itself as having; in effect, *the* G is G, without qualification. Consider, as Meinong and others have, the round square that exists; we can readily think of it, but it is false that

$(Px)(x \text{ is a round square that exists})$, i.e. $(\exists x)(x \text{ is a round square})$.

Thus (3C) is false, and with it (3): neither are tempting, once it is seen what they do; nor have they been consistently held by leading proponents of the theory of objects.

Why does Mish'alani's argument appear to carry any weight at all? Because, it seems, connecting principle (M) is true (for certain readings of "something" and "what S thinks"),[2] and because Mish'alani insists, and audiences are too ready to grant, (1) is true. And if (1) and (M) are true, so is (3), after all. It can be made to appear furthermore that the trouble arises with (M) and (3) through the admission of nonexistential quantifiers. Without (1) however the whole case collapses; and (in the *straightforward* sense) (1) is false. Furthermore the crucial part of (1) corresponding to (3C), viz.

(1C) If S thinks of *the* G, then what S thinks of is a G.

[1] Mish'alani's case for the assumptions is, however, by no means as decisive as he tries to make out. Regarding the second assumption he claims (p. 193)

> any attempt to rescue the view which is criticised in this paper by restricting the type of function which may take non-existents for arguments would be self-defeating.

This is far from true (the aim is not, in any case to provide an analysis along the lines of (1)).

The main argument in favour of the first assumption (p. 192) depends upon refusing to distinguish predicate and sentence negation; so it is worth little. However, Mish'alani's criticism of theories which reject the principles of converse is of interest, and is taken up again in chapter 7.

[2] Consider the RHS of (M) and apply ξ elimination of quantifier P. Then

(M′) (Px) (S thinks of x & $G(x)$ & …) iff S thinks of a and $G(a)$ and …,

where $a \approx \xi x$(S thinks of x & $G(x)$ & …). But a is not, of course, merely what S thinks of, but what S thinks of which is G, etc.

But what S thinks of is, by the antecedent, *the* G; hence

(1C′) If S thinks of *the* G then *the* G is a G.

Since, however, it is always true that someone can think of *the* G (by a principle of freedom of thought: see chapter 6),

(1D) *the* G is a G,

i.e. an unrestricted CP follows. Hence (1) is false. Although Mish'alani's objection may look, at the outset, as an objection based on the alleged problem of relations between nonentities and entities, on examination it turns out to be an objection relying upon use of an unqualified CP which is infiltrated by way of an explication of definite descriptions in thought contexts.

The theory of items also yields a straightforward answer to the question that generates the remainder of Mish'alani's discussion, namely:

> Is it true that:
> (2a) S thinks of a G if, and only if, $(\exists x)(S$ thinks of x, and x is a $G)$,
> provided that a G exists? (p. 194),

a question that is refined thus (p. 196):

> what should interest us is whether there is a sense of "to think of" wherein the expression "thinks of" occurs both in the antecedent and in the consequent of (2a) such that (2a) remains true.

The answer is that there is such a sense, a transparent sense (it is *not* Mish'alani's "subjective sense" which is not transparent). Let "T" symbolize such a transparent sense; and represent "a" in "a G" as "an arbitrary", i.e. "a G" is symbolised $\xi x G(x)$.

Suppose, first, $ST\xi xG(x)$. Then

$$ST\xi xG(x) \ \& \ \xi xG(x) = \xi xG(x) \ \& \ (\xi xG(x))E,$$

since by the provision, a G exists, therefore, $(\exists x)(STx \ \& \ x = \xi xG(x))$. In fact the *stronger* form which uses $G(\xi xG(x))$ – this seems to be what Mish'alani is thinking of – also results from $(\xi xG(x))E$; whence $(\exists x)(STx \ \& \ G(x))$. Suppose, conversely,

$(\exists x)(STx \ \& \ x = \xi xG(x))$. Then, for some a, $STa \ \& \ a = \xi xG(x)$, whence by transparency of T, $\xi STxG(x)$. The stronger form $(\exists x)(STx \ \& \ G(x))$ links with $ST(\text{a } G)$ for a different "a", namely "a certain". Firstly, $ST(\text{a certain } xG(x)) \supset (\exists x)(STx \ \& \ G(x))$ as above, using an existence-qualified CP for *a certain*. Conversely, for some b, $STb \ \& \ G(b)$. But if $G(b)$ then b is a certain x which is G, i.e. $b = \text{a certain } xG(x)$. Hence by transparency $ST(\text{a certain } xG(x))$.

§13. A theory of impossible objects is bound to be inconsistent: and objections based on rival theories of descriptions.

Although hostility to *possible* objects has abated somewhat[1] in recent times with the rise of modal logic semantics, and the consistent development of theories of nonexistent objects which occur in alternative possible worlds, hostility to *impossible* objects remains undiminished; and there have been continuing attempts to show that impossible objects and theories of them, are incoherent. Impossible objects have been a particular target of recent attack because it has been felt (rightly enough) that impossible objects should be impossible, and (wrongly) that what is impossible should be inconsistent and so incoherent. Impossible objects, furthermore, offer a place of maximum leverage to attack theories of nonexistent objects: such objects, if any, are likely to be serious trouble makers. Many of those who contend that there is something seriously wrong with any theory of impossible objects would like to think that there is something pretty wrong too with a theory of purely possible objects, and would like to be able to show it – if only they knew how.

Objections to impossible objects are almost invariably derived by taking over principles for possible objects and entities and showing that they lead to trouble – inconsistency or the like – when applied to impossible objects. They depend, that is, upon assuming that the logic of impossible objects is the same as, or very similar to, that of objects which are not impossible. But, as Meinong replied to Russell's objections to impossible objects, the logic is not the same; and moreover only a little reflection indicates that it would not be expected to be the same. Early objections to impossible objects, such as those of Russell, depend upon two assumptions, which though they hold for entities and

[1] With those opposed to modal logic, such as hardline empiricists, hostility has scarcely abated. But with the rise of an alternative theory, the weakness of their position has become much more evident.

for consistent situations, are separately questionable when extended to impossible objects and inconsistent situations, namely:

(1) Impossible objects are, like entities, fully assumptible, i.e. an unrestricted CP holds.

(2) Classical logic holds with respect to all objects and situations.

However, as we have seen, these assumptions both fail; and with their rejection the way is open for the development of a coherent theory of impossible objects (and in more than one direction – consistent and paraconsistent). With their rejection too Russell's case against impossible objects, and many of the later variants upon Russell's objections are removed.

These objections are not, however, simply irrelevant in the way Grossmann amazingly supposes (MNG, p. 158; similarly Gram 1970):

> Perhaps, Russell felt his objection ... that impossible objects violate the law of contradiction ... had more of a thrust because he thought of logic, not as applying to what there is, but as encompassing everything. Impossible entities show that this conception of logic is mistaken!

Grossmann's very rejection of the idea of a universal logic already applies elementary logical principles concerning impossibilia, that they are within the range of the quantifier "everything", but that they are not among what is. Moreover, Grossmann at once proceeds to reason and argue, in ways that are open to (often unfavourable) logical evaluation, about impossibilia and their properties. While Grossmann could, of course, fall back on a redefinition of "logic", in the familiar sense tied to argumentation and deductive reasoning, his thesis that impossibilia, and also possibilia, are beyond what logic encompasses is self-refuting as soon as it is defended logically.

Grossmann seems blissfully unaware how much damage his charmingly naive – and correct – claim;

> Contradictory entities ... quite obviously must violate the law of contradiction or they would not be what they are (p. 158),

does to orthodox classical reasoning, reasoning he adopts when it suits him and does not question. Let a be such an object; then for some B, $B(a)$ & $\sim B(a)$. Hence, by closure under strict (or material) implication, every statement whatsoever is true. Strictly then Grossmann's philosophical position is logically trivial:[1] but with *reconstruction* the trouble can no doubt be isolated.

The method of endeavouring to impose principles that apply to entities on impossible objects remains the underlying strategy of more recent and sophisticated objections to impossible objects, which rely upon principles concerning descriptions and identity. Two examples of this favoured strategy will be examined in some detail, an argument of Lambert (which is considered in this section and the next, i.e. §13 and §14) and a rather similar objection made by Tooley and Burdick (which is taken up in section §15).

Lambert's argument (of 1974, pp. 310–2) that "there are no impossible objects!", is summarised accurately (in his review 1973, p. 229) thus:

> To the question, under what conditions does the expression "the so and so" pick out an object? a natural answer, and one contained both in Russell's theory of definite descriptions and in modern free description theory, is this: for any object x (construed in the widest possible sense of the word "object") the so and so is x if and only if x and x only is so and so. Substitution of an inconsistent predicate for "so and so" in the condition just described, and with quantification interpreted either referentially or substitutionally, yields the verdict that there can be no such object; for example, it follows that there is no such *object* as the round square. To be sure the condition on objecthood presented above might be challenged; after all it fails in Frege's theory of definite descriptions. But then we have a right to ask what the conditions of objecthood are in Meinong's theory.

The allegedly "natural answer" to Lambert's question as to "the conditions of objecthood" – in more aseptic terms "How does one identify the *object* which is so and so?" (1974, p. 311) – is given then by none other than the principle of minimal free description theory,

[1] The fuller argument uses such textually supportable facts as that Grossmann adopts, for an important class of cases, Russell's classical theory of descriptions and its underlying logic, that, however, he extends quantifiers to range over nonentities – so it is that nonexistent objects, all of them, "have no form of being whatsoever", etc. In chapter 12 [Vol. 3 $_{—Eds.}$], another respect in which Grossmann's theory is inconsistent, and so trivial, is observed.

IP $(x)(x = (Iy)(y \text{ is so and so}) \equiv .\ x \text{ is so and so and only } x \text{ is so and so})$,

already rejected as mistaken in 1.14[1] and derived in a sharply qualified form in 1.20 [both sections reprinted in Vol. 1 —*Eds.*], the qualifications being referential in character (i.e. to existentially-loaded quantifiers and transparent predicates). In short, Lambert's answer gains its "naturalness" by importing referential assumptions; remove the referential setting – which is entirely inappropriate in the case of nonentities – and its naturalness vanishes. Yet just such removal is what the main argument depends upon: the quantifiers in IP must be interpreted so as to catch, in one way or another, impossible objects. Lambert assumes, that is,

> the formula "$(x)(\ldots x\ldots)$" is to be read as "for every *object* x … x …" and not as just "for every existing object x (or every possible object x) … x …". In other words it ranges over the nonsubsistents as well as the subsistents (p. 311). Alternatively the quantifier expression "$(x)(\ldots x\ldots)$" may be interpreted "substitutionally", that is, as being true, just in case … "… x …" is true for all singular terms in place of "x" … *provided* "$t = t$" is true for all singular terms t. (p. 312).

Thus in either interpretation the singular term "the round square", or, what is said to be the same (p. 311), $(Iy)(y$ is a round square$)$, is an admissible instantiation of the universal quantifier, and in either interpretation such terms are open to particularisation, and the corresponding particular quantifiers are similarly neutral. Hence the way is open to show that there are no impossible objects, however generous the quantifiers are taken to be.

Lambert's argument from (the neutral formulation of) IP which is supposed to "demolish the objecthood of impossible objects" (p. 312) is as follows: From IP by distributing quantifiers,

T. $(Px)(x = (Iy)Ay)) \equiv .\ (Px)A!x.$

Now let $A(x)$ represent some inconsistent wff; in Lambert's example "x is a round square". Since no object can have inconsistent features, $(Px)A!x$ is false for such A. Hence by T., $\sim(Px)(x = (Iy)A(y))$, there is *no* object of any sort that satisfied A.

[1] For the case against IP see also Routley 1976, pp. 249 and 251.

Unfortunately for Lambert, IP, so neutralised, does too much. Firstly, the same simple modelling[1] which Lambert uses to establish his "major negative thesis that Russell did *not* 'demolish' Meinong's theory of objects", "*did not prove that there are no non-existent impossible objects*", and did not show "that the world of *Aussersein* is impossible", also undermines his major positive thesis that:

> it is possible to bolster that fragment of [Russell's] argument dealing with impossible objects so that it can be seen as a very strong inclining reason against at least there being any impossible objects (p. 304).

For it follows from IP upon substituting "a round square" for "so and so".

IPC: $(Iy)(y \text{ is a round square}) = (Iy)(y \text{ is a round square}).\ \supset.$
$(Iy)(y \text{ is a round square}) \text{ is a round square.}$

But on the simple modelling (indeed by Lambert's own arguments verifying (1)–(3)) the antecedent of IPC is true but the consequent is false. Hence the simple modelling provides a countermodel to IP, and so to Lambert's argument as well as Russell's.

Secondly, inconsistency results. Were neutralised IP logically valid, since $(Iy)(y \text{ is so and so}) = (Iy)(y \text{ is so and so})$ always on Meinong's theory (and according to reputable free logic),[2] the unqualified characterisation principle

CP: $(Iy)(y \text{ is so and so}) \text{ is so and so}$

would be logically valid (contradicting (4)). And CP leads directly to inconsistency. As previously explained, the full CP is inconsistent with the validity of any logical laws,

[1] Lambert sketches in 1974 a simple modelling which shows that each of what he takes to be the key theses of Meinong's theory of objects, namely (1) every singular term stands for an object, (2) there are nonexistent objects, and (3) nonexistent objects can have characteristics, can be true while nonetheless (4) the statement form "The so and so is (a) so and so" is not logically valid. The modelling is an adaption of Carnap's method IIIb of definite descriptions (MN, p. 36) which adds to the domain of things the null thing, $*$, which does not exist and which is to serve as "the bearer of any definite description whose characteristic matrix is not true of exactly one thing in the domain of discourse, for example, to a definite description such as 'the round square'" (p. 309). The model is designed to guarantee (1), $*$ makes (2) true, and (3) holds true because the round square, for example, is self-identical, $*$ being identical with $*$. Thus the model verifies each of (1)–(3).

But nothing in the domain of discourse is both round and square. Not even $*$, that is, the round square. So, where "so and so" is "round square" , (4) is true (p. 309). Hence too Russell's objections to Meinong's theory which turn on use of a full Characterisation Postulate fail.

[2] On the strength of this parenthetical remark Lambert writes (1976, p. 252): "Though he should know better, Routley gratuitously and falsely remarks that such a principle is also adopted unanimously by free logicians" . Wrong, he did not: what he did suggest was that free logics that reject $(Ix)A = (Ix)A$ are not reputable.

including CP itself. For suppose otherwise that CP is universally valid, i.e. every definite description $(Iy)(\ldots y\ldots)$ satisfies CP. Consider (in the metalanguage) the z such that z does not satisfy CP. Then it is logically true both that the x which does not satisfy CP satisfies CP and that it does not satisfy CP, which is impossible. Therefore, by reductio, CP is not universally valid. So IP is not valid, and should be rejected.

Lambert's response is: in that case (i.e. when IP is rejected) Meinongians should

> present another *distinctive* identity criterion *à la* Quine for objects (objecta) in general, and impossible objects in particular, ... no *relevant* concrete alternative is presented by Routley (1976, p. 252).

A Meinongian may well decline to meet this "challenge": the call for such identity criteria has its extravagant side (as will emerge in the next section), and Quine's dogma "no entity without identity" is no more grounded in "common sense" than the referential assumptions that underlie it. Even so the challenge can be, and has been, met.

But first observe that Lambert's question has shifted, and his standards rise when it appears that impossible objects can meet them. Lambert's initial question as regards criteria for objecthood can be phrased: as regards object x, exactly when is $x = (Iy)A(y)$? More precisely, fill out the following wff in a nontrivial fashion:

$$x = (Iy)A(y) \equiv .\ - - -.$$

That challenge was met in the following nontrivial way (in Routley 1976, p. 249):

$$x = (Iy)A(y) \equiv .\ (\mathrm{ext}\, f)(xf \equiv ((Iy)A(y))f);$$

more generally, objects are the same iff they coincide in all extensional respects. This is a "relevant, concrete alternative"; it is quite specific; it answers Lambert's question (though using second order considerations) in the quite precise sense of filling out the schema indicated. In response Lambert, like Quine, upgraded his standards for satisfactory identity criteria – a most characteristic philosophical response.

§14. Identity again: Lambert's challenge and how Quine hits back.

Lambert has objected (1976, p. 252) that the basic identity criterion offered for impossible objects, namely coincidence in all extensional respects, does not meet the question he posed for Meinongians, "Quine's question: What *are* the nontrivial identity criteria for impossible objects?" (1974, p. 313). For firstly, "the sense of 'identity criterion' at issue is the well-known sense explained by Quine in many places, for example in [WO, chapter 6]" (1976, p. 252). Secondly,

> just as the criterion [offered] is inadequate to distinguish say, classes from attributes – both *kinds* of entities satisfy it – so it is inadequate to distinguish Meinongian objects, *qua* objecta from say, objects *qua* objectives (1976, p. 252).

This is simply false. While classes as usually conceived[1] are identical if they coincide in all extensional respects, attributes are not. To take a familiar example the property of being a featherless, comparatively hairless biped differs – for instance for modal reasons – from the property of being human, though in extensional respects – as regards the classes of entities picked out – they coincide. The bottom identity criterion for attributes is of greater than modal strength. Objecta and objectives differ similarly: identity determinates for objectives are intensionally stiffer.

While it is true that the basic identity determinable offered for impossible objects does not differ from that given for entities, that is no objection. For impossible objects are sharply distinguished from entities in *other* respects than their identity conditions (see §3 above). It is a mistake to try and rest the whole weight of the distinction between different sorts of things on their identity conditions.

Lambert's first point is harder to set to grips with. For nontrivial identity criteria for impossibilia have been given. And these will serve to meet Quine's questions (e.g. WO, p. 200) as to what counts as the same (impossible) object and what counts as another, and therewith Lambert's parallel questions (1974, p. 311), e.g. the round square and the square round are the same, the round square and the elliptical square are different. What more do they want in the way of "identity criteria"? Quine hardly explains this in the source Lambert cites (namely WO). Indeed the expression "identity criterion"

[1] Subsequently the usual conception will be questioned. Nonetheless, something of the usual contrast of classes and attributes as, respectively, extensional and intensional objects will be retained.

does not seem to figure at all. In the case of attributes, Quine simply tells us that the (rather conventional positivistic) "objections to propositions on the score of identity applies unchanged to attributes and relations" (p. 209). He does not even consider the classical account of identity of relations (the account given for instance in PM).

Fortunately Quine has recently elaborated his position in Quine 1975.[1] He puts his case against criteria for identity of objects like those offered for impossibilia in the special case of certain higher order objects, namely attributes; but he makes it plain, in the course of his further campaign against attributes, that his objections apply equally in the case of other objects than physical entities and classes constructed from these.

Quine's lead-in question (1975, p. 4) is: "What does the identity of attributes consist in?" The answer is going to be that there is no satisfactory noncircular answer, and hence that the notion of attribute is not really intelligible. For "the notion of attribute is intelligible only insofar as we already know its principle of individuation" (p. 7) – which, according to Quine, we don't. But to show the latter he has, at the very least, to dispose of straightforward answers to his lead-in question.

The "interestingly exasperating answer" Zedsky offers is that usual identity criteria suffice for attributes as for other objects: objects whether attributes or other, are identical (under a given determinate) iff they have the same features (as restricted by the indeterminate). But Quine is not interested in niceties such as identity determinates,[2] so determinates can be dropped for the purpose at hand, and suitable restriction on features understood. It is an old objection to the criterion that, so expressed, it is circular. But Russell, and others, set the definition in formalism and so showed how circularity was avoided: $x = y$ iff $(f)(xf$ iff $yf)$. Quine's initial move against Zedsky is this old-fashioned one: "Zedsky is evidently caught in a circle, individuating attributes in terms of, in the end, themselves" (p. 4). The move is met in the familiar way: by a definition which eliminates sameness in favour of quantification and biconditionality. Quine of course insists upon putting his points in terms of classes – or maybe Zedsky, eager to show how it could all be done in terms of notions Quine finds acceptable, foolishly sets the point in terms of classes. So the criterion gets transformed – not necessarily preserving equivalence – to $x = y$ iff $(z)(x \in z$ iff $y \in z)$, which makes "no mention of

[1] Lambert kindly drew my attention to this elaboration.

[2] Quine makes the characteristic referential mistake of expecting a single criterion of identity at the outset: "attributes, I have often complained, have no clear principle of individuation". (1975, p. 3).

identity" (p. 5) – nor of what Quine premisses his next assay upon, "mention of classes of attributes" (p. 5). Quine assumes that the formula $(z)(A \in z \equiv . \; B \in z)$ for identity of attributes A and B "contains only the single variable 'z' for classes of attributes". No doubt this suits Quine's argumentative purposes;[1] but Zedsky, if he sticks to his guns, can simply say that "z" is a variable for objects.[2] On the face of it, this blocks Quine's "deeper" ground:

> The real reason why the formula does not clarify the individuation of attributes is not that it mentions identity of classes of attributes [it does not], but that it mentions classes of attributes at all (p. 5).

Again it does not. Presumably Quine would restate his objection in some such terms as these: quantification over objects involves, in this case, quantification over attributes, and this supposes classes of attributes – whereupon the objection can proceed as before:

> we have an acceptable notion of class, or physical object, or attribute, or any sort of object, only insofar as we have an acceptable principle of individuation for that sort of object. ... (p. 5).

And so on: ... "no entity without identity" ... . It is important to see what the objection assumes but does *not* state. It assumes that an acceptable principle of identity for objects of a given sort cannot be provided through quantification which includes quantification over objects of that sort, *because such quantification already presupposes identity, or individuation, principles.*[3] That assumption is, it has already been argued, false. One can perfectly well talk of all objects of a given sort without there being sharp or especially clear individuation of objects of that sort: to previously cited examples such as natural objects of various kinds, e.g. clouds, mountains, waves, forests, can be added of course examples of just the kinds of theoretical objects Quine would like to dispense with, propositions, ideals, etc. Quantification does not require sharp individuation of elements, enumerability of elements, etc. (see 1.15, 1.16 [Vol. 1 —*Eds.*]). Hence identity criteria for objects of a given sort can be satisfactorily supplied – without circularity – through

[1] There are considerable advantages of course in considering doctrines of nonexistent philosophers such as Zedsky and Wyman. For positions that are perhaps hard to refute can be twisted into more malleable form, and then criticised, with the result that it can look, especially to those who have only glimpsed the original position from a limited perspective, that the original position has been equally criticised.

[2] Naturally some nonsignificance, such as "roundness belongs to 17", will have to be written off as false while the logical theory remains two-valued.

[3] There is a similar, initially more plausible looking, assumption made with regard to classes.

accounts which involve, among other things, quantification over objects of the sort in question.

Even so, objects so individuated may seem to compare invidiously with the extensionally hard sorts of objects Quine likes, notably physical objects and classes. For these objects are, so we are told, *well* individuated – physical objects by coextensiveness (which is an "impeccable principle" for *all* such objects, no matter how vague their boundaries), and classes by coextensiveness (or coincidence) of their members which are ultimately (in grounded cases) physical objects. None of this is as rock hard as has been put about. For what is coextensiveness? It is (as, e.g., OED informs as) *sameness* of extension (in space, space-time, etc.). And so meeting the old fashioned objection of circularity leads once again to quantification over objects or the like; in short, the questions Quine takes to be problems for attributes seem to be reappearing for physical objects. For consider first how sameness of extension is usually characterized in the case of classes (how Quine *does* characterise it): x and y are coextensive iff every member z of x is a member of y and conversely, i.e. $(z)(z \in x \equiv x \in y)$.[1] A comparable definition of coextensiveness for physical objects is now readily glimpsed, namely that commonly adopted in the calculus of individuals and mereology, which simply replaces membership by the part relation; that is, where $z < y$ reads "z is a part of y", $z = y$ iff $(z)(z < x \equiv z < y)$. (For such a definition see Goodman, and Quine himself). But this involves, among other things, quantification over the sorts of objects to be individuated. The escape Quine thinks *he* has from the problems he sets up for intensionalists is illusory.

Such identity criteria, as having just the same parts, are satisfied by all objects that significantly have parts (including classes and other nonentities); in particular, they are satisfied by natural objects such as clouds for which it was said clear individuation was lacking. The upshot is that, contrary to what Quine implies, identity tests such as coextensiveness are not sufficient for what would ordinarily count as clear individuation. Clouds are presumably identical iff they are coextensive, but they are hardly well individuated. The coextensiveness test does not enable objects that satisfy it to be counted for instance; yet well individuated objects should be enumerable.

Quine's unfavourable comparison of attributes with classes depends, furthermore, on a little cheating. For the impression is given that while classes can be accommodated

[1] Quine assumes existential loading of the quantifiers: but it is written not into the symbolism but the intended interpretation.

by going down through their elements attributes can only be dealt with (in a comparable extensionally admissible way) by going up, through what they belong to: hence while (grounded) classes "are well individuated" attributes are not because "classes of attributes are as badly off as attributes" (p. 5). It is plain however (what Quine says suggests it) that attributes may be distinguished through what has them. The attribute of heartedness is distinguished from that of kidneyedness by, e.g., possible (not to say actual) creatures with hearts which lack kidneys. More generally, if $(\Sigma x)(x \iota A \mathbin{\&} {\sim} x \iota B)$, i.e. some possible object instantiates A but not B, then $A \neq B$, i.e. A is distinct from B.[α] Moreover, grounded attributes will reach bottom in particulars, which satisfy identity criteria exactly like those for Quine's well-individuated physical objects: $x = y$ iff $(z)(z < x \equiv z < y)$. Venus is the same as Aphrodite because their parts (histories, etc.) coincide; Holmes is not the same as Moriarty because their parts go their separate ways.

The same points may be adopted to meet Quine's restatement of the issue in terms of open sentences (pp. 6–7): "the question of individuation of attributes [similarly classes] becomes in practice a question of how to tell whether two open sentences express the same attribute [class]". Classes are again supposed to have the advantage over attributes (not on the way up but on the way down) – the advantage that makes or breaks, and renders those that fail unintelligible:

> The desired formulation ... a satisfactory formulation which holds if and only if "Fx" and "Gx" determine the same class ... is of course immediate: it is simply "$(x)(Fx \equiv Gx)$". It does not talk of classes; it does not use class abstraction nor epsilon, and it does not presupposes classes as values of variables. It is as pure as the driven snow. Classes, whatever their foibles, are the very model of individuation on this approach (p. 7; my rearrangement).

In these times of increasing pollution even the driven snow is no longer so pure in industrialized places. The apparent purity vanishes once we start to analyze its contents. The intended content of "$(x)(Fx \equiv Gx)$" is: for every entity x, Fx iff (materially) Gx, i.e. the quantifier is restricted to entities. Thus if classes are not among the values of the variables the formulation is, even classically, inadequate, since many classes are not classes of individuals. It ceases to be evident too that we can, as Quine suggests, do

[α] Where the "Σ" quantifier is "for some possible", introduced in Vol. 1, §1.9 and defined in §1.17.3. —*Eds.*

nothing similar for attributes. Why not simply widen the scope of the quantifier to give "$(\Pi x)(Fx \equiv Gx)$"?[α] This is modally pure: we can even rewrite it "$(\underline{x})(Fx \equiv Gx)$", and only the range of the quantifier is construed differently. It too *looks* pure enough.

But before we get carried away with this enlargement on the referential criteria Quine has offered us, it would be prudent to look critically at the conventional referential wisdom on class identity we have so far been dished up. Consider, for instance, the existential coextensiveness criterion, of which the pure formulation is (usually taken to be just) a reexpression. According to that criterion the classes {Pegasus, Quine} and {Quine} are identical, and the classes {Pegasus} and {Zeus} are the same, the same as the empty set, as are all classes of numbers should it happen that numbers do not exist. This picture is, the noneist submits, entirely mistaken. Class theory, properly formulated, works in a fairly satisfactory fashion for what does not exist; indeed its major applications are in the area of the nonexistent given that mathematical objects and purely theoretical objects do not exist. It would not work correctly if the existential coextensiveness criterion were right. For similar reasons the pure test is wrong. The truth "$(\forall x)(x$ pegasizes $\equiv x$ zeusizes)", for example, has the result that "x pegasizes" and "x zeusizes" determine the same class; but they do not, one determining the singleton {Pegasus} and the other {Zeus}.

§15. Further objections based on theories of descriptions.

An almost endless series of objections to a theory of objects can result by drawing upon and varying principles of theories of descriptions tailored only to entities – evidence that there are indefinitely many ways of infiltrating the Reference Theory: description theory is certainly a favourite strategy for so doing. Objections derived from description theory take, however, two strikingly different forms.

For a curious feature of objections to theories of objects is the vacillation (sometimes even in the one critic) between saying that a theory of objects is impossible, because for instance inconsistent, and saying that all the main assignments of a theory of objects can be absorbed and explained in an extension of classical theory. But the latter situation of course excludes the former by providing a classical model for the theory of objects, thereby ensuring its consistency. This curious phenomenon is exhibited especially with

[α] Where the "Π" quantifier is "for all possible", introduced in Vol. 1, §1.9 and defined in §1.17.3. —*Eds.*

theories of descriptions; and an instructive example is provided by Tooley 1978's objections to Meinongian semantics.

On the one hand, Tooley endorses the following principle of certain classical description theories.

DP. If $a =$ (the x)$A(x)$ then $A(b)$ iff $a = b$,

which induces inconsistency in any theory of nonexistent objects with but few further assumptions. On the other hand, he (subsequently) proposes the following description scheme to account for what we are inclined to say about nonexistent objects:

TP. ((the x)$A(x)$)f iff either $(\exists!x)(A(x)\ \&\ xf) \vee .\ {\sim}(\exists!x)A\ \&\ \Box((\exists!x)A(x) \supset xf))$,

i.e. the A (is) f iff either there exists a unique x which is A and f or else there does not but it is necessary that if there exists a unique A then it is f.

TP does allow for some correct assignments to statements about nonentities, e.g. "The golden mountain is golden"; unlike DP it permits a theory of a sort about nonentities. Indeed it enables counterexamples to DP to be designed, as the argument to inconsistency from DP will reveal.

The argument[1] to inconsistency is as follows: Let "a", "b" and "c" name, respectively, these nonexistent objects: the golden mountain, the (exactly) 5000 feet high golden mountain and the 6000 feet high golden mountain. Then, using elementary principles of the theory of objects and obvious abbreviations:

(1) $a =$ (the (unique)x) $xg\ \&\ xm$

(2) $b =$ (the x) $xg\ \&\ xm\ \&\ x5000$

(3) $c =$ (the x) $xg\ \&\ xm\ \&\ x6000$

Applying the Characterisation Principle to b,

(4) $bg\ \&\ bm$

For by the CP, ((the x) $xg\ \&\ xm\ \&\ x5000)g\ \&\ \ldots$. Hence by DP, (4) and (1),

(6) $a = b$

By parallel reasoning using (3) in place of (2),

[1] The argument was suggested to Tooley by H. Burdick.

(7) $a = c$,

whence, by transitivity,

(8) $b = c$.

But the properties of b and c differ, one being 5000 feet high, the other 6000 feet high (using the CP again); so by identity principles

(9) $b \neq c$,

contradicting (8).

The conclusion would be damaging indeed – namely that any theory of nonexistent objects is inconsistent – were the premises compelling. Unfortunately for the argument, however, DP has little to recommend it. What reason does Tooley offer? That it is normally accepted. Not even this is true. It is not a principle of basic free logic for example, which rightly restricts DP to *existent* a. (Free logic yields only the universal $\forall$-closure of DP). DP does not hold, for that matter, on Russell's theory of descriptions, without important qualifications. To ascertain these qualifications, consider an attempted proof of DP on Russell's theory. The antecedent expands to $(\exists x)(A(x)\ \&\ (\forall y)(A(y) \supset .\ x = y)\ \&\ x = a)$. Suppose, firstly, $a = b$; to show $A(b)$. Since $x = a$ for some given x, $x = b$. Also $A(x)$. But to conclude $A(b)$, requires that A be *transparent*. Suppose, secondly, $A(b)$ to show $a = b$. To use the uniqueness provision, it has to be taken for granted that b *exists*. Then, by instantiation $A(b) \supset .\ a = b$, whence $a = b$. The second qualification is the crucial one. For the argument to inconsistency depends upon applying DP (twice) where b does not exist. The argument fails, that is, even on Russell's theory of descriptions.

Really DP is a totally disastrous principle. By DP, if $a = IxA$ then, if $a = b$, $A(b)$. Take b as IxA. Then, if $a = IxA$ then $A(IxA)$, by contraction. Now take a as IxA. By reflexivity $A(IxA)$. DP yields, that is, an unqualified Characterisation Principle, and thus inconsistency – irrespective of questions concerning nonexistent objects.

The other half of DP, which is the operative principle in the argument against nonentities, is only marginally better. The principle

(5) If $a = IxA(x)$ then if $A(b)$ then $a = b$,

is false, as the following sort of counterexample shows (the example is easily formalised on the theories of objects given or in other theories such as Parsons' theory): Let $a =$

the round thing, and b = the red round thing. Then b is round, but it is false that $a = b$. More generally (5) implies, what is false, that at most one object is nonexistent. Suppose there were two, b and c; $\sim bE$ and $\sim cE$. By (5) and reflexivity, if $A(b)$ then $b = IxA(x)$. Hence $b = Ix{\sim}xE = c$. (5) has at best severely restricted validity. One requisite restriction is already evident from the qualified form (5) takes in Russell's theory and on free logic: a further clause to the effect that b exists is required.[1]

Removing DP and (5) makes logical room however for Tooley's other objection from a rival theory of descriptions, that the theory of objects, though not now inconsistent, is unnecessary and the intuitions underlying it can be better accounted for through a variant reductive theory of descriptions, namely that encapsulated in TP above.

But TP is also inadequate. The following are typical objections to this Carnapian fix (a Carnapian fix being to take the Russellian theory in the unique existence case and to give an alternative where the unique existence clause fails):

1. it fails for intensional f, e.g. "was thought about by Meinong";
2. it fails for logical f, e.g. "is incomplete";
3. it sanctions ontological arguments. For example, on the analysis, the existent golden mountain exists, and, of course, the round square that exists exists.[2]

Tooley presents TP as only illustrative however of ways in which features of nonexistent objects, such as their incompleteness, may be explained in a realistically satisfying way.

> A Meinongian semantics provides no explanation of why there are incomplete objects. A realist semantics, in contrast, by analyzing sentences purportedly about nonexistent objects such as the golden mountain into sentences which are about linguistic entities, such as predicates, or about intensional entities, such as concepts, or about

[1] Strictly that is not all that is required. For principle (5) assumes a (over-) strong uniqueness requirement, so that the golden mountain, should it exist, must be one with the golden mountain further described. An interesting feature of the Tooley–Burdick argument is, in fact, that talk of entities could almost entirely replace talk of nonentities. An analogous argument with entities fails only at the very final step. Further determinations of entities do not change them (under the theory assumed); further determination of nonentities can change them – a significant logical difference.

[2] Proposals rather similar to TP appear in the literature, e.g. in Grossmann MNG, p. 161, p. 167, where it is asserted (falsely) that Meinong's claims are really about interrelations of properties; also in Dyche 1976 and in Rapaport 1978. A somewhat similar adaptation of Russell's theory of descriptions is to be found in Hintikka 1959 and in the original version of *Exploring Meinong's Jungle*, Routley 1967.

> real entities, such as properties, *can* provide an explanation of the features in question (p. 4).

It is, in the first place, just false that a Meinongian semantics provides no explanation of why some objects are incomplete, of why the golden mountain is indeterminate as to altitude. Naturally it does not provide a reduction; but not all explanations, by any means, are reductions. Moreover reductive explanations, for what they are worth, can often be parallelled by nonreductive explanation in terms of the character of nonexistent objects, e.g. that they have their characterising features but lack extensional features that do not flow from their characterisation. Secondly, there are good grounds for scepticism as to the availability of reductions of the sort required. Despite much labour none has so far resulted. Indeed there has never been such an analysis of which one could truly say: that was close to successful. And there are simple theoretical reasons for supposing that no analysis can succeed. Take, for example the special case of propositions, a sort of nonentity. Since there are nondenumerably many propositions there is no hope of individuating them in terms of finitely many actual speakers and finitely many sentences – uttered sentences. That is, a linguistic reduction is precluded. But what of reductions of objects through concepts or properties? Noneists are not going to agree that such objects exist; such reductions, were they to succeed, would be only of objecta, bottom order nonentities, to higher order objects. Thus such reductions already suppose a theory of objects: they do not offer an alternative to it.

Certainly the idea is about – it appears e.g. in Grossmann MNG and Dyche 1976 – that the distinctive truths of a theory of objects which flow from its characterisation postulates can all be somehow accounted for in terms of a picture of objects as systems of properties or natures. Such a simple picture is too simple: it is inconsistent without qualification. Moreover (as will also be seen in chapter 12 [Vol. 3 $_{—Eds.}$]), elaborating and working out the picture leads, as the development of the model through Parsons 1974 and 1978 shows, not necessarily to objections to theories of objects, but to more sophisticated and logically satisfactory theories of objects.

The reduction strategy is misconceived. Instead of trying to squeeze nonentities into preconceived moulds, which they do not fit, what is required is a phenomenological investigation – of how they behave, what features they have. Strictly the same is required of properties, since enough of their features remain obscure. A prudent course for anyone considering whether a reduction of objecta to properties or classes of properties could succeed, would be to investigate both classes first and then determine – ideally *show* – that a reductive analysis succeeded – or, more likely, failed. Reasons for suspecting failure – such as that nonentities are required to account for aspects of properties – are given in subsequent chapters (especially in chapter 12).

§16. The charge that a theory of items is unnecessary: the inadequacy of rival referential programmes.

If, however, nonentities cannot be reduced then surely they are not needed – at least in what counts, science. It is indeed a very familiar objection to theories of items that they are unnecessary. Everything one really wants to say – everything that really needs to be said it soon becomes, when it turns out that much one wants to say cannot be said on the alternative proposals – can be said without all the dubious and metaphysically objectionable apparatus of a theory of items. We have encountered this sort of objection before, and shall meet it again in subsequent chapters where it will be answered along the following lines: There is no satisfactory alternative to a theory which accepts leading theses of a theory of items if an adequate account of discourse is to be given. Alternative theories invariably leave out or distort much of what one genuinely wants to be able to say or to allow to be said. Nor is there any good reason to insist on such alternative, characteristically referential theories, once the Reference Theory and its elaborations have been seen through. For it is then apparent that though there are of course difficulties in working out how a theory of objects should be extended into new areas (where alternative theories have in any case seldom ventured), such as how the theory of items is to be appropriately combined with relevant logic, nonetheless a theory of items need involve no metaphysically dubious items. To insist that it does is to revert to referential thinking. Furthermore science is by no means all that counts philosophically. And even in the case of science nonentities are fundamental for an adequate account.

The formal mode reduction of discourse about nonentities, the Russellian reduction via a theory of descriptions, and Fregean style reductions to concepts, are only some among the many proposed ways of eliminating discourse about such "peculiar" objects in favour of honest-to-god referential discourse. Another favoured reduction programme, that undoubtedly has much appeal in the case of objects of myth, fiction and the like, is the elimination of nonentities that *do* feature in discourse by way of their (empirical) sources. The pull of such an approach Tooley[1] conveys:

> one surely feels that there is a reason why the Meinongian object, Sherlock Holmes, has the property of smoking a pipe, and lacks the property of smoking cigars. And the reason, surely, consists of certain facts about the world of existent things, facts about certain writings of Arthur Conan Doyle. But how are we to conceive of the relation between these facts about certain books, and the fact that the Meinongian object, Sherlock Holmes, has certain properties and lacks others? Is the relation a causal one? Or a logical one? Or something else? It seems natural to think of it as a logical relation, as some sort of entailment relation. But if facts about the world of existent things entail that there is a certain sort of Meinongian object, with certain properties, those facts about the existent world logically determine which fictional statements containing the name "Sherlock Holmes" are correct, and which incorrect. And in doing so, they provide a *deeper* account of the correctness and incorrectness of fictional statements – one that shows Meinongian objects to be, so to speak intervening variables.

But, firstly, this misrepresents the depth a theory of items can provide. Unless there is some intrinsic depth to referential accounts (there is not, though the impression that there is is often given), any depth of a referential approach can be at least matched by a Meinongian approach.[2] The Meinongian approach does not end – cannot satisfactorily end – with a statement of simple facts such as that a nonexistent object, Sherlock Holmes, has certain properties, e.g. he smoked a pipe but not cigars. There is always an account to be given of how nonentities acquire the properties that they have. In certain cases this will involve details of actual stories other cases theories, and so on. In some cases it may

[1] Tooley says that the "objection (1978, p. 4) was suggested by some remarks made by" Mortensen. Note that really it is not so much an objection, as a complaint that the Meinongian approach does not go "deep enough", together with a proposal for a further and perhaps rival programme.

[2] There is of course more than one measure of depth. Even so things are often said to be deep when they are not, but for instance simply difficult routes to something easily achieved in other ways. For example, a detailed Tooley–Mortensen reduction of fictional objects might be hard – very hard – to get right, but that does not offer any substantial reason to think that it would thereby be rendered *deep*. But the theory of objects, no matter how tortuous, never seems to qualify for any of the laudatory empiricist epithets, such as "deep", "solid", "sound".

involve details of merely possible or conjectural stories. For the theory should be able to take account of the deeds of unsung heroes, just as it does of unstated propositions and unformulated theories – matters that make for very serious trouble for "realist" accounts. Indeed for this sort of reason a full reduction to actual source books is impossible.

Secondly, there are problems for any reductionist programme about the relations between source books and their objects. Granted that the relation between the source book(s), and the features of Holmes is a logical one (it is in fact a semantical relation), it is *not*, it seems certain, an entailment relation: Facts about the existent world do not entail Holmes smoked a pipe, or similar. Nor can an entailment relation be dredged up – without the insertion of *some theory* (about objects) – for the following sort of reason: An empiricist-style logic such as *Principia Mathematica* (mark n model say) can incorporate all facts about the world (that an empiricist would wish to assert), but will nonetheless falsify (through the theory of descriptions) "Holmes smoked a pipe". So, there is no entailment.

Perhaps an elimination programme can succeed in other ways, e.g. through the semantical connections. These programmes, however they are elaborated, are bound to resemble programmes (to be examined subsequently), older and largely faulted programmes, for eliminating theoretical objects of science, and for eliminating discourse about universals; and they have about the same prospects of success, namely none.[1]

In conclusion, there is a general guide – it is hardly to be expected that it be fully effective – to meeting objections to the theory of items that is worth recording; namely, *give the common sense response*. The reason that such a procedure works well is that the theory of objects is, as we will came to see more clearly in subsequent chapters, *much* closer to common sense than its main opponents, the Reference Theory and its reductionistic variants. Furthermore, any opponent of the theory of objects would be well-advised, before he levels the accusation that the theory of items leads to weird results, to consider whether his own theory does not lead, *when followed out*, to conclusions that are far stranger, and further removed from common sense, than any that the theory of items entails. The main alternative theories that have been worked out in much

[1] See further chapters 10 and 11 [Vol. 3 —*Eds.*]. The programme of eliminating universals in favour of a first-order logic, for example, demonstrably fails. Where reduction programmes can succeed after a fashion, as in the case of the elimination of the intensional in favour of the extensional, explanatory power, and thereby depth, *are* sacrificed.

detail in modern times lead in such directions – consider, for example, the atomism of Russell and the early Wittgenstein; the extravagant elimination by the logical positivists of much ordinary discourse as nonsense; the scepticism about translation and the flight from intensional discourse and wipe-out of substantial parts of it by Quine. So also do other less fully worked out theories – consider, for instance, the surprises Kripke has for us; the disastrous conventionalism of Carnap and the later Wittgenstein; the underlying empiricism and nihilism of ordinary language philosophy. Well may one look upon these modern referential works and despair. At least, however, be warned before too lightly dismissing the nonreferential alternative offered by the theory of objects.

Chapter 5

Three Meinongs

The three Alexius von Meinongs, of theory of objects fame, to be considered are the unhistorical Meinong, that is the mythological Meinong of mainstream philosophical literature, the consistent Meinong, and the paraconsistent or dialectical Meinong.[1] At most one, and perhaps none, of these Meinongs actually existed, and certainly the first mythological Meinong never did exist. The primary task in what follows is not, however, the historical one of trying to determine, if it can be determined, which if any of these or other Meinongs did exist (though this is not without importance, and is attempted in section 5), but to continue the on-going case for and assessment of Meinong's theory of objects and near relatives to it, and to begin on assessing the case for the dialectical Meinongian position vis-á-vis the consistent one.

§1. The mythological Meinong again, and further Oxford and North American misrepresentation.

Meinong has been presented in the recent history of philosophy as a philosopher with weird and peculiar theories far removed from commonsense, as a philosopher whose position was to be avoided at any cost, so that mere association of a position with that of Meinong was on its own a conclusive refutation of it, as a warning to all of what would befall them should they be tempted into following his path into the jungle. Meinong's

[1] Although we shall talk about the consistent Meinong, the dialectical Meinong, and so on, really each description covers a cluster of positions. Subsequently, in chapter 12 [Vol. 3 $_{—Eds.}$], substantive issues other than the consistency and coherence of the position of the historical Meinong will be taken up, and this will lead to the consideration of various allegedly historical Meinongs. Particularly important as regards Meinong's theory of objects is the question as to how Meinong understood his objects: were some of them genuine individuals, or were they really natures, complexes of properties, or some such.

The original version of this chapter was revised. Due to a mistake by the publisher, the editor's name 'Dominic Hyde' was mistakenly appeared as a co-author in all the chapters on Springerlink and third-party websites. The correction to this chapter is available at https://doi.org/10.1007/978-3-030-26309-6_7.

D. Hyde (ed.), *Noneist Explorations I*, Synthese Library 415,
https://doi.org/10.1007/978-3-030-26309-6_4

theories encountered heavy opposition from their inception, so much so that in later works Meinong had already unduly modified and weakened some of his claims, e.g. those concerning the character of impossible objects (see e.g. EP). The opposition to Meinong's theories, and increasing misrepresentation of them, were no doubt encouraged by remarks of James and by the platonistic picture of Meinong's theories put into circulation by the later (post-1918) Russell.[1] But the really serious misrepresentation of Meinong's thought in the English-speaking world began, it seems, not with Russell but with the Oxford philosophers, and especially Ryle. It is from Oxford philosophers that the main misrepresentation of Meinong's position, and the presentation of Meinong as a figure of fun, as a philosophers' Aunt Sally, has come. This picture, according to which a central thesis of Meinong's theory of objects is that all objects have some kind of being, is very far from accurate. But it has persisted. And it has been modified only slightly by philosophers of the Bergmann school, such as Grossmann MNG, who still see Meinong as a super-platonist, as an entity-multiplier par excellence (even if a serious philosopher with some decent reasons for his claims).[2]

Major attacks on Meinong's position and substantial misrepresentation of his position, have come, not too surprisingly, from a reductionist, and usually empiricist, opposition. The misrepresentation has come primarily from philosophers who are locked into a framework of assumptions that Meinong would have rejected, chiefly facets of the Reference Theory such as the Ontological Assumption (as, for example, the Oxford philosopher Hinton whose criticism of Meinongianism was assessed in chapter 4). It is an easy step from seeing, on the basis of such a framework, a nonreductionist theory such as Meinong's as saturated with the systematically misleading, to systematic misrepresentation. And this is what has happened. Even Findlay, who has given us a fairly sympathetic overview of Meinong's work, slipped in his *later* appraisal of Meinong (written under Ryle's

[1] Though Chisholm notes (in 1972, p. 36, note 12), and Griffin argues in detail in 1977, that Russell was rather scrupulously fair to Meinong in his earlier expositions of Meinong's theories. This is not Findlay's view (1963, pp. xi–xii):

> Unfortunately Russell was far too concerned to advance from Meinong to his own notions and conclusions to bother to get Meinong quite straight, and the accounts he put into circulation of Meinongian contents as consisting of sense-data and images, and of Meinong's non-existent objects as "subsistent", are simplifying travesties of Meinong's complex opinions.

Findlay does adduce several rather convincing pre-1918 examples in the course of 1963, e.g. p. 84, p. 94. A careful reader will find other examples in Russell 1904.

[2] Grossmann's misrepresentation of Meinong will be examined elsewhere; a small beginning only on this task is made in chapter 12 [Vol. 3; cf. also chapter 4, §11 *—Eds.*].

influence) into using the very misleading terms of the opposition, into seeing Meinong as the introducer of entities (Findlay 1963, pp. 327, 333) and as having "a world too rich in forms of intensionality and overpopulated by objects" (p. 326). But the horse population of the world comprises the horses that exist. Pegasus does not exist, so Pegasus does not belong to the horse population of the world. Similarly nonexistent objects do not belong to the population of the world. The world may be overpopulated, but it is not "overpopulated" by nonexistent people.

The unhistorical criticism of Ryle (who plays the role of arch-villain in the first act of this melodrama) is more serious. Ryle's initial diagnosis of how it is that the theory of objects is fundamentally mistaken depends upon associating it with the doctrine of terms. Ryle claims (1933–34, 1971, 1972) that the theory of objects is founded on the traditional doctrine of terms, the whole structure of which is rotten. Meinong's main service to the philosophical community was a *reductio ad absurdum* of this traditional doctrine; by taking it to its absurd extreme, he showed the rottenness of the whole structure. Ryle's case however depends on several assumptions – which fail. Firstly, Ryle's case depends upon expanding the doctrine of terms, and assigning to Meinong theses from the expanded doctrine that he never held. According to the traditional doctrine of terms, *every term* stands for, or *designates, an object.* This thesis Meinong did adopt (though with qualification in the case of defective objects, see §3). It was *not* however part of the traditional doctrine of terms, or of Meinong's view, that every object so designated existed or had being of some sort (Mill makes this very plain: see 1947). But Russell and Ryle attributed such a view to Meinong (e.g., Ryle "The Platonizing Meinong", 1972, p. 7), and it has stuck. It was *not* part of the traditional doctrine of terms that the designation of a term is its meaning; it was certainly not part of the doctrine that the meaning of any expression at all was some designation. Nor was it Meinong's view: Meinong made the elementary distinction between what a term expresses and what it designates. Yet it is a favourite criticism of Ryle's that Meinong accepted, what the doctrine of terms included, the "Fido"-Fido (or reference) account of meaning (thus e.g. "the lid of ... the 'Fido'-Fido box ... was never even lifted by Meinong" Ryle 1972, p. 7). It was *not* part of the traditional doctrine of terms that every expression occupying a subject position in English is a term, in particular quantified subjects with "everything" and "something" were not traditionally accounted terms, nor were auxiliary expressions such as "the sake".

(Ryle's crack 1972, p. 14 that "Lewis Carroll would not have amused them [Meinong and Husserl], or therefore, taught them" is accordingly misdirected.)

Secondly, Meinong did not accomplish a *reductio ad absurdum* of the traditional doctrine, nor has such a *reductio* ever been accomplished without additional assumptions which go well beyond the doctrine of terms. (Nor, as logical investigations show, *can* a *reductio* be accomplished; for the main theses of the doctrine of terms can be satisfactorily modelled, in a way evident from 1.24 [Vol. 1 $_{—Eds.}$]). Meinong had no reason then to stand back and say, as Ryle believes those who were right said, "but this is ridiculous" (Ryle 1972, p. 14).

Ryle's later diagnosis of the error of Meinong's ways – included in his premature obituary note for the theory of objects – was that the main error was that of nominalisation, and the treatment of the results as logical subjects (1972, especially, p. 12), and that the solution lay in Brentano's and especially Russell's

> systematic and strategic de-nominalisation. The principle of their mutinous reaction was that wherever possible – and wherever anything was at stake – the contributions made to sentences by words and phrases must be shifted away from the "Socrates"-place and into the predicate-place (p. 12).

Ryle should have realised that the method of avoiding nominalisation by predicate paraphrase could not succeed generally. For if it worked higher-order predicate logic could be reduced to first-order logic, to take one blatant example of failure; similarly set theory could be reduced to virtual set theory. That some sample cases may be more or less paraphrased – all Ryle has ever offered by way of justification of his extensive claims – does nothing to establish a general reduction. When it comes to preserving sense, point and force of claims – preservation of which would require substitution *salve veritate* in decidedly intensional contexts – comparatively few of the nominalising operations that English affords can be discarded without loss. Ryle's reduction programme, for eliminating a wide range of "systematically misleading" nominalisations, is impossible, and not merely in isolated cases. Moreover from a Meinongian point of view the whole reduction, and systematic elimination, programme is misconceived. For it is based in large part on an assumption of ontological reduction, that by eliminative means existential commitments can be removed, since, by the Ontological Assumption, the subject places eliminated carry existential commitment. But the underlying assumption is mistaken: use of a term in a subject place carries no commitment to the existence of what the term is about.

Misrepresentation of Meinong by no means ends with the representation of Meinong as a super-ontologist or super-nominaliser who took all his (weird) objects to exist, or subsist, in some way. No less pernicious, and more serious (since platonism is not usually accounted trivial), is the very different assumption that Meinong's theory is bound to be inconsistent – indeed that *any* theory of nonexistent items is bound to be inconsistent – an objection emanating from Russell. And hardly less damaging and waylaying to attempts to systematise a theory of objects is the proposition, forcefully expounded by Quine, that nonexistent objects are incorrigible and that no worthwhile theory can contain them – a proposition that has frequently been used against Meinong's theory, without any proper account being taken of Meinong's distinctions and classifications of objects.[1]

Both these sorts of dismissal of Meinong, and of theories of objects, have been circulated from Oxford, as if furthermore the objections were entirely fatal. Thus, for example, Kenny 1968 reiterates, as insurmountable, the objections to nonentities that Quine tries to manufacture from issues as to identity and similarity of such objects (objections considered in chapter 3). More telling is the line emanating from Russell. Russell is said by Lambert (1972, p. 37) to have argued (in OD) that *any* theory which admits nonexistent objects is inconsistent. A more modest version of this thesis – qualified to: any theory respecting ordinary language – (Carnap MN, p. 65) also derives from Russell. At least in his main argument in OD Russell makes no such sweeping claims as to the inconsistency of any theory (or, what is even less likely given Russell's attitude to ordinary language, any theory respecting ordinary language) which admits nonexistent objects. Russell's claim is that "such objects, admittedly, are apt to infringe the law of contradiction" (p. 45) and that the objects of the theory of Meinong [and, strictly, his students Ameseder and Mally] do break the law. But subsequently, in his discussion (p. 54) of McCall's division of individuals into real and unreal, Russell equates *any* theory which assigns nonentities (unreal individuals) as the denotation of significant subjects (which on his view do not denote anything) with Meinong's theory, and so to be rejected as

[1] Though misrepresentation of a position and objection to the position are commonly enough quite distinct, there are other cases where there is no easily discernible line between the two.

conflicting with the law of contradiction. In short, Lambert is right (except perhaps for the bit about arguing). The invalidity of Russell's argument is striking enough once the argument is exhibited. For to derive an infringement of the law of contradiction from a theory of nonentities in Russell's way, further assumptions have to be pumped in – assumptions about assumptibility and negation, which (though they are commonly taken to be part of Meinong's theory) are not part of Meinong's developed theory, and which are *certainly not* part of alternative theories of nonentities, some of which are demonstrably consistent (as Lambert 1976 and others have explained, and as Parsons has now shown, in outline, in the case of a sophisticated theory of nonentities, with much in common with Meinong's mature theory: see his Parsons 1978, IV).

But the myth that any theory which admits nonexistent objects is inconsistent or absurd persists, at least at Oxford, as Dummett's work reveals:

> ... absurdities result from adopting the view of Meinong – and early Russell – that there are objects which do not exist, objects which are not actual, but merely possible, ... (Dummett 1973, p. 197).[1]

Furthermore,

> ... as Meinong's experience showed, we run into uncomfortable antinomies when we try to lay down the truth-conditions for statements about possible objects (1973, p. 386).[2]

What is the argument for these claims, from which major logical conclusions are drawn? Sandwiched in between the claims (with intervals of about 100 pages either side), we find (1973, pp. 279–80) a repetition of Russell's argument (discussed in Findlay 1963, p. 104ff.) that Meinong's theory enabled novel applications of the ontological proof:

> The most telling objection to the thesis that "exists" is a predicate which applies to some things but not to others derives from the ex-

[1] It is important to observe that Dummett does *not* make similar objections to platonism, and that similar objections are not usually made to platonism, since it is widely recognized that (properly) qualified platonism is at least a consistent doctrine.

There is then a tension between the representation of Meinong as inevitably embroiled in an inconsistency and of Meinong as a super-platonist, that some critics have failed to observe.

[2] Unless the second claim is little more than a repetition of the first, it is farfetched. Meinong's experience did not appear to include anything that would be counted as "laying down truth-conditions" in the modern sense.

In any case, truth-conditions for elementary theories of nonentities can be laid down without encountering antinomies of any sort.

> perience of Meinong in his ill-fated attempt to maintain that thesis.[1] The thesis goes, of course, with taking our expressions of generality as ranging, not only over actual, but also over merely possible (and, in Meinong's case, also over impossible) objects, and with the taking of names which lack an actual referent as standing for a merely possible (or, again, perhaps an impossible) one.[2] In that case, we are faced with the problem when a predicate is to be held to apply to a non-existent object: and *the answer which forces itself on us is that it must apply whenever its doing so follows from the mere sense of the name of the object.* It is precisely from making just this move that ... the ontological argument starts. But now, if "exists" is a predicate like any other, there is no reason why the requirement of satisfying this predicate may not be incorporated into the sense of some complex name; indeed, there is nothing to prevent us inserting at the beginning of any definite description, the word "existent": and it will then appear that we are committed to the thesis that true sentences result from putting any such complex proper name into the argument place of "ξ exists', i.e. that the referents of all these complex names exist [emphasis added].

Thus, to use Russell's examples, the existent present King of France exists (and also does not exist), and the existent round square really exists. The fundamental flaw in Dummett's argument is exhibited in the italicised assumption (an unqualified version of the Characterisation Postulate). Such an answer did not force itself upon Meinong, or upon earlier or more recent investigators of theories of non-existent objects; and there are excellent reasons for not giving such an answer. It follows from Meinong's developed views that we cannot make existence, or any other sort of genuine being (i.e. being carrying the modal moment), part of the nature of an object (cf. Findlay 1963, p. 105). The argument from the sense of "the existent ξ" to the truth of "the existent ξ exists" accordingly fails, because existence is not a characterising, or nuclear, feature, and so not assumptible. Dummett has done nothing to refute theories of nonexistent objects, other than the very

[1] The inaccuracy of this attribution emerges from Findlay 1963, p. 105 and p. 176.

[2] These connections are not necessary, or even universal. Free logics counter some of the connections, the taking of existence as a universal property another, one of Meinong's positions another.

naive ones Russell had already disposed of, and nothing to demolish Meinong's considered views; indeed Dummett has not bothered to take account of these views.

It is not, of course, being suggested that misrepresentations of Meinong's position now emanate only from Oxford. The picture of Meinong as a super-platonist has had much philosophical coverage in many places, particularly in the American mid-West,[1] and the idea that Meinong was bound to accept, and did accept, an unqualified Characterisation Postulate is also alarmingly widespread. The more common style of misrepres-

[1] Examples abound, especially in publications of the Bergmann school.

A recent striking example is Lycan 1979, where the paradigm of Relentless Meinongianism, which is equated with Meinongian realism, is Lewis's Democritean theory of alternative existing worlds, and where Meinong is portrayed as a theorist whose ontology included all sorts of nonexistent objects. Relentless Meinongianism is intended (like Hinton's resolute Meinongianism) to be an irreducible form of Meinongianism, which (unlike Hinton's Meinongianism) takes as primitive non-standard Meinongian quantification and a nonuniversal existence or actuality predicate. Unfortunately for Lycan's unconvincing attempt to assimilate Meinong and Lewis, as relentless Meinongians separated by mere terminological differences, even the primitive logical apparatus is interpreted very differently (not to mention other things). Lewis's quantification is standard, not neutral; his objects of quantification are existent beings which are consistent and complete; and his actuality predicate simply effects a partition of these entities which separates out those of one world. Meinong's neutral quantifiers *are* non-standard, the objects of quantification include many objects which have no existence or being at all, and very many of which are incomplete or inconsistent (and not fully assumptible) and so could not exist. Furthermore all of Meinong's objects are in the object-domain of the actual world, and true unmodalised statements can be made about all of them. It is not just that the objections made (in chapter 4) to attempts to convict Meinong's theory of platonism apply against Lycan's "terminological" transformations – transformations which fail to preserve philosophical positions, and which commit the cardinal ontological sin of converting "is an object" into "exists" – but that Lewis's world-partitioned universe, the logical theory of which may be classically formulated, differs drastically from anything genuinely Meinongian.

Because the terminological transformation fails, some of the drawbacks to relentless Meinongianism which Lycan enumerates – which are objections to Lewis's theory – fail to apply to Meinongian theories. And others are easily met. For example, nonentities cannot be (classically) met by entities (or counterparts encountered by their real life originals) because of the Brentano condition, that entities cannot stand in entire physical relations to nonentities; and knowledge of nonentities is obtained by reasoning from their respective source books. Which leaves just one objection, which Lycan repeats over and over, namely that what amounts to neutral quantification is taken as primitive, is not reducible (to referential quantification) and (so) is literally unintelligible. While it certainly is referentially irreducible – it can be defined, e.g. in the lambda theory, but only presumably in ways that would generate parallel objections – unintelligibility is not a consequence. For analogous objections could be made to referential quantifiers, which are ultimately explained in terms of natural language quantifiers. But nonreferential quantifiers can be similarly explained, e.g. in terms of "every" and "some". This reveals that the unintelligibility of neutral quantifiers is not the real issue, but only symptomatic of the opacity to reference theorists of the Meinongian notion of *object* or *thing*. For the neutral quantifiers can be explained exactly as Quine explains his quantifiers (which Lycan finds intelligible): "(Px)" reads "for some thing x" and "(Ux)" "for every object x" (cf. Quine, e.g. 1959, pp. 83–85). All that differs is the range of objects encompassed (Quine's being restricted to referents, i.e. transparent entities, or in more old-fashioned terms, clear and distinct entities). So Lycan's objection boils down to an objection to the neutral notion of "object": but such objections have already been dismissed in chapter 4.

entation is exemplified by the following invalid argument – a *some* to *all* argument – of Linsky (1977, p. 33: similarly Lycan 1979):

> Meinong uses the example of the gold mountain which both does not exist and is made of gold. This commits Meinong to the universal validity of the formula $\phi(\imath x)(\phi x)$.

No such thing: all the premiss establishes is that Meinong accepted $\phi(\imath x)(\phi x)$ for some ϕ. Linsky's other argument that Meinong was committed to the universal validity of the principle is different but no less fallacious: it is (p. 34) that the insistence that "$(\imath x)(\phi x)$" denotes $(\imath x)(\phi x)$ "entails that $\phi(\imath x)(\phi x)$ is always true for any choice of ϕ". Again no such thing, without the further naive, and mistaken, assumption that the object denoted is, because presented as the object which is ϕ, ϕ. What does require more work to counter is Lambert's contention (1976, p. 253) that Meinong imposed no restrictions on the Postulate, and indeed that there is abundant textual evidence that this is so. This appears misinformed. Lambert rashly relies upon work Russell was supposed to have had available; but part of Russell's case against Meinong – the argument concerning the existent King of France – rested not on Meinong's theory but on what one of his pupils had said, and Russell's early work of course took no account of Meinong's mature thought. There is, moreover, clearcut evidence from Meinong's later work that the Characterisation Postulate was restricted (cf. also 4.7).

The evidence comes firstly from Meinong's doctrine of the modal moment (explained in Findlay 1963, p. 102ff.),[1] in terms of which Russell's objection that the theory of objects generates ontological proofs (e.g. novel proofs that impossible objects exist) is met:

> ... the pure objectum is indifferent both to being and non-being; these distinctions come to it from outside, and are only to be found in objectives which concern it. It follows that, strictly speaking, we cannot make existence, or any other sort of being, part of the nature of an objectum. ... genuine existence does not belong to the sphere of so-being (1963, p. 105).

Thus, for example, the existent round square (though it has, so Meinong allows, suppositious existence) does not exist because its existence lacks the modal moment. Furthermore

[1] The doctrine is presented in Mög, chapter 37; but Findlay's presentation is more accessible.

> we *cannot*, by means of a judgement or an assumption, attribute the modal moment to an objective [or object] which does not possess it (1963, p. 107).

Put differently, predicates carrying ontological-style commitment (e.g. existence or factuality) are not assumptible, the properties they specify are not characterising, not part of so-being.

The difference indicated, between properties which can be part of the nature of an object and those which cannot be (but which are, for instance, founded on the nature of the object), is consolidated in Mally's and Meinong's distinction between nuclear and extranuclear properties. Extranuclear properties, such as existence, determinateness and simplicity, are not, to put it bluntly, assumptible: the Characterisation Postulate does not apply without important restriction where extranuclear properties figure. As far as restrictions on the Postulate are concerned, the second piece of evidence deriving from Meinong's deployment of the nuclear–extranuclear distinction (on which see Findlay 1963, c. p. 176), supersedes the first piece deriving from the modal moment since, logically important though the modal moment is, the property distinction alone, properly applied, is enough to meet all objections to theories of objects based on illegitimate appeals to the Characterisation Postulates. The Meinong whose theory includes an unrestricted Characterisation Postulate is accordingly, like Meinong the super-platonist, a mythological Meinong.

§2. The Characterisation Postulate further considered, and some drawbacks of the consistent position.

The separation of the consistent and paraconsistent Meinong also comes with the Characterisation, or Assumption, Postulate, according to which an object has all its characterising features, i.e. in the form which will be of main concern, $(\tau x\ xf)f$, where f is a characterising (or nuclear) property and τ an appropriate singular descriptor. The Characterisation Postulate certainly has to be properly qualified if inconsistency, and also triviality, are to be avoided; indeed the Postulate has been perhaps *the* main problem in working out a consistent Meinongian theory, since other postulates, e.g. those of neutral quantification theory, all admit rather obviously of elementary finite modelling. The problem, in determining requisite qualifications, consists primarily in working out what counts as a characterising feature (this was, as we have seen, one of Meinong's later problems also).

It might be thought that the Characterisation Postulate is simply a special problem for Meinongian theories and that the Postulate and its difficulties can be done away with entirely in classical theories. This is not so. The Ontological Argument and its variants remain to vex classical logical theory. Moreover, there is no question of doing away with the Characterisation Postulate classically: it occurs in an important form in classical theories, e.g. in Hilbert's ϵ-scheme $(\exists x)(f(x) \supset f(\epsilon x f(x))$, in Fregean description theories (see, e.g. Kalish and Montague 1964, pp. 242–3), and in the scheme $E!\ \imath x f(x) \equiv f(\imath x f(x))$ of Russell's theory of descriptions (cf. PM *14.22). Thus according to Russell it is true that the man who wrote *Waverley* wrote *Waverley* but false that the man who squared the circle squared the circle. It is *very* doubtful that untutored logical intuition would agree, but the matter has perhaps ceased to be one of hard data. Nor is the question of the restrictions on the Characterisation Postulate without difficulty classically. The case of sets is enough to establish this: for set-theoretically the restrictions on the Characterisation Postulate amount classically to the restrictions on the set abstraction axiom, i.e. on set existence, as we shall shortly see.

An easy way out of difficulties might seem to be to simply follow the classical lead and not impose any characterisation postulates for nonexistent particulars. This would, however, amount to abandoning the Meinongian enterprise, to giving away the claim that the golden mountain is golden, and the round square both round and square, and more important to abandoning the seemingly correct thesis that nonentities have more or less definite natures and that different ones have distinctive characters. More important, it would be to renege on the commonsense basis on which the theory of objects is built, according to which nonentities can be distinguished and do have properties. And given the defensible commonsense basis, it emerges by transcendental arguments that limited characterisation postulates hold for nonentities (see chapter 6).

Such transcendental arguments yield, however, only limited information regarding properties of nonentities. They do not indicate how properties other than those supplied by characterisation are arrived at; they do not indicate *which* of the extensional properties that a nonentity says it has in its own description, for example, it has. Which such properties nonentities do have, which properties are *characterising*, will have to be determined or delimited, by somewhat independent arguments;[1] and on this issue theories of objects split. A most important split is into consistent and dialectical theories.

Let us consider first consistent theories, and the sorts of restrictions imposed on characterisation postulates in such theories. The first stage of the logical theory of the consistent Meinong can be obtained by a fairly simple reinterpretation and enlargement of classical logic. The logic of (first-order) quantificational theory can remain exactly the familiar syntax, though to emphasize the fundamental semantical changes, and because the usual quantifiers are wanted for their usual (though now defined) role, it pays to adopt new notation for the neutral quantifiers. In place then of the referential quantifiers $\forall$ and $\exists$, let us use the neutral quantifiers U (for every) and P (for some). Neither ordinary language quantifiers nor those of quantification logic need be restricted by the existence and identity requirements that classical logicians have tried to impose in their regimentation of discourse, to suit the demands of an empiricist philosophy. The removal of classical regimentation greatly increases the scope of application of the logic, what can be said using it, and the arguments that can be formalised with it, and several problems simply disappear with the reinterpretation.

A conspicuous difficulty for what we are generously calling the "consistent theory" is to make such apparently[2] inconsistent assertions as that the round square is round and also not round, because square, emerge from the Characterisation Postulate. A solution to this problem, and thereby a consistent theory, is obtained by making, in one way or another, a distinction between sentence and predicate negation (in the modelling of Parsons 1974, for example, predicate negation is taken as property negation). The distinction – reflected in symbolism naturally enough as the difference between $\sim xf$ and $x{\sim}f$ – enables such objects as *the round square* and *the round nonround* to be incorporated in a consistent theory. For consider the round nonround, i.e. in obvious notation $\imath x(xr \;\&\; x{\sim}r)$. By the Characterisation Postulate, the round nonround is round and the round nonround is not round, i.e. $(\imath x(xr \;\&\; x{\sim}r))r \;\&\; (\imath x(xr \;\&\; x{\sim}r)){\sim}r$. No violation of consistency results because $y{\sim}r$ does not imply $\sim yr$ when y is inconsistent, i.e. predicate negation cannot be converted generally into sentence negation. "The round nonround is not round" does not imply, and so does not mean, "It is not the case (it is

[1] Those mentioned below for various different cases – for abstraction principles, for the case of entities, and for the cases of nonentities (cf. 1.21 [Vol. 1 $_{—Eds.}$]).

[2] Someone may doubt that the consistent theory really includes impossible objects. In one good sense it does (that the existence of the objects would yield a contradiction), in another good sense it does not.

false) that the round nonround is round". What – one soon asks or, should one neglect to do so, the opposition soon enough asks – about the object that is round and is such that it is not the case that it is round, i.e. more clearly in symbols, $\imath x(xr \mathbin{\&} {\sim}xr)$? According to the consistent theory the Characterisation Postulate does not apply in such cases: if it did inconsistency would of course result with $zr \mathbin{\&} {\sim}zr$ where $z = \imath x(xr \mathbin{\&} {\sim}xr)$.[1]

Such a restriction on the Characterisation Postulate may appear arbitrary, but it is not. It is based on the old idea that an item cannot be (properly) characterised negatively, in terms of what is absent from it; that genuine characterisation should be positively achieved in terms of what properties or their opposites items have. In these terms, predicate negation, which introduces opposites, is a positive operation along with conjunction.

There are other ways of arriving at similar restrictions on the Characterisation Postulate. One way emerges from the extensionality restriction (soon to be considered): looked at one way negation is not a fully extensional matter, so sentence negation should be excluded in characterising an object.[2] More important are the general reasons already alluded to that we shall come back to in arguing for the general qualifications that go with characterisation postulates: that one is only free to determine what is open to determination, and that consistency like other less global features of the actual world is not so open to determination.

These sorts of explanation – though of a similar order of merit to that given in explaining why the Characterisation Postulate is restricted classically to existent objects – are not entirely satisfying. It looks a bit as if Meinong's theory has been rendered consistent by formal chicanery, when its natural tendency is perhaps, as Russell thought, to inconsistency.[3] Is there really so much difference between $\imath x(xr \mathbin{\&} x{\sim}r)$ and $\imath x(xr \mathbin{\&} {\sim}xr)$, so that the first object is round and not round and the second is, at best, round, and its inconsistency shown only indirectly (e.g. by a *reductio* on the supposition that it exists, or by a qualified Characterisation Postulate for possibilia)? Consistency is (as always) purchased at the price of restrictions on what can be said, to the detriment of what

[1] In the more restrictive setting of Parsons 1974's modelling, such objects as z cannot be formed. Since such objects can readily be introduced using natural language this is a limitation of the modelling, if an advantage from the point of view of ensuring consistency.

[2] The detailed argument depends on defining extensionality semantically in terms of no world shift (even to reverse worlds), and observing that negation evaluation in relevant logic involves a shift to a reverse world [by way of the Routley-$*$ operation used in defining relevant negation; see RLR —*Eds.*].

[3] Perhaps the natural tendency of an *adequate* theory is to inconsistency?

ought to be truly sayable. The cost of these restrictions appears in a sharper way with paradoxical or defective (as Meinong called them) objects, such as Russell's class and the statement that this very statement is false, than with inconsistent items such as z. In the case of many paradoxical items, most of us, before instruction anyway, do want to apply the Characterisation Postulate, to obtain from the Liar statement $\imath p(p \leftrightarrow Fp)$, the equivalence $T(\imath p(p \leftrightarrow Fp)) \leftrightarrow F(\imath p(p \leftrightarrow Fp))$, whence (by Excluded Middle, and the Tarski biconditionals, $Tp \leftrightarrow p$ and $Fp \leftrightarrow {\sim}p$), p_1 & ${\sim}p_1$, where $p_1 = \imath p(p \leftrightarrow Fp)$. What is more we can argue rather convincingly – by the paradox arguments themselves – for the contradictory conclusion reached by application of the Characterisation Postulate. So why restrict the Characterisation Postulate to exclude the result of these apparently valid arguments (which are not however really independent of the Characterisation Postulate)? And why not attempt a uniform treatment of paradoxical and impossible objects? Indeed why not. But before answering the questions, it is of historical interest to (try to) see where Meinong stood.

§3. Interlude on the historical Meinong: evidence that Meinong intended his theory to be a consistent one, and some counter-evidence.

It is at least clear that Meinong's theory of objects is unclear and indeterminate in several crucial logical respects, most notably in the matter of eventual restrictions on the CP, but also as to issues such as identity theory (see further 12.2 [Vol. 3 $_{—Eds.}$]). Is it that the historical Meinong is also indeterminate between the consistent Meinong and the paraconsistent Meinong? Apparently not, but it is hard to be quite sure.

The chief evidence that Meinong was prepared to endorse contradictions as true is sometimes said to come from a frequently-cited passage where Meinong is replying to Russell's objection that impossible objects such as the round square are apt to infringe the law of contradiction. According to Meinong,

> the principle of contradiction is to be applied by no-one to anything but reality and possibility (Stell, p. 16).

But if a contradiction were true, af and ${\sim}af$ were both true for some statement (which can be given the form af), then reality would not remain unscathed, inconsistent statements would both be true. Thus if the principle of (non)contradiction, in this form, applied, without qualification, to reality a contradiction could not be true. This was not however what Meinong meant. Meinong was not denying non-contradiction in the semantical (or in the propositional) form, he was not contending that some propositions are both true and false. The logic of objectives, Meinong's replacement for propositions, was two-valued: objectives either obtained or they did not, and no objectives both obtained and did not. This is important evidence that Meinong had not contemplated, or really allowed for, paraconsistency.

In the given context "reality" concerns what exists, not what is true, and is better replaced (as in Findlay's rendition 1963, p. 104) by "actuality". What Meinong meant, so it was previously argued, was that the traditional principle, asserted of objects, that no object both has and lacks some one property, applies at most to existent and possible objects. Hence his subsequent elaboration of the point, that exceptions to logical principles that are confined to impossible objects are not important limitations (Mög, p. 278). Hence too his choice of examples (in Stell and Mög), of objects such as the round square, which are simultaneously round and not round, i.e. which satisfy predicates and their negations.

In defence of Meinong's claim, it is worth reiterating that the typical, and Aristotelian, applications of such logical principles, and also standard defences of them, occur in settings where existential presuppositions are made and where restrictions to entities are commonly assumed. However, to a large extent, Russell and Meinong were at cross purposes. Russell's rejoinder to Meinong (Russell 1905, p. 439) was that Non-Contradiction is asserted, *not* of subjects, *but* of propositions; but this evades part of the issue. For Meinong was concerned with the traditional formulation in terms of subjects, which had wide currency at the time Russell was writing. Moreover, Russell's own theory allows violations of such traditional forms, violations of Excluded Middle in the case of nonentities (e.g. the present King of France is neither bald nor not bald), and of Non-Contradiction in the case of classes, violations that are quickly said to be *apparent* because they disappear when class and object talk are translated into entity talk. Meinong could allow for impossible objects in a similar way – except that logical violations are not merely apparent – in effect, through a theory of predicate negation, the rudiments of which are to be found in his work. The Law of (Non)Contradiction that no one would expect to hold for impossible objects was a predicate negation formulation of the principle.

In sum, the evidence does little to establish the alleged conclusion, to show that Meinong had moved outside the intended framework of a consistent theory of objects, but if anything helps to confirm the contrary impression.

Other major evidence comes from Meinong's examination of paradoxes. Meinong considered the logico-semantical paradoxes (EP, p. 10ff.), but he never really contemplated a paraconsistent assimilation of these paradoxes.[1] Thus he saw the Russell paradox as a dilemma to be resolved (p. 11 bottom), and as "easily resolved if a rigorous interpretation of the situation is given" (p. 13). Even if Meinong's method of resolution of the Russell paradox is rather opaque (it is a curious blend of elements later worked out by Lesniewski, by Ryle, and by others), the upshot of the resolution for paraconsistency is clear. Nothing self-contradictory is left; contradiction is dissolved by the time-honoured method of locating a difference of respect:

> all that it means, in the end, is that two objects may be similar in one respect and dissimilar in others. Black and white are different from each other; neither of them can be predicated of red or blue; in this respect they are similar.

It is the same with some of the semantical paradoxes which do force Meinong a little closer to a paraconsistent position, and really into a difficult corner. Consider what Meinong says of Mally's problem: whether a thought about a thought which is not about itself is about itself.

> A thought which is at once about itself and not about itself is indeed peculiar. But these two adequacy-relationships, i.e. the thought being at once about itself and not about itself, can coexist as long as they relate to different foundations. There is *no more incompatibility* here than in the analogous coexistence of exact likeness and unlikeness (as obtained in the case of colours) (p. 14; my italics).

Meinong realises that this sort of strategy is hardly going to succeed with every paradox.

> ... there are (peculiar) circumstances which might prevent us from allowing this kind of tolerance, and which would make the charge of "meaninglessness" [Mally's charge] plausible and significant (p. 14).

[1] It is only fair to say that the simple technical apparatus on which a coherent paraconsistent approach is based was not available in Meinong's time.

Note that all page references in this section without further citation are to EP.

Consider "what I apprehend is false" where my thought points to nothing beyond this falsity.

> In the case of such an incomplete expression, as in the analogous incomplete expression "What I apprehend is correct" one is confronted with a peculiar defectiveness in the object of thought, which always becomes evident when an apprehending experience tries to refer to itself as immediate object (p. 15).

The defectiveness lies in a failure of reference and, more generally, of a reference chain, to have an ultimate ground, to closed reference loops. Just as "no relations can be based exclusively on relations as *inferiora*", reference chains "must have an end and that end must consist in an object that is not itself an apprehension". Defective objects violate this requirement: they are defectively incomplete objects, to which Mally's expression "meaningless" may appropriately apply.

> [If] the defective object is itself apprehended ..., then one is confronted with defective objects which lack even *Aussersein*, though this expression is indeed peculiar. In this case one is not really confronted with an object, and experiences of apprehension in this instance lack a proper object.

With this admission the fabric of Meinong's theory is beginning to break down: defective "objects" are not objects, though the objects of such presentations; and there is no excluding such presentations, since their occurrence is a phenomenological fact, and every presentation must have an object. Nor, as Meinong realises, will it honestly do to say, in *every* defective case, that the presentation is of something other than what it seems to be.

In fact Meinong does not settle the status of defective – or as we may call them *paradoxical* – objects, and leaves it open that defective objects might somehow be counted as objects. He considers the following objection to the proposal that defective objects are not proper objects and lack *Aussersein* (p. 20):

> If *Aussersein* cannot be denied to the round square, how can it be denied to defective objects, which in some respects pose fewer difficulties?

Meinong emphasizes that he has found no sufficient grounds to revise his claims concerning impossible objects, despite doubts repeatedly expressed about them; but he also indicates that he could be prepared to revise what he has said about defective objects.

> It would not be surprising if further research in this peculiar, unfamiliar field of objects should produce unexpected results; and it might be a mark in favour of future resolution of present controversies if the objection to the theory of defective objects just set forth should be shown to have something right about it which has erroneously been taken to be an argument against the *Aussersein* of impossible objects (p. 20).

By way of partial summary: Meinong discerns a class of defective or paradoxical objects and he also diagnoses the cause of the defectiveness, namely closed reference looping (or in more picturesque terms, failure of namely riders). He is also clearly aware that as a result of contingent circumstances perfectly satisfactory objects (statements) may be rendered defective, e.g. of thoughts, wishes, and so on, which are not normally paradoxical, which become paradoxical owing to contingent circumstances (cf. Prior's family of paradoxes in 1961). He tentatively excludes these paradoxical objects from the theory of objects proper; in particular they are not open to quantification (p. 19), and they are absurd. But he also leaves the way open for an alternative treatment; and certainly a paraconsistent treatment could be adopted which would save his theory from breaking down at critical points. However Meinong appears completely unaware that the paradoxes might be handled in a paraconsistent fashion, and rules such an approach out in the case of the Russell paradox.

My conclusion, then, also tentative, is that Meinong saw his theory of impossible objects as part of a consistent framework, a framework that would be shattered by the full admission of paradoxical objects. If this is right the paraconsistent Meinong is not the historical Meinong: Meinong intended his position to be a consistent one. But though he aimed for a consistent theory, he did little to show that he had arrived at such a theory. Thus Findlay's assertion (1963, p. 327) that Meinong certainly showed us how to talk consistently about objects including numbers, classes and propositions, is a vast overstatement. There is certainly no demonstration, nor even a discussion of the question of demonstration, to be found in Meinong. And of course, for all our techno-logical advances, we remain pitifully supplied with demonstrations of consistency, even for formalised parts of the theory of objects such as bits and pieces of mathematics.

Was Meinong consistent, in accord with his intentions? It is difficult to be sure. But it seems that he probably failed, restrictions on the CP not being tightly enough secured. However, the theory cannot be simply dismissed because of this, especially as relatively simple modifications would appear to render important parts consistent. Moreover most productive philosophers are probably not consistent. At most only parts of their theories can be dismissed on that basis.

The exciting idea, fashionable in Brazilian logical circles, of reconstructing (the central part of) Meinong's theory of objects within a paraconsistent framework, is then unhistorical. Even so, the paraconsistent approach has much to recommend it, and it appears to be a more promising way of elaborating the theory of objects than the consistent route Meinong tried to follow.

§4. The paraconsistent position, and forms of the Characterisation Postulate in the case of abstract objects.

The questions at the end of §2 take us directly to the paraconsistent Meinong. Before waving goodbye to the consistent Meinong, it is worth taking stock of some of the very real advantages that are lost by going paraconsistent. A major disadvantage of the loss of consistency is that analogues of the classical theory and arguments cannot be simply taken over, appropriately reinterpreted and enlarged, e.g. by predicate negation. As explained in detail in RLR, classical logic fails outside consistent situations; specifically Disjunctive Syllogism and the corresponding rule of Material Detachment, i.e. (γ) A, $\sim A \vee B \rightarrow B$, both fail.

In short, paraconsistent interpretations of Meinong exclude the general use of classical logic (reinterpreted as in chapter 1 [Vol. 1 —*Eds.*] and elsewhere), though classical logic can be used in a restricted class of cases, e.g. those where consistency can be proved or, more likely, is assumed. Going paraconsistent requires then some far-reaching logical changes (of the sort already outlined elsewhere, e.g. UL [reprinted in Vol. 4 —*Eds.*]), whose ultimate foundational satisfactoriness remains, at this stage, a somewhat unknown quantity and in too large a measure a matter of conjecture. Firstly, it is not known whether various paraconsistent theories of interest, such as paraconsistent set theories of modest strength, are nontrivial.[1] Secondly, it is not known how much such paraconsistent theories are good for, e.g. how much of classical mathematics can be recovered within them (without further, and unguaranteed, assumptions).[2] Classically, one would expect a trade-off like that between consistency and completeness, only now between nontriviality and adequacy; but the classical arguments for such an exclusion depend really on the paradoxes and cease to apply outside a classical framework (cf. UL).

These uncertainties do not, however, have much bearing on the philosophical adequacy of paraconsistent theories, unless, for example, it is erroneously taken that classical mathematics as a whole is some sort of revealed gospel which stands without any addition in the way of consistency assumptions. But the philosophical adequacy of paraconsistent positions is certainly in question on other grounds, not least the amount of damage it does to traditionally accepted modes of argument, such as the rejection of inconsistent theories without further ado. Such modes of argument are not as generally applicable as has been supposed (cf. Routley 1980). But before taking up these issues it is worth asking whether going paraconsistent, or more accurately going dialectical, helps solve the problem of determining the scope of validity of the Characterisation Postulation, and whether, in particular, restriction on the principle can be removed in a paraconsistent setting.

The paraconsistent shift certainly removes the anomaly in the negation case, between the round non-round [invoking predicate negation —*Eds.*] and the round which is not (sententially) round, and it enables a more uniform and appealing treatment of paradoxical items. However the CP does not emerge unscathed in an unqualified form – nor would one expect it to – even in paraconsistent theories, unless the theories are not closed under Modus Ponens or are trivial (and so not really paraconsistent). For consider, e.g., the statement r_0 which is true and entails every statement (or entails the false). By an unrestricted CP, r_0 is true and entails every statement, hence, using just the Modus Ponens rule (for entailment), the theory is trivial, since every statement is true. That is, the Characterisation Postulate yields, *without* an intermediate argument through logical principles such as Contraction (of antecedents), versions of the Curry–Moh–Shaw–Kwei

[1] Postscript: But see now point 1 of the Preface to "Ultralogic as universal?" [reprinted in Vol. 4, where it is claimed that "the nontriviality of dialectical set theories of interest has now been established" —*Eds.*].

[2] The notion of recovery is not uniquely determined. There are trivial ways in which what is called classical mathematics may be "recovered", e.g. by simply adding enough assumptions, or by invoking limited consistency assumptions, presupposed classically anyway.

paradoxes. Thus the Characterisation Postulate is a more powerful principle than the various abstraction principles which it implies,[1] since in weak logics the latter principles do not trivialise.

The reasons for failure of unrestricted forms of the CP are obvious enough. Whether a statement is true, like whether an object exists, is not really up to it, but a matter of external circumstances, of factual matters beyond its control. A statement can no more determine its own logical status directly, e.g. by announcing its own truth or what it entails, than an object can directly determine its own ontological status and its own circumstances, whether it exists, and to what actual things it relates. These evident features show how characterisation postulates, and more generally postulation, should be controlled.

It is not just with objects of higher order that an uncontrolled CP causes havoc, at worst triviality. Consider to illustrate a method which applies to both individuals and numbers, the number which is identical with 7 and which multiplied by 2 gives 16; then for this number n_1, $n_1 = 7$ and $n_1 = 8$, so $7 = 8$ by transitivity. By iteration of the method, there is only one number, and one thing – a simple proof of monism. But there are also many things and infinitely many numbers, so monism is combined in an inconsistent union with pluralism.

A shift to paraconsistency does not give then complete absolution to an uncontrolled CP, though it renders determination of the forms of the CP *much* easier, especially in the case of higher order objects, and the results more satisfying in bottom order cases (by removing "anomalies" such as the exclusion of sentence negation in CPs). The question, even for the paraconsistent Meinongian, is: what are the bounds on postulation? More narrowly, what are the correct forms of the CP? Enter the Idiosyncratic Platitude of commonsense philosophy (cf. chapter 6): every sort of item has its own sort of logic. In the case of entities, a definitive form of the CP has already been investigated (in 1.21 [Vol. 1 —*Eds.*]). In the case of important objects of higher order the question is not as difficult to answer (within limits) as might be expected. Consider, for example, sets: the cases for properties and relations are similar, and those for propositions and propositional functions have much in common.

[1] It may be that abstraction principles yield all the information that is required concerning objects of higher order, and that the CP vanishes into such principles at the higher order. This proposal is followed up shortly.

What is the set of elements which are A, i.e. $\{x : A(x)\}$? First, it is an object. This is especially obvious in the Meinongian case, since it is an object of thought, of predication, and so on. Furthermore there is little difficulty in saying which object it is. It is the object comprising exactly those objects which are A. Cashing *comprising* in terms of membership and *exactly* in terms of a biconditional, $\{x : A(x)\} = \imath w(Uy)(y \in w \leftrightarrow A(y))$. Note further that this set $\{x : A(x)\}$ is an object specified in the form $\imath wB(w)$, i.e. $\{x : A(x)\} = \imath wB(w)$. Hence, applying the CP, $B(\imath wB(w))$; i.e. $(Uy)(y \in \{x : A(x)\} \leftrightarrow A(y))$, whence particularising $(Pz)(Uy)(y \in z \leftrightarrow A(y))$, i.e. the familiar comprehension principle follows.[1]

This sole application of the CP, which can be independently argued for, yields practically all, perhaps all, of the features of sets. Accordingly it is tempting to suppose that the CP can be replaced entirely in the case of sets by one of its proper consequences, an independently defensible principle which gives the correct conditions on the CP for sets.

This is, in effect, just the restriction that *was* assumed classically. For the classical CP delivers a classical (naive) abstraction axiom, and the classical abstraction axiom yields in return the classical CP for sets. The classical connections are these:

$$E!(\imath xB(x)) \equiv B(\imath xB(x))$$
$$\{x : A(x)\} = \imath x(\forall y)(y \in x \equiv A(y)), \text{ i.e. } \imath xB(x) \text{ for short.}$$

Hence,

$$E!\{x : A(X)\} \equiv (\forall y)(y \in \{x : A(x)\} \equiv A(y)).$$

Thus, given the abstraction principle, $E!\{x : A(x)\}$, i.e. $E!(\imath xB(x))$, whence the CP, $B(\imath xB(x))$ follows; and given the CP, set existence follows, whence the abstraction axiom results as before.

For abstract objects of other varieties a similar replacement of the CP by independently defensible abstraction principles appears equally promising: the independent defence is just that sometimes given for "naive" abstraction principles. For example, the

[1] The argument is correct classically on the hypothesis that $E(\{x : A(x)\})$ i.e. that the class $\{x : A(x)\}$ exists (uniqueness being guaranteed by extensionality).
Strictly z is subject to conditions required for correct particularisation, viz. z is not free in $A(y)$.

relational CP is simply $(Py)(x_1 \ldots x_n)((x_1 \ldots x_n)\ \iota\ y \leftrightarrow A)$, subject perhaps to the provision that y is not free in wff A. Where $n = 1$ the property CP results, where $n = 0$ the propositional CP. All this suggests that *in the case of abstract objects, characterisation postulates may be replaced by abstraction postulates*, i.e. abstraction characterises abstract objects. Once classical existence and possibility assumptions are got rid of this suggestion appears to have much merit.[1]

§5. The bottom order Characterisation Postulate again, and triviality arguments.

It remains to control the CP in the case of particulars or objects of bottom order where, because the objects are not abstract, no abstraction principles are available to undertake the talk. The general principle restricting postulational methods in general and the CP in particular we have already observed: one can only successfully determine what is open to determination, and very much is not open to determination so one is not free to determine it by postulation, e.g. whether Capetown is north of Tangiers, who was the father of John Stuart Mill, whether the aether exists, how old Ivan Illich is, whether Peano arithmetic is consistent. In particular, then, an item cannot determine through its own characterisation, by describing itself appropriately, matters that are otherwise determined, it cannot settle its own ontological or modal status or its relations to actual things where these relations induce new relations between actual things. What goes wrong if the CP can determine matters that are not open to determination, in a way incompatible with how things actually are we can see by examining a variant of Parsons' "version of Meinongian ontology" (in 1974 and 1975),[2] call it the *naive* variant. Parsons' 1974 theory is based on the following principle (1975, p. 75):

[1] Of course classical connections such as $A(\imath x B(x)) \supset B(\imath x B(x))$ have to be rejected too.
Whether the entire postulational method for higher order objects of mathematics can be reduced to abstraction axioms or, more generally to characterisation postulates (classically then to existence assumptions), it is unnecessary to settle. This is a technical issue as to a general class of reductions (assuming of course that choice principles and the like can be accommodated in some suitable way). The success hitherto of reductions, e.g. in reducing numbers (which are still at first-order stage characterised through induction), analysis etc., does not guarantee future success with respect to objects introduced by esoteric postulation procedures – unless postulation procedures can themselves, as has sometimes been supposed, be exhaustively catalogued.

[2] The term "ontology" is Parsons': although such uses of the term have wide currency, especially in North America, they are decidedly misleading, something I believe Meinong would have thought also. Such uses typically presuppose the Ontological Assumption.

PP. For every object x and every nuclear property ρ, x has ρ iff ρ belongs to the set correlated with x.

The principle is

> true in the case of real objects, because this just was the defining condition for what the correlation was in that case. It's to be true in all other cases as well – by edict (p. 75).

As before, determining which predicates are nuclear (Parsons' version of characterising) and which are extranuclear is obviously crucial to the enterprise. For consider the set of properties {goldenness, mountainhood, existence}. If existence were nuclear then the object correlated with this set, the existent golden mountain, would exist, contradicting the facts. "Exists" is accordingly extranuclear. Among nuclear predicates, which he variously describes as ordinary predicates and as predicates which stand for properties of individuals, Parsons lists these (p. 76) "is blue", "is tall", "kicked Socrates", "was kicked by Socrates", "kicked somebody".[1] Extranuclear predicates include ontological and modal predicates, intensional predicates, and logical ones such as "is complete" and "is determinate". Indeed the division is very similar to (but *not* the same as) that (of 1.21 [Vol. 1 $_{—Eds.}$]) for ch and non-ch predicates. The reasons for the similarity are (we should like to think) that the division reflects a natural classification. But it is also true, as Parsons likewise observes, that development of a theory of objects of a Meinongian cast tends to force such a classification, as well as clarifying and sharpening it.

The following more problematic sorts of predicates are nuclear as well under the accounts Parsons gives: "is taller than Socrates", "is father of Socrates" , "lived off Baker Street". But with this admission (necessary for an adequate theory, so it was argued in 1.21) *the naive Parsons theory is inconsistent with the facts.*[2] For consider objects like the following:

(1) the Greek who is taller than Aristotle and shorter than Socrates,

[1] Parsons is not specific about the sort of quantifier, especially whether its range is restricted to entities. If not, then ancient puzzles can arise, such as what happens when the irresistable force meets the immovable body, or when the universal solvent is applied to the insoluble element.

[2] Inconsistency, and even total theoretical destruction, is the fallout from going excessively nuclear. The naive Parsons' theory is not alone in suffering such a fate. Castañeda's presentation in 1974 of a "Meinongian" theory appears (the theory is not sharply enough articulated to make this completely certain) to encounter a similar disaster (see 12.4 [Vol. 3 $_{—Eds.}$]).

(2) the man who is father of St. Thomas Aquinas and son of Bertrand Russell, i.e. the object correlated with the set {human, the property of being father of St. Thomas Aquinas, the property of being son of Bertrand Russell}.

Then by PP and (1), the Greek in question is taller than Aristotle and shorter than Socrates, whence, by the logic of relations, Aristotle is shorter than Socrates, which is false. Similarly from (2) and PP – upon defining *grandfather* so that, if for some x, x is son of y and x is father of z then y is grandfather of z – Bertrand Russell is grandfather of St. Thomas Aquinas!

Two possible ways out of this trouble – neither so easy if one is attracted by the splendid simplicity of Parsons' original theory, before, that is, one sights the theory of relations – are to redefine *nuclear* or to amend PP. Neither course is satisfactory. PP is a consequence of the Characterisation Postulate HCP[α], already argued for, by principles of the logic upon letting χ be the set correlated with x and taking $A(f)$ as $f \in \chi$ and allowing HCP to generate every object. To abandon PP would be to leave the theory without a Characterisation Postulate and so any way of effecting, what is essential, the assignment of extensional features to nonentities (as to the reasons for this and its importance see the transcendental argument of 6.4). However modifications to PP are not thereby excluded and are feasible. One such, designed to ensure that nonentities accord with familiar deductive practice and so are closed under consequence, is as follows: for every object and every nuclear ρ, x has ρ iff ρ is entailed by properties in the set correlated with x. But firstly this is but a special case of PP, obtained by taking the properties in the set correlated with x to be closed under entailment; and secondly it is not appropriately general, since many nonentities (e.g. those of visions and dreams) are not so closed under deducibility. The same will be found of other seeming more plausible modifications. The full strength of PP *at least* is wanted if the rich variety of nonentities of theory and thought, imagination and experience, are to be included in the theory.

A repair by excluding relational predicates from among nuclear predicates has already been excluded as a live option. Such a course would have the very damaging effect of preventing nonentities from entering into relations of a range of sorts. Yet many nonentities certainly stand in relations. Fictional items can be grandfathers, taller than others, live in various cities, etc. But though relational predicates cannot be ruled out

[α] HCP. $(Px)(chf)(xf \equiv A)$, with x not free in A. Cf. Vol. 1, 1.21. —*Eds.*

(without so weakening the theory that it cannot account for the data), there is an intermediate course: namely to exclude from the class of nuclear (relational) predicates those which state relations to entities. This is not so restrictive as may at first appear, since *duplicates* of entities can be used, e.g. in place of "lived in London", "lived in Dickens' London", and similarly "Shakespeare's England", "that England", "merry England". Duplicates may be used because *they* do not interfere with what is independently, and perhaps differently, determined, e.g. by force of historical circumstance. *Such* duplicates are characterised, and criticised, in chapter 7: the theory is nothing if not generous with respect to the range of objects it allows, as it should be in accord with its charter (p. 2 ff.).

While there is nothing theoretically amiss with duplicate objects, and indeed (unlike say sense data) there is a place for them, they are not adequate to the data to be theoretically reflected, either in fiction and cognate topics, or as regards isolated and homeless "philosophical" objects. For example, the difference between the queen of France and the queen of England lies in the relations of the queens respectively to France and England, not to duplicates or to a duplicate in one case but not in the other (irrespective of the problem of defining duplicates in such cases, where source books tend almost to vanishing point). Likewise the tension of the queen of France's features with the historical circumstances, that France has no queen, is precisely due to the relation to France, not to a tension-removing duplicate. And so on (see 7.7).[1]

Given that relational predicates are in, as nuclear, the viable course for a consistent theory of objects (that adopted both in Parsons' far from naive theory 1974 and in 1.21) is to *modify*:

[1] Since the logic of relations can be rendered much more (classical and) familiar by weakening to duplicates, the weakened alternative theory thereby indicated, with characterising relational predicates restricted to those with suitable duplicate terms, is no doubt worth considering. But there are grounds for suspecting that such a theory incurs serious new problems e.g. as to the interrelations of more and more defective duplicates with their originals. A fine taste of how things might go is afforded by D. Lewis' counterpart theory, since counterparts resemble platonised duplicates.

Ruling out relations to entities from among characterising properties would remove part of the case for excluding intensional properties from among characterising or nuclear properties; e.g. difficulties with the objects out the window seen by you that you did not see (since they would only be seen by your duplicate, or *one* of them). Does it remove the case entirely? There are other reasons for thinking it does not. One is that characterisation aims to give something like the nature of the object characterised and it is commonly assumed that nothing, not even nonentities, can be intensionally characterised. That isn't really so obvious. Another reason is this: just as a set is a certain object, so an individual a is a certain object, namely the a-ish object, or in modified Russellian, $a = \imath x(x = a)$. Now the requisite criterion for identity is coincidence of extensional features. Accordingly it would seem (but the argument can be defeated) that a is characterised by none but extensional features, a conclusion already independently reached.

(i) the classical theory of relations,[1] and

(ii) the classical theory of definition.

It is pleasing to discover that these modifications involve absolutely no work, at least formally. For the classical theories are always imposed on top of the underlying logic. The theory of definition, where presented, is a separate section in the *meta*logic, while the theory of relations requires many additional axioms that are not part of the basic theory. None of these additional schemes are part, for instance, of quantification theory, e.g. it is not a thesis, but an additional postulate, that *taller than* is transitive, and that *taller than* contradicts *shorter than.*

A *neutral theory of relations* – to give the rival an appropriate name – simply does not accept *all* the classically-imposed relations and conditions.[2] It would be quite fallacious to conclude however that it accepts *none.* What the neutral theory does is to impose *duly qualified* forms of the classical axioms; e.g., certainly (significance aside) $(\forall x)(\forall y)(x < y \;\&\; y < z \to. \; x < z)$ where $<$ represents "is less than", and certainly such a restriction to entities though sufficient is not necessary, since similar transitivity holds, for instance, for all ordinal numbers. But it does not hold even for all numbers, as examples like (1) restated with numbers show. Consider

(1′) the object [number] less than 5 and than which 7 is less.

Then $n_c < 5$ and $7 < n_c$, for that inconsistent number n_c, whence by transitivity for all numbers $7 < 5$. That does not however impugn transitivity for all natural numbers or for other well-behaved nonentities. Thus the neutral theory of relations does not, as a null theory of relations for nonentities would, cripple what Meinong saw to be among the most important applications of the theory of objects, those in accounting for mathematics and significant parts of the theoretical sciences, in treating mathematics as a *daseinfrei* science. For mathematics is substantially based (as PM has demonstrated) on the logic of relations.

---

[1] Similarly Chisholm 1972, p. 36. In a way the *basic strategy* for a theory of objects is to break vulnerable classical connections derived from the logic of entities, e.g. such imposed relations as $a{\sim}$round $\to$ $\sim a$ round. In like manner the theories of Castañeda 1974 and Rapaport 1978 break classical predication assumptions (but wrongly: see 12.4 [Vol. 3 —*Eds.*]).

[2] The conditions modified include extensionality conditions: see 7.7, subsection 2.

§6. Characterising predicates and elementary and atomic propositional functions, and the arguments for consistency and nontriviality of theory.

As repeatedly seen, nonentities do not have, by any means, all extensional features their descriptions or (more generally) source books present them as having. Nonentities cannot settle their own ontological status, their own identity, nor can they by their mere sayso impose relations and conditions on actual things and places. But a great many features remain that they do have, (*s*-features and) features that are reliably presented, characterising features.[α]

It is enough to begin with, and for a distinctive logical theory of nonentities, to indicate with *some* precision two moderately comprehensive classes of predicates, first those that definitely are characterising, and second those that definitely are not. It does not matter, for the time being at least, if the classification is not exhaustive, if questions are left open for further, later, determination (cf. deductive theories which are commonly incomplete in this sort of way; and recall intuitionistic mathematics on further determination). It would aid the noneist cause, by helping to reduce the charge of arbitrariness in the division of predicates adopted (that of 1.21 [Vol. 1 $_{—Eds.}$]) if the division can be found in, and has a solid-looking basis in, classical orthodoxy, and fortunately this is the case. The first two primitive ideas in *Principia Mathematica* *1, are those of *elementary proposition* and *elementary propositional function*; and these will serve. The primitive notions are explained – in none too sharp a fashion – as follows:

> By an "elementary proposition" we mean one which does not involve any variables or, in other language, one which does not involve such words as "all", "some" , "the" or equivalents for such words. A proposition such as "this is red", where "this" is something given in sensation will be elementary. Any combination of given elementary propositions by means of negation, disjunction or conjunction will be elementary.[1]

[α] On the supposition operator s see Vol. 1, §1.21(8), (c): "given the predicate f, the operation s yields the predicate $s(f)$ – also written sf – read 'presents itself as f' or 'says of itself that it fs' or 'has supposititious f-ness'." $_{—Eds.}$

[1] Elsewhere Russell says, more succinctly, "A proposition containing no apparent variables we will call an *elementary* proposition"; it can be safely assumed that the *variable binding test* applies to wff in *primitive* form.

Because Parsons' nuclear predicates may include quantified terms, as does "kicked somebody" (1975, p. 76), nuclear predicates do not coincide with elementary (or atomic) propositional functions, or (so it

A "consistent" noneist would of course modify the final recursion clause, ruling out negation and disjunction, and perhaps conjunction also, but in a paraconsistent theory such qualifications are unnecessary, and undesirable. Elementary predicates are determined in terms of elementary sentences in the expected way (PM, p. 92):

> By an "elementary propositional function" we shall mean an expression containing an undetermined constituent, i.e. a variable, or several such constituents, and such that, when the undetermined constituent or constituents are determined, i.e. when values are assigned to the variable or variables, the resulting value of the expression in question is an elementary proposition.

Examples Whitehead and Russell give of propositions which are *not* elementary are (PM, p. 93): "Every individual is identical with itself" and "There are individuals". Other examples of sentences which are not elementary in the sense given are "This is a universal solvent" (an example due to M. Tooley), "This is self-identical" and "This exists", all being excluded because they involve implicitly quantification. Whitehead and Russell would have excluded sentences involving the predicates "identical" and "exists" on precisely the same grounds. Identity is explicitly defined in quantificational terms (*13). Consider "exists", $E!$ in the notation of PM. According to PM (pp. 174–5), $E!z$ is only significant (and well-defined) where z is a description. Hence judgments of the form $E!z$ involve implicit quantification (which elimination of descriptions explicitly shows), and so are not elementary.[1]

Certainly too Whitehead and Russell would have accounted intensional predicates nonelementary, since in PM extensionality is always a matter of functions of propositional functions, and so in one obvious respect – though it does not conform to the letter of the characterisation given – is nonelementary. To obtain the requisite definition of "elementary" which does exclude intensional predicates (as seen by Whitehead and Russell) it is enough to turn to the second edition of PM (PM2). There (PM2, pp. xv–xvii) elementary propositions are all truth-functional compounds of atomic propositions, where

> atomic propositions may be defined negatively as propositions that contain no parts that are propositions, and not containing the notions

will now be made plain) with characterising predicates. The classes also differ in other ways as regards relational predicates (see again 1.21).

[1] Elementary wff of second-order logic 2Q were precisely defined in 1.18 [Vol. 1, p. 293 —*Eds.*].

"all" or "some". Thus "this is red", "this is earlier than that", are atomic propositions.

Atomic propositional functions accordingly provide pretty much the class of characterising predicates (for putatively consistent noneism). Where the match is not perfect it can be made so by but trifling adjustment.

For the arguments which follow, however, a perfect match is not required and the predicates concerned may be a rather more sweeping class. Let us call them, in keeping with the popular particle analogies, *fundamental* predicates. It is intended that all characterising predicates are fundamental, and where precision is required fundamental predicates can be defined as the smallest extension of atomic predicates that includes ch predicates and *ipso facto* ch-forming operations thereon (i.e. fundamental predicates result from the closure of atomic predicates under ch-forming operations, such as predicate negation, the s-operation, etc.). Thus the predicate negates of atomic predicates are fundamental; and in the paraconsistent theory the sentence negates are also, i.e. in paraconsistent theory all elementary predicates are fundamental. Naturally fundamental predicates will conform to logical postulates, e.g. to the postulates of the carrier logic and predicate negation. What is important for nontriviality is that fundamental predicates have *no special constraints* imposed upon them, that is they satisfy no further conditions beyond these, such as axiom schemes interrelating them or definitional conditions. In particular then, neither the theory of relations nor the theory of definitions places any constraints on these predicates. As usual in nontriviality arguments definitional issues can be removed by working always with primitive notation.[1] The damaging effect the theory of relations could have is evident enough: if, for instance, it yielded connections of the form $(x)(xf \supset {\sim}xg)$ for f and g fundamental (e.g. x is taller than Tom materially implies that it is not the case that x is shorter than Tom), then f and g would not be jointly satisfiable in any consistent theory, and contradiction would result from application of FCP to $\xi x(xf \;\&\; xg)$.

The nontriviality problem for *first* and *second-order* theories of items is approached by stages. The carrier logics are consistent, as may be shown by either semantical or syntactical arguments, e.g. by a finite domain modelling of modalised type theories (as of

[1] Thus identity determinates are replaced in second-order logic $\Box$2Q+ [cf. Vol. 1, 1.18, subsection 1 —*Eds.*] through their defining clauses. No postulational conditions remain, *except* those on the predicate ext, which is noncharacterising, being logical.

the logics of Bressan 1972 and of S5-modalisation of the type theories of Church 1940 and Henkin 1950). Such modellings show that the logics are consistent, and the modellings can be syntactically presented as finitary consistency proofs (as in Church 1956 for second-order logic). The same applies to relevant type theories, since they can be viewed as subtheories of modalised type theories. Also shown consistent in the same standard fashion are theories with many of the additions of □2Q+ (and its relevantisation), e.g. abstraction schemes, λ-conversions, predicate identity, additional predicates, etc. There are in fact only two serious sources for doubt at the second-order, namely Characterisation Postulates for bottom order objects, and adding to this, s-predicates and their logical conditions. The issues are treated in stages. The stages consist in considering step by step additions of CPs for bottom order objects to first-order and then second-order logics. The point of such a treatment is to reveal some of the different methods that are available, and also because the treatment is not fully worked out or definitive.

FCP and a "first-order" stage: The stage is first-order in a liberal sense, that it admits such logical apparatus as ξ used in formulating FCP, as follows in the paraconsistent case:

> FCP. $A(\tau x A(x))$ where $A(x)$ is an elementary predicate containing just x free and τ is an appropriate descriptor, e.g. "an arbitrary", "the".

This form of the Characterisation Postulate, which is not far removed from Meinong's eventual form, has the substantial advantage that a straightforward proof of its nontriviality can be given within a paraconsistent neutral quantification logic, thereby putting an end to a considerable class of doubts as to the coherence of any such alternative common-sense logic (many other doubts remain, needless to say, to keep noneists in business for many years). Consider the logic LQCP obtained by adding FCP, with specific descriptor ξ, to the logic LQ of relevant quantified entailment with neutral quantifiers, *with the elementary predicates in* FCP *consisting of atomic one-place predicates* of LQ *and* appropriate (same variable) *truth functional compounds thereof*: LQ may be any one of a range of quantified relevant logics (See 1.23 [Vol. 1 $_{\textit{—Eds.}}$]). Thus LQCP has as well as the connectives &, $\vee$, $\sim$, $\rightarrow$ and neutral quantifiers U and P of LQ, the term-forming descriptor ξ, one-place elementary (i.e. atomic) predicates, and also some sentential parameters. A key feature of the semantics of LQ (as given in 1.23) is that for any zero-degree (or classical) wff there are situations where it holds and situations where it fails. This sort of feature can be exploited in showing that LQCP is nontrivial. The argument conveniently takes a semantical form, though, like the argument for the consistency of quantification logic which it resembles, it could no doubt be rewritten syntactically: note that, like the semantical consistency argument, only soundness of the semantics is required. Interpret LQCP over a domain $D = \{\text{the round square}\}$ consisting of one object d (= the round square). Assign every atomic one-place predicate so that it holds of d at T and does not hold of d at $T^* \neq T$, e.g. for f atomic and t a term, $I(f(t), T) = 1$ iff $I(t) \in I(f, T)$, i.e. by the specification, iff $d \in \{d\}$; and $I(f(t), T^*)$ iff $I(t) \in I(f, T^*)$, i.e. iff $d \notin \{d\}$. Interpret $\xi x A$ as d, i.e. $I(\xi x A) = d$ always. And finally for some sentential parameter r (understood as, say, "The theory of objects is not true"), assign r not true, i.e. r fails at T, i.e. $I(r, T) \neq 1$, i.e. the theory of objects is true. Then, by induction, every elementary predicate wff holds at T, i.e. is true, so FCP is true. But r is not true, so the theory is not trivial.

An analogous argument using strict-implication in place of relevant implication, will show that a version of FCP in which negation does not occur can be added consistently. That is, a (weakly admissible) *positive* characterisation postulate does not render neutral logic inconsistent. The arguments can also be extended to accommodate predicate negation (construed as a "positive" operation). Similar arguments also extend to second-order, and they can be used to show the relative, nontriviality (or positively restricted, the consistency) of stronger forms of FCP than that adopted, in particular the predicative form of FCP (which admits "first-order" quantification as well as sentential operations).

Although the logics of nonentities for which nontriviality arguments have been sketched can hardly be criticised from a comparable classical perspective as too weak, the logics are too weak for various expected and intended purposes; in particular FCP, in whatever form, does not fulfil the important negative role of excluding features from characterisations of nonentities. To obtain a more balanced logic of nonentities, it is necessary to strengthen FCP, as in 1.21 [Vol. 1 $_{—Eds.}$], to HCP.

The HCP stage in an enlarged second-order setting: The setting is 2Q+ [or a relevant analogue] with all the additions that were made in advance of characterisation postulates – so excluding both HCP and further classes of characterising predicates, such as *s*-predicates, hyphenation and the like. Consistency [or nontriviality] of the background logic (symbolised BL) is established as before by rather standard methods, so the main question at this stage is whether consistency can be extended to include HCP. The procedure of *method 1* is to (try to) show that in certain models among those for BL, HCP holds.

To satisfy HCP it is enough to satisfy the principle:

HCP′. For every set χ of ch predicates, there is some x for which $xf \equiv f \in \chi$.

For $A(f)$ of HCP determines a set $\chi' = \{f : A(f)\}$ of predicates and will be satisfied if this substitute is; and HCP only looks at characterising predicates in χ', so χ' can be restricted to χ. Now it is rather easy to find models of extended second-order (and higher-order) logics which satisfy at least one half of HCP′. For

FS. Every set of fundamental (one-place) predicates is simultaneously satisfiable wrt some model.

FS can be trivially guaranteed by equating every fundamental predicate with some one atomic predicate (taking predicate negation as an identity operation, etc.). Less trivially, the methods of a Skolem–Lowenheim argument can be applied (cf. Church 1956, exercise 54.5, p. 317). FS is like the result of the first stage, and depends similarly upon the lack of special constraints on atomic predicates (hence the importance of excluding logical predicates such as = and ∈ from among them). Principle FS yields only one half of

HCF′, that for some x, where $f \in \chi$, xf.

To ensure the converse negative half as well FS is strengthened to

SFS. For some model, for any set χ of fundamental predicates, there is an item which satisfies just the predicates in χ.

The argument involves doubling-up on the elements of the domain of the model M′ for FS. Let χ be any set of fundamental predicates. Suppose in M′, applying FS, z' satisfies every element in χ. Let z, the double of z', satisfy just those predicates z' satisfies that are in χ. That this specification is in order, and a model results, derives again from the absence of constraints on fundamental predicates.

To establish the consistency of an extended second-order logic which includes HCP, select a model for which SFS holds. In short, proof of SFS for a BL model suffices to establish the consistency of BL + HCP.

There are ways of modelling HCP in second-order theories which appear to yield more information and which deliver syntactical proofs of consistency (though presumably the argument outlined can be recast syntactically). As we have seen, in order to model HCP in a given logical framework it is enough to ensure that any set of characterising properties exactly determines an object which has them. One obvious way to guarantee such a condition is simply to represent objects as sets of such properties. This is the genesis of *method 2* and of *Parsons models* for theories of objects.[1] Models of this type admit a reductionistic strategy, which explains their popularity. For the properties can (it has seemed) be taken as those that existing objects have, i.e. as properties of entities. Then of course nonentities can be "reduced to" entities; they "amount to" sets of properties of entities. One glaring defect, however, of such reduction proposals is that nonentities may well have properties that no entity ever has; other glaring defects are considered later (in chapters 8 and 12 [Vol. 3 $_{—Eds.}$]). A Parsons model presupposes no such reduction; though it is sometimes technically advantageous to restrict the properties considered to those of entities (e.g. in modelling *s*-predicate theory), such a restriction is mostly not imposed in what follows.

The charm of method 2 is readily seen if HCP is reexpressed using restricted characterising variable $\underline{f}$, thus:

HCP^{r}. $(Px)(\underline{f})(x\underline{f} \equiv B(\underline{f}))$, where x is not free in $B(\underline{f})$ and $B(\underline{f})$ (i.e. $A^{\mathrm{r}}(\underline{f})$) is the restricted rewrite of A.

Now when an object is represented as a set of properties – so x translates to χ – an object's having a characterising property is represented as the properties belonging to the set – so $x\underline{f}$ translates to $\underline{f} \in \chi$. Thus HCP^{r} translates to

HCP^{t}. $(P\chi)(\underline{f})(\underline{f} \in \chi \equiv B^{\mathrm{t}}(\underline{f}))$, with χ not free in $B^{\mathrm{t}}(\underline{f})$;

that is HCP^{r} *translates into* nothing other than *an abstraction scheme*, and so is automatically vindicated in a suitable translating logic. Thus if the translation can be extended to the whole of BL + HCP, i.e. BLCP, and suitable logical features (such as logical op-

[1] Though the idea of such a representation – or as often, but mistakenly, a reduction – of theories of objects has occurred to several workers, and in principle goes back to Locke (see 12.4 [Vol. 3 $_{—Eds.}$]), there are two distinctive features of Parsons' use of the representation that justify calling the model a "Parsons model"; firstly, the (essential) restriction to nuclear features (though this appears also in Mally's theory), and secondly the detailed presentation (in 1978) of the resulting models, and application of them in establishing metamathematical results, notably consistency.

erations) preserved under translation, consistency of BLCP will follow from that of the translating logic.

There are two features of Parsons modellings that determine the framework for the remainder of the detailed modelling or translation adopted. From the perspective of type theory, which supplies suitable abstraction schemes such as $\mathrm{HCP^t}$, a double type lift is involved: objects are represented as items of order 3 (properties, or sets, of properties), which can be taken to correspond in certain cases, when they exist, to items of order 1, entities. Thus since quantification over objects has to be accounted for, a translating logic of at least fourth-order is required. But so long as it includes the apparatus of fourth-order logic, and ideally of BL, the translating logic, TL, can be any of a variety of sorts. For example, a part of ZF set theory without the axiom of infinity could be used. A very convenient translating logic however is an S5-modalisation of type theory formulated with ξ (e.g. a variation on Henkin's formulation in 1950 of Church's simple theory of types); nor is modalisation really required, since $\Box$ can be translated into an identity functor (thus $m(\Box A) = m(A)$ where m is the mapping function). Since such type theories have finitary consistency proofs, TL can be assumed consistent. Secondly, a type inversion occurs, in effect, in ascribing a characterising predicate to an object: for on translation the object is attributed to the predicate, i.e.

$m(xf) = m(f) \in m(x)$, in the set-theoretic modelling,

or $= m(x)(m(f))$, on a pure type (property-theoretic) modelling.

All extensional operators translate into themselves, e.g. $m({\sim}) = {\sim}$, $m(\&) = \&$. Hence $m(A \mathbin{\&} {\sim}A) = m(A) \mathbin{\&} {\sim}m(A)$; thus if BLCP is inconsistent, so is TL. Variables $x, y, \ldots$ of BLCP translate into predicates of predicates, say $X, Y, \ldots$ for convenience of representation, i.e. $X = m(x)$, etc. Similarly predicates $f, g, \ldots$ translate into predicates, say $F, G, \ldots$ of the former predicates. Accordingly, $(x_1, \ldots, x_n)f$ translates to $F(X_1, \ldots, X_n)$. It is with the translation of ch$f$ that inversion occurs: ch$f$ translates to $(UY)(F(Y) \equiv Y(f))$. (Strictly it should be rendered: $(PF)(UY)(F(Y) \equiv Y(f))$, but the translation can be chosen so that F correlates with f.)

The background logic BL remains intact under translation, i.e. theorems translate into theorems. To verify HCP after translation it has to be shown that $(PX)(UF)$ $((UY)(F(Y) \equiv Y(f)) \supset .\ F(X) \equiv mA(f))$. By an abstraction scheme of TL, for some Z, say Z_1, $Z_1(f) \equiv mA(f)$. Since $(UY)(F(Y) \equiv Y(f)) \supset .\ F(Z_1) \equiv Z_1(f)$, $(UF)((UY)(F(Y) \equiv Y(f)) \supset .\ F(Z_1) \equiv mA(f))$, whence the classical result upon particularising on Z_1.

Working out method 2 suggests a simple translation which requires no type lift, *method 3.* Translate not just all extensional operators, but all primitive variables as themselves in (appropriate) TL, where TL, as before, contains at least the logic of fourth-order type theory. In order to guarantee HCP it is enough to translate ch$f$ suitably, for example as $(U\alpha)(Py)(yf \equiv \alpha(f))$, where α is a predicate of predicates of TL. Then HCP translates to $(Px)(Uf)\,.\ (U\alpha)(Py)(yf \equiv \alpha(f)) \supset .\ xf \equiv A(f)$, which is derivable, since $(U\alpha)(Py)(yf \equiv \alpha(f)) \supset_f .\ y_1 f \equiv A(f)$, by wff substitution upon α (the equivalent of abstraction) and particularizing y.

To complete this outline of consistency proofs, it remains to include firstly ECP and secondly, what takes us to the next state, KCP$'$ and s-predicates.[α] ECP is straightforward, provided that TL contains the expected principle $(\xi x A)E \supset .\ (\exists y)A$. For then ECP translates, under method 3, with E translating to itself, to an expression of the form $(\xi x A')E \supset A'(\xi x A')$. This follows however from the following consequence of the ξ axiom scheme: $(\exists y)A \supset A(\xi x A)$. Type theories usually do not contain E (PM is an exception), in which event a simple move, if it can be got away with,[1] is to translate $(\xi x A)E$ to $(\exists y)A'$ where A' translates A. Again ECP is guaranteed under translation by the ξ-scheme.

The KCP and KCP$'$ stage: In this final stage s-predicates have to be taken care of, so there is not merely the problem of satisfying KCP, for example, but also of meeting the axiomatic conditions on s-predicates. One of the latter conditions makes it easy to show that KCP can be satisfied when HCP is, as it of course it can be; namely, the condition: if ch$f$ then $f \approx sf$. For define sf partially by cases: when ch$f$ then $f \approx sf$; when $\sim$ch$f$ then … (to be specified). Then *KCP may be derived* as follows: Let B abbreviate ch$f \supset .\ xf \equiv A$, C ch$f \supset .\ xsf \equiv A$, and D $\sim$ch$f \supset .\ xsf \equiv A$. By HCP, $(Px)(Uf)B$. Since $B \supset .\ ($ch$f \supset .\ xf \equiv A)\ \&\ (chf \supset .\ f \approx sf)$, $B \supset .\ chf \supset .\ (xf \equiv A\ \&.\ f \approx sf)$, whence $B \supset C$ upon replacement. Hence generalising and distributing

[α] ECP. $(\xi x A(x))E \twoheadrightarrow A(\xi x A(x))$; KCP$'$. $(Uf)(z_o sf \equiv A)$ where $z_o = \xi x(g)(xg \equiv A)$ and x is not free in A; for full explanation, cf. Vol. 1, 1.21. —*Eds.*

[1] Care has to be taken with respect to other conditions on E. For example, guaranteeing $(x)xE$ – as the equivalence $xE \equiv (U\,\text{ext}\,f)(x{\sim}f \equiv {\sim}xf)$ would, if predicate negation is translated into sentence negation – could have disastrous results, viz. UCP.

quantifiers $(Px)(Uf)C$. But $C \supset .\ C \vee D$ and $C \vee D \supset .\ xsf \equiv A$, so $C \supset .\ xsf \equiv A$. Hence generalising and distributing again, $(Px)(Uf)(xsf \equiv A)$, i.e. KCP.

Guaranteeing KCP′ – that is, $(Ug)(z_0sg \equiv A(g))$ where $z_0 = \xi x(Ug)(xg \equiv A(g))$, that the object as such which is A is precisely sA – comes down to choosing models appropriately, models for TL and so for the full second-order logic of items. There are two cases to consider according as the condition C, i.e. $(Ug)(xg \equiv A(g))$ with x not free in $A(g)$, is satisfied or not.

Case 1. Condition C is not satisfied. Then z_0 can be arbitrarily chosen among the domain. Choose z_0 as an element satisfying $(Ug)(xsg \equiv A(y))$. Since KCP holds in the model some elements are such, so the choice is legitimate.

Case 2. C is satisfied. Then $(Px)(Ug)(xg \equiv A(g))$, whence $(Ug)(z_0y \equiv A(g))$. Hence, as in the derivation of KCP, $(Ug)(\text{ch}g \supset .\ z_0sg \equiv A(g))$, i.e. HCP holds with respect to z_0. But now $(Ug)(z_0sg \equiv A(g))$ may be derived in exactly the way KCP was. In sum, neutralised type theory supplies specialised models in which all object axioms hold good.

Catering for further axiomatic conditions concerning s-predicates is partly a matter of completing the definition of sf, that is defining sf in case f is not ch, and partly a matter of ensuring that the translating theory satisfies certain conditions, namely that for some g, effectively sf, ch$g$ and $(Ux)(xE\ \&\ \text{ext}\, f \supset .\ xg \equiv xsf)$ for ~ch$f$. It is rather easy to satisfy chsf: simply identify sf (where ~ch$f$) with an arbitrary characterising predicate. Although the axiom $xE\ \&\ \text{ext}\, f \supset .\ xsf \equiv xf$ is more demanding it is easy to satisfy it vacuously, at the same time, by design of models where nothing exists (set $\sim f$ distinct from f always). A less trivial way of guaranteeing s axioms, in accord with method 1, is to elaborate the modelling by duplicating predicates (in the Henkin construction), each one-place predicate being duplicated by another elementary predicate. If f is ch, then its duplicate $f^{\#}$ is equated with f itself; while if f is ~ch then $f^{\#}$ is equated with $\lambda x(\sim x \sim f)$. The additions cannot disturb consistency; by virtue of the equations any inconsistency would already have occurred before the duplication. Now for f noncharacterising, define $sf = f^{\#}$. Then chsf, since $f^{\#}$ is elementary; and $xE\ \&$ ext $f \supset .\ xsf \equiv xf$, by the definition of existence. Similar procedures can be combined with methods 2 and 3, which admit however interesting alternatives. The alternative approach is to make sure that the translating logic supplies a subtheory of entities over which certain predicates are characterising. This can be straightforwardly achieved under method 2 by restricting bottom objects, individuals, to entities and assuming that their one-place predicates are characterising. Then it is a matter of translating sf so that it is tantamount to a predicate of individuals (namely, to that predicate of individuals that is equivalent to f's translation applying to properties of the same individuals).

Thus far the representation for special principles, HCP and beyond, of the theory of items has been, in effect, a consistency modelling for the consistent theory. The same sorts of procedures can be reapplied, however, to show nontriviality of the paraconsistent theory by replacing translating logic TL by a relevant logic of types. Such a logic is obtained (cf. PLO) from the simple theory of types, e.g. as formulated by Henkin 1950, by replacing classical sentential and quantification axioms by those of a suitable relevant logic, and by upgrading main connectives in other axioms (such as $A\xi$) to corresponding relevant connectives.[1] Then most of the argument goes through as before. The working logic of items is nontrivial.[2]

Since every consistent theory admits – if only as regards demonstration by remaining relatively weak – of consistent extension, and likewise every nontrivial theory admits of nontrivial extension, in each case, to accommodate further logical notions, it is evident that fatal objections to the theory of objects which show inconsistency or triviality are not going to be achieved except by importing (referential) assumptions, no matter how natural or plausible they are presented as being, which the theory should not accept. In short, logics of nonentities are here to stay. They are no longer going to be defeated by demonstrations of inconsistency or triviality.

[1] Only the extensionality axioms cause trouble, and for the most part those are best dropped. HCP is formulated with a relevant biconditional, and corresponds to a similarly formulated abstraction scheme.

[2] However larger nontriviality questions remain outstanding at present, notably those for CPs for higher order objects: see 6.4(2).

Chapter 6

The theory of objects as commonsense

It is beginning to be appreciated that the Meinong of the mainstream philosophical literature is a mythological figure, that Meinong's philosophy has in fact been presented in an unfair fashion (perhaps even by largely sympathetic expositors such as Findlay 1963), and that the theory of objects in particular has been either widely misunderstood or else deliberately misrepresented. What has yet not been much appreciated is that Meinong's theory of objects represents an important alternative to standard (Russellian) logical theory.[1] Whereas the entrenched theory is both reductionist and pragmatico-empiricist in spirit,[2] the alternative is nonreductionist, antiverificationist, and commonsense. Since the theory of objects has often – there are important exceptions – been taken to be the very antithesis of commonsense, there is some explaining to be done. The problems are compounded by the fact that it is not at all easy to say what commonsense amounts to, and even more difficult to show that a philosophical theory is a commonsense one.

---

[1] There need be no apology for calling modern, standard, orthodox, "nondeviant", "classical" logic "Russellian". The orthodox logic of the textbooks consists essentially of variations and improvements (or sometimes the reverse) on the logical theory devised in large measure by Russell, building on the work of Peano and others, and worked out in collaboration with Whitehead in *Principia Mathematica.* Certainly there have been important additions by Hilbert, Wittgenstein, Tarski, Gentzen and others but these do not affect the general claim. In these terms influential modern logical theories, such as those of Quine FLP, are but variations on a theme of Russell's. And they share the reductionist empiricist assumptions of Russell's logical theory.

[2] The linkage of classical logic and empiricism is much more rigorous than this suggests. The linkage is through the Reference Theory, which classical logic encapsulates, and of which empiricism is the epistemological correlate given an account of existence which empiricism dictates. The linkage will be elaborated in later chapters.

The original version of this chapter was revised. Due to a mistake by the publisher, the editor's name 'Dominic Hyde' was mistakenly appeared as a co-author in all the chapters on Springerlink and third-party websites. The correction to this chapter is available at https://doi.org/10.1007/978-3-030-26309-6_7.

D. Hyde (ed.), *Noneist Explorations I*, Synthese Library 415,
https://doi.org/10.1007/978-3-030-26309-6_5

§1. Nonreductionism and the Idiosyncratic Platitude.

Philosophical theories may be roughly divided into reductionist and nonreductionist theories. Wisdom put the difference in this way (1953, p. 51):

> The Verification Principle is the generalisation of a very large class of metaphysical theories, namely all naturalistic empirical positivistic theories [and also of idealisms and conceptualisms]. While its opposite, which I venture to call the Idiosyncratic Platitude, is a generalisation of all commonsense, realist, transcendental theories. The verification theory is the generalisation of such theories as: A cherry is nothing but sensations and possibilities of more; A mind is nothing but a pattern of behaviour; … .

The difference is strikingly enough exhibited in the difference between the earlier and later Wittgenstein, between the *Tractatus*, and the *Philosophical Investigations* (though Wittgenstein never quite shook off underlying assumptions of his earlier work, such as the Reference Theory).[1]

In a number of respects Wisdom's way of putting the difference is, however, less than satisfactory. Though reductionist positions are very commonly motivated by the Verification Principle in one form or another – those of idealisms and intuitionisms as well as of empiricisms – they need not be. More seriously astray is what Wisdom has to say about the(?) nonreductionist position. It is not the case that nonreductionism is adequately stated by the Idiosyncratic Platitude in the form "Everything is what it is and not another thing". It is not the case that nonreductionism admits no analysis or logical representation of discourse – analysis is, as the chemical comparison suggests

[1] What Wittgenstein did demolish is the substantially stronger thesis, the Meaning-is-Naming Theory, according to which the meaning of every word is the referent it names, in Wittgenstein's terms (1953, p. 2) on

> this picture of language …
> Every word has a meaning. The meaning is correlated with the word. It is the object for which the word stands.

This theory implies the Reference Theory; for that the meaning of every word is its referent implies that meaning is a function of reference. But the converse does not hold. Although some philosophers (Davidson for example) are committed to the Reference Theory, given their assumptions that truth is a function of reference and meaning a function of truth, they are not committed to, and do not maintain, the Meaning-is-Naming Theory (cf. the discussion in 1.3 [Vol. 1 $_{—Eds.}$]). Furthermore some philosophers (for instance Quine), who accept the Reference Theory in its important truth formulation, explicitly reject the Meaning-is-Naming Theory while looking askance at the whole notion of meaning.

clearly enough, not always reduction. It is not the case that all interesting and exciting philosophy is reductionist in spirit and also not the case that (correct) nonreductionism is trivial and platitudinous. Illuminating though Wisdom's general philosophy of philosophy in terms of the interplay of paradoxes and platitudes may be, it too is paradoxical in his generous sense of "paradoxical" – as Wisdom himself may have been happy to agree.

According to the Idiosyncratic Platitude, in a more satisfactory form, also stated by Wisdom, *every different sort of item has its own sort of logic, and such different sorts are not in general in need of reduction.* Natural language is sufficiently in order as it is: it is not crammed full of "systematically misleading" expressions, it is not in need of complete reform and restructuring beginning with a levels-of-language restructuring and the elimination of descriptions and all ordinary names, and ending, characteristically enough, with the rejection of substantial slabs of familiar discourse as logically incoherent.

Classical logics, and also many non-classical logics, do however lead in these reductive directions, inevitably in the case of classical logic and its main modifications (see the argument of Part II, chapter 1 [Vol. 1 —*Eds.*]).[1] As Findlay remarks in his appraisal of Meinong (1963, p. 324):

> We cannot treat [Meinong], as did Russell, as affronting our "robust sense of reality", and as requiring to be exorcized by an elaborate Theory of Descriptions and a general method of "logical constructions". We all know where this theory and this method ultimately led. It led to the construction of physical things and minds out of infinite classes of sensed and unsensed sensibilia, or rather their dismemberment into the latter ... .

Most of us know that the logic of *Principia Mathematica* led directly to the philosophy of logical atomism, we know where the extensionalized form of that logic in the *Tractatus* led, and we are beginning to see where more recent revivals of these ventures such as the logic of *Word and Object* (i.e. Quine's WO) lead. It is less evident, but becomes clear upon applying or attempting to apply the underlying theories to the analysis of discourse, that such theories lead *inevitably* in these sorts of directions, to reductionisms such as atomism and away from realism and commonsense. It is commonly supposed that logic,

[1] So do reductionist accounts which reflect classical logic, such as Ryle's account (in 1971) of systematically misleading expressions. These accounts certainly do not regard things as in order as they are.

being reductive analysis, is bound to lead in these sorts of directions (cf. Wittgenstein 1953, ILT, Blanshard 1939, Marcuse 1964). This is the upshot however of far too narrow a view of logic, of logic as necessarily being, for example, extensional, contextless, without categorial distinctions, and so on. None of these features are limitations on modern logic, as recent developments in alternative nonclassical logic have shown (see especially the argument of Slog, chapter 4). More and more can be handled logically without reductive distortion, even if logics of ordinary discourse are still some way from being satisfactorily articulated.

Meinong's theory of objects yields a logical theory – or rather, as we have started to see, logical theories – of the latter less reductive kind.[1] Meinong's theory is, essentially, a rationalist, nonreductionist, commonsense theory, a beginning on working out the features of intensionality and non-existence, in particular, without reduction. As already noted, it will strike many as pretty extraordinary to claim that Meinong's theory of objects is, at bottom, a commonsense nonreductionist theory, a theory falling under the Idiosyncratic Platitude. For Ryle's account (in 1933–34) of Meinong as "the supreme entity-multiplier in the history of philosophy" has been influential, and even in Passmore's sober judgement

> ... objectivity has been preserved at a considerable cost. The Universe, it would appear, is populated by a variety of entities with the most surprising properties (1957, p. 186).

These assessments, especially Ryle's, are (as explained) misleading. Meinong did not take his objects to exist, to be elements of the Universe, which is what the assessments suppose. Meinong's theory did not include the surprises for commonsense that these assessments suppose.

Evidence that Meinong's theory, his theory of objects in particular, is strongly nonreductionist is not difficult to locate. Almost every sort of object is assigned, on Meinong's theory, its own sort of logic, and objects are investigated in meticulous phenomenological detail as they are, without reduction. But some important examples of nonreduction are enough for present purposes. Firstly, discourse about nonentities is irreducible according to the theory of objects to discourse about entities; and secondly

[1] Main logical theses of the theory of objects include those set out on pp. 3–4 [Vol. 1 $_{—Eds.}$]. In addition to nonreductionist theories of objects of both consistent and paraconsistent varieties, there are reductionist theories of objects such as those of Castañeda 1974 and Parsons 1974, which partially explicate Meinong's theory while sacrificing however some fundamental theses (see 12.4 [Vol. 3 $_{—Eds.}$]).

discourse about higher order objects is not reducible to discourse about ground floor objects. This does not imply of course, that isolated sentences cannot be paraphrased out, as e.g. "Pegasus does not exist" may be rendered, preserving truth,[1] as "it is not the case that there exists an entity identical with Pegasus". What is claimed is that the whole class of sentences of a given type is not replaceable by a class of another type: thus philosophers' attempts to establish reducibility theses (e.g. Ryle's in 1971 to show that talk of universals is replaceable by talk of individuals) by showing that a few sample sentences can be paraphrased, are quite inadequate.

Evidence that the theory of objects is antiverificationist, and liable to damage empiricist assumptions, is likewise not hard to come by. Truths about nonentities are not readily, or at all, verified or tested by experience. However this issue gets smudged once theories are introduced, the hypothetical deductive method applied, and empiricism rendered less falsifiable.

Evidence that Meinong's theory is, at least in its broad initial outline, a conmon-sense one may be reckoned a little harder to assemble but it can be found, especially if we take a little time off to get clearer as to what a commonsense theory is like. That the theory of objects is nonreductionistic does not show that it is commonsense, it simply indicates that the theory is not excluded as a commonsense one since commonsense is typically antireductionist (partly because reductions tend to run foul of hard data).[2] But a nonreductionist position that defends its claims by appeal to various special ways of knowing, for example, is not a commonsense one.

That the distinctive theses of the theory of objects, which are often taken to show that the theory is far removed from commonsense, do not show anything of the sort has been maintained, e.g. by Findlay (though he offers little in the way of argument for the claim, and later on contradicts it):

> Meinong's most famous and characteristic doctrine, that of an unbounded realm of objects which are *daseinsfrei*, indifferent to the antithesis between being and non-being, and his frank espousal of the anti-Parmenidean position that *what is not* is as much the object of

[1] Though *not* force, or thus replacement in more highly intensional frames.

[2] At this point we begin to get into all the, considerable, difficulties as to what a commonsense philosophy is, difficulties brought out in Grave 1967; for example, Berkeley considered his metaphysics complied with commonsense.

> significant reference and valid examination as *what is*, might seem to prove Meinong's extravagance and unsoundness, his wide exceeding of the bounds of commonsense. The doctrine, however, is eminently arguable at a commonsense level, and was once even justified by Russell on the basis of "perception" (Findlay 1963, p. x).

Even major problems for a theory of objects – the status of the Characterisation Postulate, the postulate according to which, roughly speaking, objects, whether existent or not, have their characterising features, and the matter of restrictions on the application of the Postulate, the issue which separates consistent and paraconsistent forms of the theory – are problems which arise for nonreductionist philosophical theories that are solidly grounded in commonsense: *they arise*, so to speak, *like other philosophical problems, out of commonsense taken as a theory.* The preliminary task this gives rise to, is to try to bring out how it is that, despite the apparently, or allegedly, extravagant features of his position, Meinong is essentially a philosopher of commonsense in the same tradition as Reid, and also as Moore,[1] and how the problems for the theory of objects, which is a refinement of a nonreductionist commonsense position, arise from warranted commonsense assumptions.

§2. The structure of commonsense theories and commonsense philosophy.

A *theory*, in a comprehensive logical sense, is[2] a class of statements closed under certain (logical) operations. The commonsense theory is the class of statements of commonsense closed at least under such operations as simplification and adjunction (i.e. A & B belongs to the theory iff A and B each belong to the theory). Commonsense is not a *logical* theory; for it is not, it would seem, closed under logical consequence. The logical working out of commonsense may well lead away from commonsense. Commonsense is rather a set of beliefs, (beliefs are represented propositionally in a theory) and like other theories comprising propositions of belief, not all consequences of what is believed are believed.

[1] The two main works in English on Meinong, Findlay 1963 and Grossmann 1974, both compare Meinong with Moore. Only Findlay considers (and that in a superficial fashion) the commonsense bases of the philosophies compared. And both accounts are inaccurate and unsatisfactory in other respects, e.g. Findlay's because of the simplistic claim that "each is responsible for one great idea" and the ideas listed, and Grossmann's because of his very misleading view of Meinong's philosophy primarily in terms of the development "over the years from a sparse ontology into an ample one", and because of the differences he alleges in general characterisation between the development of Moore's views and Meinong's.

[2] In the loose sense common in reductionistic mathematics. Strictly a theory is given by, i.e. is a certain function of, a certain class of statements.

In refining commonsense into a coherent and logical theory one of the first measures to be taken then is to close the theory under logical consequence. This already suggests one corollary, that refined commonsense has its problems, which arise when the consequences of commonsense may diverge from apparent commonsense, as happens with paradoxes. It may be unkindly said that the working out of the theory of objects as a development of commonsense provides adequate confirmation for this corollary.

Closure under consequence is only the first of several refinements that philosophers of commonsense would presumably wish to impose on properly refined commonsense. Much more difficult to formulate sharply are other sorts of refinement. One is the winnowing out of prejudice and superstition from the initially given beliefs of commonsense. Commonsense itself has usually suggested a distinction between the propositions that are basic and stand in need of no further argument, and those that are not so basic but belong to theoretical superstructure. The proposition that God exists, for example, was not regarded as basic in this sense even at times, such as those of Reid's Scotland, when it was a commonplace doctrine.[1] Argument was still required, if only for the Fool. More difficult to separate out than the overt theses of a theory of nature, or the universe, or of religion or other life-directing theories, are doctrines as to man's place in nature, his relations to other animals, his entitlement with respect to natural objects, and so forth. One way perhaps to remove prejudices of this type is to require that genuine principles of commonsense be transcultural. Such a requirement also removes the existence of a god of this sort or that sort from commonsense propositions. But the most satisfactory way (I can see) to avoid problems of all these types is to adopt the route Moore took, to list, insofar as required, basic assumptions of commonsense: this is like listing the axioms of a theory.[2] Although practically everything in Moore's splendid list of truisms

[1] According to Grave (1960, pp. 146–7):

> The question of God's existence they hold to be almost answered by commonsense ... the existence of God is inferred ... the premisses (the two which Reid requires) are supplied by commonsense.

The main principle of the inference to the existence of God is that "design or intelligence in the cause may be inferred with certainty from the marks or signs of it in the effect." This principle Reid ascribed to commonsense (*Intellectual Powers*, Vol. 1, p. vi), thereby revealing the extent to which his commonsense philosophy was a theory which went rather far beyond the initial harder data supplied by commonsense. Moore stretched "commonsense" almost as far, commonsense leading to the propositions of sense data theory (in Moore's famous "Defence of Commonsense", in 1959). In showing that the theory of objects is commonsense no such testings of the notion of commonsense are required.

[2] The resemblance of the principles of commonsense to mathematical axioms was pointed out long ago by Stewart: cf. Grave 1960, p. 149.

can be retained – there is a world external to us, there are material things that exist independently of us, with quite ordinary properties such as being brown, solid, and of definite size, there are other people, with thoughts of their own – other propositions need to be added. Furthermore the list can be open-ended, open to extension, and presumably to some revision, and some statements in the list may have less certainty than others, though any that are *un*certain will be rejected. Refined commonsense theories will not then be uniquely determined.

The axioms of commonsense, though commonly regarded as uncontroversially true, do not have to be sceptic-proof, philosopher-proof, or incapable of being criticized or doubted. No statements occupy the latter position; for just as any statement *can* be assumed, so any statement *can* be doubted even if in no way in doubt, even if self-evident (cf. Routley[2] 1975). This meets one of two serious difficulties Grave raises (in 1967) for Reid's defence of commonsense against philosophical scepticism, and (in 1960, p. 119) generally for commonsense, namely if the truths of commonsense are self-evident, how can they be denied, or how can there be dissent from them. People, philosophers for instance, can dissent from what is self-evident. Grave's other difficulty – namely if these truths are self-evident how can they be made evident when denied – can be met, for example in Moore's way: even the self-evident can be argued for, and its evidence elucidated.

It is important to be somewhat more specific as to the character of the axioms, since these are what distinguish a commonsense theory. The axioms are first truths: according to Buffier first truths have these characteristic marks:

> No attack upon them, and no attempt to prove them can operate from premisses that surpass them in clarity or evidence. They are and always have been, acknowledged by the vast majority of mankind. Those who imagine they reject them act like men in conformity with them (Grave 1967, p. 156).

These marks have their problems, but with some qualification they can be accepted; and they can be added to. For example, the initial truths, or axioms, should be transcultural in character.

A commonsense theory is restrained not only through what it accepts, as it must include sufficiently many basic commonsense assumptions, but also through what it rejects.

Such a theory cannot reject basic commonsense assumptions, nor can it systematically reject hard data claims as to what is so.

In summary, *a refined commonsense theory c* is represented by a class of statements:

(i) containing sufficiently many basic commonsense assumptions;

(ii) closed under deducibility, and under certain other logical operations such as adjunction;

(iii) excluding the negations of basic commonsense assumptions; and

(iv) excluding the negation of hard data claims.[1]

Then a *commonsense philosophy* is a theory which is a constrained extension of a refined commonsense theory, i.e. it is an extension, thus taking commonsense assumptions of (i) to be true, which excludes what the commonsense theory excludes under (iii) and (iv). A little more explicitly formally, c is represented by a pair $\langle \Delta, \Theta \rangle$ with Δ giving what the theory includes, i.e. (i) as closed under (ii), and Θ giving what the theory excludes, Θ being disjoint from Δ in the case of a satisfactory theory. Then a commonsense philosophy is given by an extension $\langle \Delta', \Theta' \rangle$ of $c = \langle \Delta, \Theta \rangle$ i.e. $\Delta \subseteq \Delta'$ and $\Theta \subseteq \Theta'$.

It should be evident that a commonsense philosophy may diverge drastically from what is sometimes called the philosophy of commonsense, i.e. (according to the *Concise Oxford English Dictionary*) "accepting primary beliefs of mankind as ultimate criterion of truth". Moore's philosophy was certainly not part of the philosophy of commonsense in this – rather revolting – sense, since although Moore took commonsense perception claims as certainly true he thought they stood in need of analysis and were capable of a deeper explanation. Nor did Reid, that paradigmatic philosopher of commonsense, fit under such a philosophy of commonsense rubric: nor any longer do most other commonsense philosophers. Although, according to Grave (Grave 1967, p. 156)

> philosophers ... who have argued from commonsense and for its beliefs have often thought of commonsense in this way [i.e. as an intu-

[1] There is indeterminacy, of course, at several points in the characterisation, and perhaps some of the claims could be strengthened, e.g. in the tradition of Moore the assumptions of (i) would be reckoned not only true but certain.

A refined commonsense theory differs from *critical* commonsense, as explained in Grave 1967, p. 157ff., in being much more theoretically structured. Critical commonsense concerns only the character of the axioms, their being open to revision and modification.

> itively based *common consent*], they have ... as often thought of it in a more ordinary way, as the commonsense that is opposed ... to high and obvious paradox.

A refined commonsense theory may be seriously incomplete and underdetermined. It may lack even quite obvious connections between theses, it may lack expected syntheses, it may lack any explanation of its claims, it may give no answers to a great many questions that a philosophy would be expected to pronounce upon. Most important, a commonsense theory may give no account of how what it asserts is true or possibly can be so: these are gaps that an extension may fill by transcendental arguments. It is elaboration of these matters, the filling of these or some of these gaps that leads from a refined commonsense theory to a commonsense philosophy. As a result of the extension too, a commonsense philosophy may assert things, especially things in more technical terminology, that sound strange to commonsense ears not attuned to such constrained extensions of commonsense. Just this happens with Meinong's theory of objects, simple variants of which are, so I am going to contend, (part of) a commonsense philosophy. The basic reason for this is that the theories only add constrainedly to commonsense connections and linkages which fill it out – with a view to obtaining such features as coherence, explanation, synthesis, and so on.

Not any extension of refined commonsense will do. A commonsense philosophy should meet certain conditions of adequacy. An important, but far from sufficient requirement, is that of nontriviality: not every statement should hold on the philosophy, some separation of falsehood from truth should be effected. In fact if the closure of basic commonsense is nontrivial, as was assumed – perhaps without sufficient warrant – then a commonsense philosophy, since a constrained extension of the closure, will also be nontrivial. Other piecemeal conditions of adequacy are not too difficult to locate; more sweeping criteria, such as those that go under headings such as coherence, are much more difficult to nail down.

§3. Axioms of commonsense, and major theses.

To reveal the theory of objects as a commonsense doctrine, as part of a commonsense philosophy, it is important to enlarge upon Moore's truisms, or axioms, of commonsense (in 1959, pp. 33–4). Recall that these axioms included such propositions as the following:

The earth has existed for many years; its inhabitants have been variously in contact with, or at different distances from, one another and other things. In particular (p. 34),

> ... I have had expectations with regard to the future, and many beliefs of other kinds both true and false; I have thought of imaginary things and persons and incidents, in the reality of which I did not believe ... .

It is but a very small step from Moore's generalisations of those truisms (under heading (2), p. 34–5) to two further important commonsense axioms, not explicitly cited by Moore but representative of axioms of metaphysical relevance:

Axiom M1. We can and commonly do make true statements, of kinds that can be readily indicated, about what does not exist. We can think correctly and talk truly about what does not exist.

Let us call the items concerned, items which do not exist, *nonentities.* These nonentities will be of various different sorts, as the next axiom begins to assure us.

Axiom M2. Not all nonentities are the same. Nonentities are of many sorts, in particular some are possible and some are impossible.

Both assumptions were taken to be merest commonsense by Reid:

> Indeed I know no truth more evident to the commonsense and to the experience of mankind [than that] men may barely conceive things that never existed ... they know that they can conceive a thousand things that never existed ... [a man may] conceive a centaur, he may have a distinct conception of this object, though no centaur ever existed (Reid 1895, pp. 368–9, my rearrangement).

Just because these propositions are cited as axioms – in the colloquial sense – it is not to be taken that they are merely self-evident and cannot be defended or even deduced from more aseptic principles. For example, it follows from the fact that the Icosohedron is different from the Dodecahedron or that Pegasus is different from Chiron that not all nonentities are the same. That is, the axioms may be defended by appeal to the hard data.

It will be protested that in stating M1 and M2 in the way given, the issue has been prejudiced against reductionist theories. M1, for example, which is, it is said, far too theoretical for commonsense, should be restated at least in the form:

Axiom M1′. We sometimes make true statements purportedly about what does not exist,

and preferably in more aseptic examples. But M1, not M1′, is what commonsense assures us of. Moreover as far as the sorts of arguments to be advanced are concerned changes of this kind can be admitted. The arguments will take us back to M1.

There are two major theses to be argued:

(A) A theory of objects fits with commonsense, in the sense made more precise, that it is a constrained extension of commonsense.

(B) Only a theory with central assumptions in common with a theory of objects, in particular rejection of the Ontological Assumption, can fit with commonsense.

Eventually one might hope to prove these propositions, in particular (B) for which it is only necessary to know some of the axioms of commonsense and some hard data which rival reductionist theories violate. Proof of (A) may look a much more difficult business because not only is a sufficient axiomatisation of commonsense apparently required, but more of the character of the hard data has to be known as well. In fact comparatively weak underdetermined theories stand a much better chance of fitting with commonsense, as the following sort of arguments reveals. Let $c = \langle \Delta, \Theta \rangle$ be a refined commonsense theory which includes axioms M1 and M2, and let d be the extension of c by subject variables, the universal predicate "is an object", the truth-functional connectives "&", "$\vee$", "$\sim$" and the quantifiers "for every object" ("U") and "for some object" ("P") subject to the axioms of quantificational logic (with the rule (γ) of material detachment restricted in its application to the imposed logical theory and consistent extensions thereof) and to the axiom "for every x, x is an object". Then d is certainly an extension of c, it contains a weak logical theory of objects, and the only hard question is whether it is a constrained extension. But if the underlying theory c does not include the logical operators, as it may well not, then because there is no interaction of the extension with the underlying theory, the extensions would be constrained. All this, even if right, only shows that a quite weak theory of objects fits with commonsense, not that a more full-blooded theory like Meinong's does. But it is a useful start on which we can build by more indirect and circumstantial considerations.

There are other less theoretical bases from which a start can also be made. One, which won't however bear *too* much weight, is by direct appeal to the commonsense phenomenological character of Meinong's investigations. Findlay 1973 has brought out the extent to which Meinong was a phenomenologist and how in important respects his work was purer than that of Husserl; however the reductionist anti-commonsense directions in which phenomenology has led are familiar enough. A better start is by appeal to the sort of commonsense data which Meinong expected his theory of objects to respect.

> Meinong [and] ... Moore ... share ... the same sensitiveness and deference to what people actually believe and commonly say ... and ..., though many would question it, the same common sense, in the sense of an unwillingness to abandon what we plainly understand and know, and what forms the firm foundation of our discourse, at the behest of theories much less lucid and indubitable (Findlay 1963, p. ix).

This may be confirmed not only by detailed examination of Meinong's work; it may also be confirmed by such circumstantial evidence for the commonsense character of the theory of objects as that it gives something like the naive approach to the solution of many fundamental logical problems (see, especially, Castañeda's argument in 1974), and that it solves all, or almost all, the philosophical puzzles over naming and denoting in a natural way (as Linsky in effect remarks in 1977).

Another good start, which again strictly calls for much documentation, is with Reid's philosophy. Reid was a philosopher of commonsense *par excellence*, and yet Reid's philosophy, a clear example of a nonreductionist philosophy conforming in leading respects to the Idiosyncratic Platitude, incorporates essential features of a theory of objects, e.g. the Independence Principle (1895, p. 403), nonexistential quantification (e.g. p. 368), and most important rejection of the Ideal Hypothesis or Theory of Ideas, an epistemological version of the Reference Theory.

Indeed features of Meinong's theory of objects which offended the "robust sense of reality" of Russell and others were already an integral part of Reid's commonsense philosophy: this "robust sense of reality" is little more than a penchant for applying the Ontological Assumption, a prejudice Reid explicitly rejected. Thus if Meinong's work remains a scandal, a very similar commonsense theory of objects can be built up from the philosophy of Reid and other commonsense philosophers. Finally, there is evidence now coming in from defence of theories of objects, for example of this sort: if the commonsense answer is given to new difficulties the theory encounters, then this turns out almost invariably to be the answer that the theory delivers, or should deliver.

Defence of (B) is also part of the case. The argument for (B) is, in initial form, simply that no analysis or reduction of common true statements purportedly about nonentities to true statements about entities can be given which does not violate the hard data. It is known that all the standard theories of descriptions, which were designed to undertake this sort of job, break down on quite firm data (see, e.g., chapter 1 [Vol. 1 $_{—Eds.}$]); and a more general argument that all such theories are bound to break down can be designed (it is given in chapter 8 [Vol. 3 $_{—Eds.}$]). The conclusion is that no adequate reduction of talk about nonentities to talk about entities can be effected. That is, a reasonably comprehensive commonsense philosophy will contain at least the rudiments of a theory of objects. The next task is to build on the rudiments.

§4. No limitation theses, sorts of Characterisation Postulates, and proofs of commonsense.

Meinong's Independence Thesis – according to which *Sosein* is independent of *Sein*, according to which objects have definite characteristics even though they do not exist – in its various and fuller formulations combines, or connects, several different, commonsense, principles which it is important in logical investigations to separate out.

1. *No limitation (or Freedom) theses.* These theses come in several forms, for assumption, for consideration, for imagination, and so forth (for wishing, thinking, doubting ...). According to the Unlimited Assumption Thesis there are no limitations on what one can assume; one can assume anything one likes, no matter how bizarre, inconsistent or paradoxical.[1] A direct analogue of the Unlimited Assumption Thesis is the Unrestricted Imagination Thesis fundamental to a comprehensive theory of fiction, that there are no

[1] The Thesis is a purification of Meinong's principle of freedom of assumption, for which see, e.g. Mög, p. 282ff.

restrictions on what is imaginable (even the unimaginable is imaginable). There is, as in all these cases, only one major catch: what is assumed may not in fact have all the properties it is assumed to have. The reason for this catch, if not obvious, will become so: one may consider, or wish for, a round square which exists, but that won't lift the object out of *Aussersein.*

Now assumption is propositional: a creature assumes that so and so, where the so and so is typically specified by a sentence. Unlimited assumption is reflected in a logic through no limitations on the propositions it can consider – any statement at all can be represented in its notation – and no limitations on the propositions that can be assumed in hypothetical arguments and proofs. Most logics now allow for these things, up to a point: up to the impossible, but not into the paradoxical. One can assume, and in *reductio* proofs characteristically does assume, what is impossible. But one is barred, by stratification theories of one sort or another, e.g. order theories, levels of languages theories, from assuming what is paradoxical, e.g. from assuming that what one is assuming is not the case. This classical distinction in the propositional case between the impossible and the paradoxical or defective we find in Meinong in the object case: and really, since propositions are objects (of higher order), the classical case is just a special case of Meinong's more general separation.

In terms of the freedom thesis there is no basis for this classical restriction to the nonparadoxical. Nor does commonsense support the restriction: on the contrary natural languages, which tend to mirror commonsense, are conspicuously free of stratification, and provide no warrant at all for levels-of-language impositions or the like. The admission of paradoxical assertions into a logic – into what remains of the object language – though an important step in the direction of dialectical logics, does not of course settle the logic of the assertions admitted: *this* depends on the issue (under 6.4(2) below) of how these propositional objects are characterised.

Let us generalise then from propositions to all objects.[1] According to the Unlimited Contemplation Thesis there are no limitations on what one can consider or contemplate: one can consider any item at all, no matter whether or not it exists, or whether it is

[1] It is worth emphasizing that propositions offer a valuable model for objects: for arguments for the objectivity and irreducibility of propositions, though not undisputed, are well known and widely accepted. Compare also worlds, objects which offer, though perhaps to a lesser extent, somewhat analogous advantages in explaining the situation with respect to objects in general.

possible or not, or whether it is paradoxical or not. Once again, however, these objects will not have all the features they present themselves as having. The objects which present themselves as, respectively, the perfect existent being and the class of all classes not members of themselves will be regarded with some suspicion these days, much as the representative at the door is not accepted except by the unduly naive as the person his introduction card says he is (certainly not without a good deal of independent investigation in many cases). But with objects other than those we've been taught to beware of many of us are like the unduly naive; for we've been brought up philosophically with the Reference Theory and with regular existent objects where assumption never goes wrong. It is not that nonexistent objects are devious or less trustworthy: it is that one mistakes their logic if one assumes that, like entities, they have all the properties they present themselves as having. Consider, to get the feel of the direction we shall have to travel, the following object: the oil rig that is 2 miles due north of Tangiers and 2 miles due south of Capetown. If this oil rig's credentials were accepted, it would follow that Tangiers is 4 miles south of Capetown. But there is *nothing* to stop us considering the object.

These considerations answer the question: why can't we think of objects that violate the logical laws, any logical law at all? The answer is that we can – only in a coherent theory, the objects won't have all the properties they are thought of as having, that they present themselves as having.[1] It is here that the separation between logic and the axiomatic disciplines, such as geometry, appears to come; for one can have objects which genuinely violate their axiomatic principles (at least as they are usually formulated).

Unlimited consideration should be reflected in a logic, and in a satisfactory language, through no limitations on the objects it can consider or describe – any object at all should be representable in its notation (Hilbert's theory, Hilbert and Bernays 1934–39, fails here) – and no limitations on the objects that can be reasoned about, generalised or particularised about, and so on (all logics except neutral logics fail here). Yet surely what is good enough for impossible propositions, e.g. quantification over them, should be admissible also for other sorts of impossible objects.

Once again the freedom thesis leads to the removal of classical and other restrictions (including those of so-called "free logics" which are much less free than the medieval free cities). Quantification logic, duly reinterpreted over domains of objects, extends to *all*

[1] Hardline referentialists are properly obliged to object to both.

objects – and there is nothing at all to prevent *pure* quantificational logic so extending to all objects (to reiterate, important changes from classical *formalism* only begin with identity and description theories). And once again the removal of classical interpretational restrictions is supported by commonsense: classical restrictions are not incorporated in natural languages, but only in the classical canonical strait-jackets philosophers characteristically endeavour to force natural language into.

The way the initial clauses of the semantics of neutral quantification logic go are very straightforward. To illustrate: "Pegasus is winged" is true iff the item "Pegasus" is about, namely Pegasus, has the property specified by "is winged", i.e. that of being winged, i.e. iff Pegasus is winged. Such clauses, which are admittedly low in informational content, form a much more comprehensive class than classically. For there is no restriction on well-formed subjects to logically proper names: any ordinary (or extraordinary) names, or for that matter any singular subjects or descriptions, can be admitted whether they are about what exists or not. Thus singular descriptions can serve as direct subjects and as substitution values of variables. Accordingly a great many subject-predicate expressions of English, rejected as misleading in classical logical analysis, are in order as they are and admit of direct treatment in neutral logic. In short, neutral logic, unlike classical, is an appropriate vehicle for the formulation of freedom assumptions congenial to commonsense.

2. *Characterisation (or Assumption) Postulates.* These postulates also come in several forms – assumption or reduction postulates in the case of individual objects, abstraction or comprehension postulates in the case of classes, attributes and propositions. But in their completely naive, and theory-trivialising, form they assert that an object has all the properties it presents itself as having: an object which is f is f, the object which is f is f, and so on. In order to stop things going completely haywire, this assumption postulate naively transferred from the logic of entities – where it holds – to the logic of nonentities, which platonists (and others) mistakenly take to be the same – where it certainly fails – has to be appropriately qualified.

The Characterisation Postulates for the logic of entities, i.e. genuinely existing things, are comparatively straightforward, and take the following forms for different descriptors:

– *an* cases, for both "an" meaning "an arbitrary" and "an" meaning "a certain": if an object which is A exists then an object which is A is A, i.e., in symbols:

$$E(\kappa x A(x)) \supset A(\kappa x A(x)), \tag{CPi}$$

– *the* case: if the object which is A uniquely exists then the object which is A is A, i.e.:

$$E!(\imath x A(x)) \supset A(\imath x A(x)). \tag{CPii}$$

These postulates do not extend however even to possibilia, as simple cases of objects determining their own ontological status indicates. Consider, for instance, the actual fountain of youth to be found in Garema Place, assuming that to be a logical possibility. Then a full characterisation postulate for possibilia would make it true that there exists a fountain of youth in Garema Place. The conclusion is not that there is something wrong with possibilia, but that the logic of possibilia is not the same as the logic of entities.

The classical position is really that nonentities have none but a barely minimal logic, and, more precisely that characterisation postulates begin and *end* with entities, that (CPi) and (CPii) can be strengthened to biconditionals, that *only* entities are so assumptible. This goes along with the thesis that nonentities have no properties in their own right, they have none but derivative properties, these deriving from features of entities. The classical position is entirely rejected by Meinong, by noneism, and by commonsense, according to all of which nonentities have distinct, irreducible properties, from which it will emerge, by a transcendental argument, that limited characterisation postulates hold for nonentities.

Findlay argues that incomplete and impossible objects "must have a definite internal makeup if only in order *not* to exist or *not* to subsist anywhere" (foreword to EP, p. xiii). This is a little swift. An item with no properties and no definite makeup, a nothing, will certainly not exist. There is no valid argument by Findlay's route to the assignment of characterising properties to nonentities. There are other arguments, however, to the conclusion that nonentities do have extensional properties (enough often to yield a definite makeup) and that some qualified form of the Characterisation Postulate holds. An important transcendental argument takes the following form:

(1) Some statements of identity and difference are true (apparently) of nonentities, e.g. Pegasus is distinct from Cerberus, mermaids are distinct from unicorns.

(2) The truth of identity and distinctness claims about nonentities can *only* be adequately explained by supposing that the items themselves have properties. The same goes for likeness and unlikeness claims.

Given that "Pegasus" and "Cerberus" are about what they seem to be about, namely Pegasus and Cerberus respectively, (2) is immediate from (1). For if Pegasus is distinct from Cerberus some features must distinguish them, i.e. they must differ as to properties (or at least as to what is true of them). Nor can this difference be a merely intensional one. I cannot make Venus distinct from Aphrodite by mistakenly *thinking* that one is beautiful and the other is not.

But it may be denied (e.g. by those who espouse familiar concept or linguistic positions), that "Pegasus" is about what it seems to be about. Contrary to common supposition, however, mere location of differences in associated concepts or the like (e.g. senses) – worse still in associated names – will not do. While (to repeat pp. 50-51 [Vol. 1 —*Eds.*]) we might be able to explain the truth of a distinctness statement such as "Unicorns are distinct from mermaids" by reference to the distinctness of the concepts *unicorn* and *mermaid* or the difference in the senses of expressions "unicorns" and "mermaids", we cannot similarly explain the truth of a contingent identity statement such as "What I am thinking about is identical with a unicorn" by reference to the sameness of the concepts or senses involved, because they are not the same. And to explain the truth of the identity statement by identity of reference, by saying that the concepts apply to or the expressions refer to the same items, is to push the responsibility for the truth of the identity back to the items themselves, and therefore to admit that the items must have properties. Yet unless some other *entities* can be produced whose identity or difference can explain such contingent identity statements, we will have to fall back on the identity or difference of the items themselves, which implies that they have properties.

(3) The properties of nonentities cannot be purely intensional, but must include some extensional features. Firstly, we would not be able to assert the contingent identities we do unless some of the properties were extensional. Secondly, if all properties of nonentities were intensional there would be nothing for these intensional properties to focus upon. there would be nothing, for example, to distinguish thinking about Pegasus and thinking about Cerberus: certainly it could not be done by other thinkers thinking without a

vicious circularity being induced. Not all objects, certainly not all commonplace public nonentities, can be characterised purely intensionally.

(4) Nonentities have the extensional properties which serve to characterise them. By (3) nonentities have extensional properties: how can they have these properties? Only in the end in virtue of their being what they are, by their characterisation. And this is the way we do in fact come to know *many* features – not *all* features – of many nonentities, not by observation but by assumption and deduction, as consideration of examples such as the Triangle reveals.

The transcendental argument only gives a limited amount of information as to the properties of nonentities. It does not indicate how properties other than those supplied by characterisation are arrived at. It does not indicate *which* of the extensional properties that a nonentity says it has in its own description, for example, it has. Which such properties nonentities do have, which properties are *characterising* will have to be determined, or delimited, by somewhat independent arguments, which impose appropriate conditions on characterisation postulates (some of the moves have been gone through in chapter 5).

It is in part through characterisation principles that the Idiosyncratic Platitude gets satisfied by the theory of objects. Every sort of item does have its own sort of logic: this is a result primarily of assumptibility, not simply, or even at all – as is often thought these days – of identity criteria. The identity conditions for entities and nonentities are exactly the same but the characterisation principles are very different. Some of the objects and their corresponding characterisation principles (argued for in the previous chapter) are as follows:

Sorts of object	**Corresponding CP**
entities	unrestricted CP
nonentities (of bottom order)	severely restricted CP, restricted e.g. to elementary characteristics
sets	set abstraction postulate
properties and relations	attribute abstraction postulates
propositions (and objectives)	propositional abstraction

The list is incomplete for there are of course other sorts of objects; and there may be further distinctions to be made. For example, there may be separate characterisation postulates for possibilia which do not hold for other nonentities. But, in any case, the logic of possibilia can be separated from that of impossibilia in other ways, e.g. through the fact that characterisations of impossibilia lead to contradictions, and the effect of such inconsistency elsewhere in the logical theory. That is, other logical features also contribute to the satisfaction of the Idiosyncratic Platitude.[1]

Whether the Characterisation Postulates which emerge render the theory of objects a paraconsistent or a consistent theory is an important matter, not merely for the assessment of the theory, but also for claims of the theory to be commonsensical. For surely commonsense does not accept inconsistency! A little reflection reveals however that the claim a commonsense theory cannot be paraconsistent is not quite so clearcut, especially when common reactions to the paradoxes are taken into account, e.g. the reaction "Yes, that's true; yes, it's contradictory". There is no need however, to introduce such controversial claims as "commonsense involves true contradictions" in order to defend even the paraconsistent theory of objects as commonsense. It is *enough* that the paraconsistent theory conservatively extends a refined, consistent, commonsense basis.

What remains to be argued, in order to make out a *prima facie* case for the theory of objects as commonsense, is

(C) CPs are either commonsense, or

(D) CPs can be made part of a constrained extension of commonsense.

Although examples of CPs have wide ordinary appeal, they are scarcely truths of the uncontroversial type accounted truths of commonsense. The precisely qualified forms of CPs are even further removed from commonsense. And the Abstraction Axiom for sets is doubtfully a first truth of commonsense – even if it is of mathematics. So it appears that the more difficult proposition (D) will have to be argued. But establishing (D) in detail is a very large order, larger than that hitherto required for any accepted philosophical theory. For it requires, as a part, a nontriviality-proof for paraconsistent set theory, since the Characterisation Postulate in the case of sets is just the Abstraction Axiom. Some important elementary bottom order cases of (D) can be argued (by applying the results of earlier chapters, especially 5.4); and these cases can be combined with a proof of the

[1] The Platitude may seem vacuous, but it is not entirely so owing to its nonreductionist bias. In particular, on rival classical theories, nonentities really have no separate logic, and so do *not* have their own sort of logic. Nonentities may appear to have something of a logic, but that is an illusion: on closer inspection this residual logic vanishes into the logic of entities.

nontriviality of extensional paraconsistent set theory (as given in Brady and Routley 1989). But the larger issue, as to whether the theory of objects appropriately formulated with Characterisation Principles for each sort of object is a commonsense theory, i.e. a *constrained* extension of commonsense, is an open question.[1] My conjecture is, of course, that the theory of objects is a constrained extension of conmonsense, that is, is a (and perhaps *the*) commonsense philosophy.

[1] Postscript: See now point 1, "Preface to the appendix" in UL [reprinted in Vol. 4 —*Eds.*].

Chapter 7

The problems of fiction and fictions

> The people were much bothered by the tricks of the Bullim-Boukan. The Boukan were two mischievous spirits in one. They were plural and indivisible, and only looked like one. That is why they were called Bullim, the Two-Boukan (Massola 1968, p. 76).

Literary phenomena provide a severe testing ground for logical and semantical theories; and certainly (at least until recently?) the phenomena have regularly revealed serious weaknesses in each new theory proposed. But typically the phenomena are not seriously considered; typically issues of much literary interest such as metaphor, simile, transference of sense, irony, satire, allegory, fictionalisation, and so on, are set aside at the beginning of logical and semantical studies, where the emphasis has always been on literal meaning and reportative discourse, to the cost of most of the remainder of discourse.[1] The literary phenomena set aside are at best given perfunctory treatment after the important work of dealing with literal reportative discourse has been accomplished – to the very limited extent that even this discourse can be analysed on the main semantical theories so far proposed.

The serious neglect of literature, and its important and theory testing (and often defeating) phenomena, in modern semantics can be ascribed largely to the almost exclusive concentration of semantical investigations on what is taken to be the language of science and on its everyday linguistic counterparts concerned with chronicling reality,

[1] In this study too, it has to be confessed, most of these issues – fictionalisation and the like excepted – are neglected. The universal semantical theory does, however, provide the apparatus for analysing such phenomena as transference of sense and metaphor. The key devices are those of context-induced world shift (much as with fictional statements) and worlds where semantic assignments result in extraordinary senses of certain expressions in these worlds.

The original version of this chapter was revised. Due to a mistake by the publisher, the editor's name 'Dominic Hyde' was mistakenly appeared as a co-author in all the chapters on Springerlink and third-party websites. The correction to this chapter is available at https://doi.org/10.1007/978-3-030-26309-6_7.

D. Hyde (ed.), *Noneist Explorations I*, Synthese Library 415,
https://doi.org/10.1007/978-3-030-26309-6_6

such as "The cat sits on the mat". Such discourse, which it was mistakenly supposed was, or need be, only literal, reportative, extensional and about things that actually exist, encompassed all that was of real philosophical importance; the rest, insofar as it could not be conveniently reduced to the this "clear and distinct" classical basis, was dismissed with various grades of abuse, ranging from "nonsense" down to "not really worth bothering about". Such was the position of the scientific semantics which obtained its roughest and toughest formulation with logical positivism and which persists, sometimes in almost unrefined form, in modern empiricism and its variants, such as pragmatism. Since it is empiricism which has inspired much philosophical analysis and practically all formal semantical investigation, positivist preconceptions about what is clear and distinct, what is basic and requires no further analysis, and what is fundamental and important, persist in most modern semantical work, superficially softened perhaps but largely unquestioned. Because of the reality fixation, not to say science worship, characteristic of empiricism, a narrow sample of scientific discourse has been the model in semantics, to the detriment again of non-reportive discourse.

Semantical difficulties concerning certain literary phenomena, those of fiction and fictions in particular, are not however so easily set aside or separated from general semantical puzzles – puzzles firstly over talk about nonexistent things and about intensional discourse, especially the language of invention and of imagination and the language of scientific discovery, conjecture, and speculation; and puzzles secondly about scientific theories, especially false theories, which after all predominate among theories. As to the first, the logical theory, logic and semantics, of fiction and fictions is a special, and very instructive, case of any general logical theory of nonexistence and intensionality. As to the second, scientific theories, especially mistaken ones, closely resemble[1] fiction – much more closely than has usually been allowed by accounts other than fictionalism – but the main characters are commonly not human ones. Studies in ethology with animal characters as the main objects of study provide a bridge between fiction of the typical human chauvinist variety with humans and their often humdrum problems dominating the landscape, and the "fiction" of science where the main figures may be one or other elementary particles.

[1] Nonetheless the resemblance is only a resemblance (contrary to a main tenet of fictionalism, cf. chapter 11 [Vol. 3 $_{—Eds.}$]). Differences between theoretical objects and fictional objects will be brought out in what follows (§10).

There have been attempts (e.g. in Woods 1974) to sharply distinguish the cases here being assimilated, to make a distinction between nonentities (of fiction) and nonesuches (e.g. the phlogistons of science and the kings of France and Lilliput of philosophy); but the distinction, as made out, will not bear much weight. To the extent that the distinction (important in Woods' work) is explained by Woods at all, it is explained as follows (1974, p. 29):

> ... even though Holmes does not exist, we know who he is. He is a non-entity who is a somebody. The present king of France is a nonesuch [a nobody?]. Nonesuches do not fall within the values of ... bound variables. There are no nonesuches.

But wherein lies a difference of logical import? Nonesuches are not just *like* nonentities, they are nonentities, i.e. objects that do not exist; and Sherlock Holmes is a nonesuch in that there exists no such person. The suggestion is that fictional objects are somethings, objects that are quantifiable over and possessing properties; whereas nonesuches are none of these. However "nonesuches" are all of these, e.g. phlogiston is something, namely a heat substance with some fairly specific properties, about which we can truly say quantificationally that there exists no such substance, nonesuch. The present king of France, though hardly an object of modern scientific theory, is likewise quantifiable over, is a something in this sense, and also in the sense of having properties, e.g. he is a king and, less controversially, he differs from the present king of Australia, is distinct from Meinong's round square, and so on.[1] The differences between the objects reduce then to differences in source (a literary fiction having its source in works of fiction, and a scientific fiction in some group's theory); but these differences are removable and accordingly no bar to a general logical theory which comprehends them all.[2] For one thing, it is easy to arrange to have a theory presented as fiction, and conversely fiction often presents theories. In this way depauperate objects such as present king of France can be seen as limiting cases of fictional items: as they have no source book, apart from what is given by their descriptions, their only features are those of their descriptions together perhaps with what follows therefrom. But such an assimilation of fictional objects with the nonentities of scientific theories and of erroneous history (purportedly real objects) cuts two ways: for though it means that the problems of fictions are not so readily evaded even in the

[1] He is rather the object of an untold pseudo-history which extends to the present.

[2] This is not to say that classifications of objects that do not exist into various sorts, e.g. fictional objects, theoretical objects, pseudo-historical objects, etc., are not of logical significance: quite the contrary.

semantic analysis of scientific discourse, it also means that a theory of real objects is by no means independent of a theory of fictional objects, and hence that any properly comprehensive theory of objects (or descriptions) is obliged to supply a theory of fictions.

§1. Fiction, and some of its distinctive semantical features.

The failure (soon to be perfunctorily observed) of most marketed philosophical and semantical theories to cope with just one group of literary phenomena, that of fiction and fictions, is a minor theme in what follows: the major theme is the more vexing one of trying to locate minimal adjustments to comparatively naive, and commonsensical, theories of fictions – at least to the extent of clarifying when, and how fictional statements are true – and to devise a semantical analysis, fitting into the general framework of universal semantics (of chapter 1, §24 [Vol. 1 $_{-Eds.}$]), for the resulting adjusted account. Since the availability of a universal semantics, a semantical analysis for every language, induces the reasonable belief that the semantics of fictional discourse, whatever it is, can be accommodated in the general theory, the enterprise of determining the semantical structure of fiction reduces to an exercise in applying the semantics to the data, if only the literary data can be fixed. But the availability of a general framework makes the determination of the semantics of fictional discourse, even if rather more of an applied venture, only a few degrees less difficult.

Delimiting the field of enquiry and the region from which literary data are to be drawn, though not without its difficulties, is not a serious obstacle to semantical investigations. Standard dictionary accounts of *fiction* serve to define the field sharply enough: *literary* fiction comes in both spoken and written forms, in spoken forms as legends, sagas, epics, stories and tales, as contrasted for example with spoken reports, statements, descriptions, summaries of facts, briefs; and in written form as works of fiction, novels, short stories, shockers, written tales, fables, as contrasted with reportive narrative, history, memoirs, written reports. A piece, or work, of fiction typically has an author or set of authors (perhaps unknown or anonymous) and is intended to reach some audience: thus it is a *communication*, or message, in the wide sense. More generally, (a piece of) *fiction* is authored discourse (whether written or spoken) or communication which consists of imagined or invented statements or narrative, which conveys a *story* as contrasted with factual or reportative discourse. A work of fiction will generally be about fictional items or creatures (i.e. *characters*), that is the story is about *fictions* in another important sense of that term.

The more general communicational account (or something like it, e.g. an account in terms of theories of certain sorts) is required because a piece of fiction or of make-believe may involve both much more than discourse and much less than discourse. For example, a film or television show, unlike a bedside story or fireside tale, may aim to absorb the audience visually as well as auditorily, and evidently the film experience could be expanded to involve other senses as well. Since a film may be silent and contain no script, discourse in the usual sense, though characteristic of fiction, is not essential for nonliterary fiction: potential communication or *portrayal of an imaginary situation* through some medium is. It is also open to question whether the characteristic features of communication, a sender and a receiver or, in the case at hand, an author and an audience, are required. Certainly in many cases no audience (other than the author) is reached, or need be reached, by a work of fiction, and exceptional circumstances can be envisaged where an unauthored piece of fiction turns up, an *objêt trouvé*. Suspending such characteristic (though not invariable) distinguishing marks of fiction enables us to see the extent to which *a work of fiction is*, semantically at least, like *a theory*.

The bounds of fiction are those of imagination, together with the resources of the language used. Naturally not all imagined worlds – certainly not all imaginable worlds – are recorded in fiction. But the author of a work of fiction can try to set down whatever he imagines; and the free range of imagination is virtually unlimited.[1] It is not bounded by laws of nature or even by laws of logic: it is not restricted to possible worlds. Nor is there a problem in portraying or envisaging scenes where violations of natural and logical laws occur, or in describing characters (like the Bullim-Boukan) who violate these "laws". Let us imagine a further James Bond novel, for example, in which Bond proves

[1] The logic of fiction, science fiction in particular, has much in common with the logic of the ultramodal functor "It is imagined that", the semantical evaluation of which calls for worlds far beyond those of modal logic semantics. Likewise fantasy can deploy worlds remote from the natural order of things.

The account given of fiction is intended to be general and to cover all the varieties of fiction in the main modern sense, including (with at most minor variations) drama, opera, soap opera, etc. Subsequently some other varieties of fiction will be noted, and fiction (in a narrower literary sense) distinguished from myth, legend, and fantasy.

The extended (theory-laden) senses of "fiction", where fiction includes theoretical objects or legal objects or, quite generally, every object that does not exist will also be considered subsequently – and rejected. For they write in assumptions as to the similarity of, e.g., fictional objects and theoretical objects that it is important to reject.

that $E = NR$: of course we don't, and can't, get or expect full details of the proof, since such a proof is impossible (because $\Box(E \neq NR)$). This one example thus displays two features of fiction that put it beyond the reach of most going semantical theories, namely possible inconsistency and invariable incompleteness (both of the world portrayed and of objects portrayed). It is time to try to bring out these features in a more systematic way.

A work of fiction, e.g. a novel or film or series of such, N can be regarded as depicting a (dynamic) world,[1] the world of the story. The world $a(N)$ is given (in a canonical form) by the class of statements that hold according to N; it has as domain $d(N)$ the items of N, e.g. the characters and things which it is about. There are some difficulties in determining, and some unproblematic elasticity in, *what holds according to N*. For what holds according to N does not coincide with what is presented by, e.g. written in, N. Much as the source book for a character of fiction does not simply coincide with what is said in the authentic books or works depicting the character, so the world of a work or series of works does not simply coincide with the sum of assertions of the work(s). How does the world $a(N)$ differ from what N asserts, or is taken to assert? There are two possibilities:

(1) Through addition of further statements not made in N;

(2) Through subtraction of statements from those (apparently) made in N.

The first, addition, way is most important; but it is doubtful that the second, deletion, way should be admitted, and it certainly makes for a clearer and more assessible theory if it is not. For then the statemental representation of $a(N)$ is just an extension of what is asserted in N. Furthermore, admission of subtraction is a most serious obstacle to the formulation of a satisfactory theory of how $a(N)$ is determined from N, and so also in the way of a theory of interpretation. For interpreting a work N overlaps the matter of working out $a(N)$. Is there good reason for allowing deletion? For special purposes there may be, e.g. if the object is to determine only part of what work N conveys, e.g. the political theme, a minor plot, etc.: but such reasons evaporate when the object is to determine world $a(N)$. For example, the natural and cosmic forces at work in (almost any) one of Hardy's West Country novels may not figure in recounts of the plot of the work (though they are relevant to what transpires), but they are very definitely part of

[1] Or part of a world, or a system of worlds. But each subworld or world-system can be taken as a world; compare the situation with source books, chapter 1, §25 [Vol. 1 —*Eds.*]. $a(N)$ is the *source world* for N.

the world portrayed, and should not be deleted. More questionable as regards inclusion are *ir*relevant moralising included in a work or general observations on the state of the world, human psychology, etc. which have (as they may have in a *bad* work) little or nothing to do with setting the tone, mood, and background effects for the carrying of the story. Such material, while it may not add to the definition of the world concerned, does not (because of its irrelevance) significantly interfere with it either, and can simply be carried; that is, subtraction is again not required.

What has made the admission of deletion appear compulsory is the modal account of *worlds*. But while deletion accounts *are* forced by classical and modal theories, relevant and ultramodal theories can avoid them. The modal problems arise thus: once addition to worlds or enlargement of worlds is considered, as it often has to be, e.g. in treating fiction, in assessing counterfactuals, subtraction has also to be admitted, since addition can destroy properties of modal worlds such as possibility and/or completeness; and subtraction from worlds is very difficult to control technically. With the larger class of worlds ultramodal theories include, these problems vanish: addition does not have to be coupled with subtraction.[1] Accordingly it can be safely assumed that subtraction does not apply, that $a(N)$ results from what N says by addition principles only.[2]

It remains, then, in indicating how $a(N)$ is determined to indicate how and where addition is legitimate. The principles underlying addition may be divided into two groups, *material elaboration* conditions, and *formal closure* conditions. The material additions include factual or quasi-factual material, e.g. descriptive geographical or historical material, alluded to in setting the scenes of N, that may be made use of in N, but which is not explicitly spelt out in N. The material – which consists primarily of relatively common knowledge of parts of the current world and its history – may include, usually in very sketchy form, such geographical information[3] as the arrangements of certain streets of London, of the upper reaches of the river Thames, of a minor rail network and its stations, of the climate of a certain region; such historical details as the prime minister of a certain government, parts of a genealogical tree of an "important" family; such sci-

[1] This is the genesis of R. and V. Routley's impossible theory of counterfactuals, on which see RLR. It also underlies the way in which ultramodal theories are able to avoid crippling total evidence requirements of modal theories: see UL [reprinted in Vol. 4 —*Eds.*].

[2] Since it is what N states that is the starting point, syntactical errors such as mistaken grammar, misprints, etc. in a script of N are automatically eliminated.

[3] When such details are not generally known to intended audiences, or – differently – are invented, sketch maps are sometimes supplied, giving requisite information.

entific information as that normal physical forces, like gravity, operate on characters[1] and that characters are in main psychological and sociological respects like sometime-actual humans of such and such a place and time; knowledge of other works of fiction or the like in the tradition of which the work is a part; information as to social customs; etc. How much material of this sort goes into $a(N)$? Since the limitations of modal semantics have been removed there is no requirement to *complete* fictional worlds. Rather the opposite of a maximal principle of the type completion leads to seems to be wanted, namely *a principle*, if not *of minimal* required *additions*, at least of a very conservative type. Roughly, the additions should be enough to encompass the material the work alludes to or relies upon but does not set down explicitly, enough to enable the story to be understood and avert certain misunderstandings, and no more. It should *not* include empirical, logical, or other information extraneous to the work and unnecessary to its the comprehension (all this is suspended).

The closure principles are also much more restrictive than is often assumed, definitely more restrictive than modal accounts, which allow closure under strict implication, suppose. For a fictional world may well not, and generally *will not*, include all necessary truths. Closure under a *good* logical consequence relation is certainly not sufficient to account for additions admitted, in virtue of material principles; nor (as we shall see in more detail) is it necessary. Commonly, however, what holds according to N will include the entailmental closure of what is stated in N; i.e. if $A_1, \ldots, A_n$ hold in N and $A_1 \ \& \ \cdots \ \& \ A_n$ entails B then B holds in N. But in exceptional circumstances such closure can be varied, by switch to an idiosyncratic entailment connection (determined by N), or waived altogether. It may be suggested that, on the contrary, closure under entailment of $a(N)$ is not even commonly required.

The questionable assumptions that seem to underlie this suggestion are worth exposing. They are, firstly, a modal-based assumption that the admission of inconsistency of fictional worlds, even if it cannot be ruled out entirely, is to be avoided, in particular

[1] Not the general scientific laws, but relevant instances of them.

The account given of material principles of addition leaves much to be desired; it is (inevitably?) loose, open-ended and awkwardly intensional; and it can almost certainly be improved upon to some extent.

This issue may look rather unimportant for ultramodal noneism, since worlds not entailmentally closed are easily included under the semantical theory. But the matter will turn out to be of larger import, and to involve the question of whether $\underline{\underline{T}}$ belongs to K, i.e. whether the real world is logically normal, or at least commonly so. [For a discussion of this notation, see Vol. 1, §1.17: $\underline{\underline{T}}$ is the factual (actual) world, the actual world or factual model, including nonexistent items. —*Eds.*]

amputatable remote inconsistencies are to be excluded, and, secondly, that $a(N)$ somehow represents what the audience understands from N. The argument against entailment closure of $a(N)$ runs as follows: works of fiction, such as N, can be envisaged which are such that if the consequences of (widely scattered) statements of N were drawn out inconsistent statements would result, i.e. N leads to remote inconsistencies; but these inconsistencies are not part of the (ordinary?, intended?) audiences' understanding of N. The reply is that while $a(N)$ may include all the basic information necessary to understanding N (insofar as N is intelligible), it need not include just that, and often does not. In short, what is (part of) a necessary condition on N has mistakenly been inflated into a necessary and sufficient one. In fact a smarter than ordinary member of the audience may eventually detect the inconsistency; so presumably it should be included. In any case an inconsistency in $a(N)$ does not matter and there is no need to exclude it. There are other reasons for imposing closure under inference as a common requirement. For instance, certain works expect the audience to make inferences, draw conclusions, etc. It is evident that closure under entailment is only one of the principles commonly presupposed. For each work N there is, then, a set of rules, depending on the sort of work N is, under which $a(N)$ is closed.

The strategy here adopted of determining worlds from works of fiction is the reverse of the move in Jeffrey 1965, who tries to explain possible worlds in terms of a class of novels, namely consistent and complete ones. The inadequacies in Jeffrey's move begin to reveal the nature and range of the worlds of fiction. Firstly, no novels are complete in Jeffrey's sense of including at least one of the statements A and not-A for every statement A, and indeed where the domain of items considered in a novel is not finite a complete novel is impossible given that all novels are of finite length. Secondly, some novels are not consistent, in that they include statements of the form A and also of the form not-A for some statement A.

It is evident enough in fact that the worlds of fiction may be very bizarre and deviate enormously from the actual world $\underline{\underline{T}}$,[α] which can be represented by the class of true statements.[1] (Only realistic and historical novels perhaps are intended to mirror,

[α] Recall Vol. 1, §1.17: $\underline{\underline{T}}$ is the factual (actual) world, the actual world or factual model, including nonexistent items. —*Eds.*

[1] Though a world can be *represented* by (and is isomorphic to) a class of statements, it is not, in general, identical with such a class of statements, as the case of worlds portrayed by silent films reveals (an important point made by C. Mortensen).

in small part, the real evolving world $\underline{\underline{T}}$.) The extent of the deviation is important both for the semantical analysis of fiction and for determining conditions of adequacy for a theory of fictions. Firstly, as with what, and what worlds, can be imagined, so with the worlds of works of fiction: there are in principle *no* restrictions as to kind. There can thus be no restriction of the worlds required for a semantical analysis of works of fiction, of fictional discourse, to the possible worlds of most modern semantics: very many of the worlds of a universal semantics, which go far beyond the possible, are required to account for the potential range of works of fiction. Not only the class of worlds goes beyond the possible, their structure far exceeds what is tolerated by modally determined semantics. In particular, fictional worlds can be combined to yield other worlds, their unions; for example, the worlds of *The Hound of the Baskervilles* and of *The Sign of Four* and also of the other Conan Doyle Sherlock Holmes' stories can be integrated to yield the world of Sherlock Holmes, as represented in all the Doyle stories (or, differently, the Doyle stories and stories of the Holmes' tradition). The union of complete possible worlds (i.e. modal worlds) is not in general such a world; in the Sherlock Holmes' case it is certainly not, because of incompleteness and of internal inconsistency between the Holmes stories.

Even so, very many of the worlds of universal semantics are not fictional worlds, for two reasons: They have no author or set of authors; they lack structural requirements of coherence, continuity, organisation, and so forth that distinguish fictional worlds, even those of rather bad works of fiction, from other worlds; and often they satisfy requirements that fictional worlds do not, e.g. they are normal, complete, etc. In short, then, a *fictional* world is a world (of the universal semantics) which typically is authored, which satisfies structural requirements,[1] and which fails to fulfil other requirements, such as those for modal worlds.

A second upshot is that the thoroughly naive theory according to which fictional statements are always true on the author's sayso has to be rejected – or at least substantially qualified, e.g by distinguishing separate classes of "fictional truths" different from author ratified statements (in which case the theory will tend to converge with the supplemented contextual theory to be advanced subsequently). There are three reasons for this. Firstly, the world of a work of fiction may diverge from what an author says.

[1] The precise nature of structural requirements, like the criteria of organisation that distinguish a good work of fiction, though important, are not a present concern. Forster 1949 is a well-known and still valuable discussion of the organisational requirements on a novel, e.g. on plot, theme, character, etc.

There is commonly more than an author explicitly ratifies to go upon, and sometimes there is perhaps less than the author asserts in the work or (especially) later, when s/he may become just one more interpreter of the work. Secondly, among the truths of $\underline{\underline{T}}$ are at least the logical truths (i.e. $\underline{\underline{T}}$ is logically normal), but there are fictional worlds in which any given logical truth (or theorem) fails to hold or is violated. Where c is such a world, c is not included in $\underline{\underline{T}}$. Consider the new James Bond story again and let the world of this story be d. Since $\underline{\underline{T}}$ is closed under entailment and x proved that A entails A, it would be the case that if James Bond proved that $E = NR$ were true, i.e. belonged to $\underline{\underline{T}}$, $E = NR$ would also belong to $\underline{\underline{T}}$; but it does not, so contraposing, the statement that James Bond proved that $E = NR$ does not belong to $\underline{\underline{T}}$. Hence also d differs from $\underline{\underline{T}}$.[1] Thirdly, it is evident on other grounds that there are bound to be exceptions to authors' saysos as to what is true. An author's sayso cannot, for example, guarantee the actual existence of his characters, the flatness of the earth, and so on: the empirical truths of $\underline{\underline{T}}$ are determined by factors beyond the control of authors.

These sorts of objections destroy the *absolutely naive* theory of fiction according to which whatever an author asserts, especially as regards his characters, is true without qualification, i.e. according to which an author can really ratify whatever he pleases. But, damaging though they are, the objections do not totally demolish modified versions the absolutely naive theory, which can be kept logically alive (just?) by letting $\underline{\underline{T}}$ be sufficiently queer, i.e. by letting $\underline{\underline{T}}$ be radically inconsistent, so that both character x exists (in $\underline{\underline{T}}$) and also x does not exist, and no longer closed under expected entailments. But any such theory would deviate far too far from the controlling intuitive data as to what is true and what is not to be at all acceptable.

If the theory developed in what follows is correct the character of the actual world $\underline{\underline{T}}^{\dagger}$ of the absolutely naive theory, and modifications of the theory away from absolute naivety, can be worked out. For $\underline{\underline{T}}^{\dagger}$ is just the union of $\underline{\underline{T}}$ with all the worlds of fiction. A first, and *serious*, problem is to argue that $\underline{\underline{T}}^{\dagger}$ is not trivial. This will depend on suitably *restricting* the class N of works of fiction – something at odds with the main motivation for the absolutely naive theory. For suppose some author of a work asserts: everything

[1] Closely allied to the failure of expected inferences when fictional statements are literally true (and thus fail classically expected logical requirements upon $\underline{\underline{T}}$), are transparency failures considered below – under the heading "fictional paradoxes", e.g. the fact that xRa may hold on author sayso but not xRb though $a = b$.

is true. On his sayso his work is trivial, and accordingly not of much interest; but if the work belongs to N then $\underline{\underline{T}}^\dagger$ is trivial. Nor it is enough to exclude merely such works from N, to limit authors' sayso only in this sort of respect. N will have to be restricted to (something like) the class of all actual works of fiction, or the class of all actual works of fiction over all time. For if N is, as it should be, the class of all works of fiction, actual or not, then as every statement is included in some work of fiction or other – whatever coherence requirements are imposed on a work of fiction – $\underline{\underline{T}}^\dagger$ is again trivial.

In sum, the obvious absolutely naive theory has to be rejected; but a modified, and so no longer "naive", theory can be kept alive by restricting the class N – though there remain major difficulties in so restricting N in a non-ad-hoc way. And however these are resolved, the modified theory should also be rejected. For it places unwarranted restrictions on an author's freedom to create whatever he will within the bounds of his art. If an author's sayso is a paramount determinant of truth for his worlds, then $\underline{\underline{T}}^\dagger$ should be trivial.

Thirdly, as is apparent from the new Bond story, fictional worlds are generally incomplete, seriously incomplete. In the case of Bond's proof details are logically bound to be incomplete or fallacious, assuming that any at all are supplied. The incompleteness of fictional worlds, which more commonly derives from the finitude of books, tapes, and the like, and the limited details they supply and issues they settle, also separates them from the actual world and the usual possible worlds, which are complete: in this respect fictional worlds resemble some of the incomplete worlds of relevant logics.

Just as the worlds of fiction are typically incomplete, sometimes inconsistent, and sometimes physically impossible, so are the objects of these worlds, including the central objects, the characters (in a generous sense), of each N. Physical impossibility is evident from the Faust stories and appears explicitly in much science fiction, e.g. Zap gets into his machine and zooms off at a superluminal speed or into time, in a way defying relativistic laws. Incompleteness and indeterminacy are, however, more pervasive features of fictional objects than inconsistency of one sort or another. To illustrate the point, Mr. Pickwick is incomplete (indeed ω-incomplete) with respect to stature; for although Mr. Pickwick, as a man, presumably has some height, he is not 5'6" and not 5'7" and not 5'8" and not any other specific height you care to nominate. There is a similar incompleteness as regards other physical characteristics, e.g. girth; for although Mr. Pickwick was a plump man no exact tape measurement specifies his waistline. There are, moreover, logical difficulties in the way of specifying fictional objects so as to ensure completeness of characterisation. Since complete physical objects tend to be infinitely complex, since for example they can be viewed from endlessly many angles at endlessly many times, no finite listing of properties will suffice; only some general recipe can provide a full characterisation. Satisfactory recipes are, however, not so easily obtained. The apparently simple procedure of comparing a fictional character with an historical person quickly encounters problems. It will not do to say that x is *exactly like* Winston Churchill, because a fictional character must differ at least in intensional respects, owing to their fictional locale, from historical figures. Nor can x be exactly like Winston Churchill in extensional respects or his exploits could not be separated, he would be extensionally identical with Churchill, and the work would be not fiction but biography. If, however, it is said that x is (exactly) like Winston Churchill extensionally *except* in this and that respect, then there are new problems; firstly Churchill's life is bound together in such a way that isolated parts of it are hard to vary without encountering breakdown of lawlike connections; but then, secondly, there is no guarantee that x so described is appropriately complete, that the features he has that differ from those of Churchill are consistently or completely described, since some differences will lead to infinitely many differences (under predicate compounding). Such considerations suggest, but do not show, that complete fictional characterisation is impossible: purely fictional characters can only be more or less rounded out and fully characterised, with matters left open in a way that defeats extensional completeness. And such is one of the assumptions behind the definition of existence provisionally advanced in chapter 1 [Vol. 1 $_{\text{—}Eds.}$] and 9 [Vol. 3 $_{\text{—}Eds.}$], according to which an object exists when it is complete and consistent in all extensional respects.[1] If the definition were to stand up,[2] no consistent purely fictional object can be complete, since such fictional objects do not exist.

Thus just as the worlds of a semantical analysis adequate to model the structure of fictional discourse will have to include worlds that are inconsistent, logically and also physically, and worlds that are incomplete (or else somehow simulate such worlds), so the objects of these worlds will have to include objects that are inconsistent and objects

[1] Consistency and completeness were assumed to guarantee lawlikeness. For suppose that physical object x violates universal law L, i.e. $\sim L(x)$; but as L is universal it holds for all physical objects, so also $L(x)$. Hence x is not consistent. But the argument is not without its shortcomings.

[2] The limitations of the definition were brought up in chapter 3 and are discussed again in chapter 9.

that are incomplete, *given* that the objects have the characteristics ascribed them in the worlds in question.

§2. Statemental logics of fiction: initial inadequacies in orthodoxy again.

The same considerations which begin to determine the semantics of fiction also help shape the logic of fiction and fictions. In particular, these considerations reveal that the general logic of fictional worlds cannot be classical, and hence that the logic of fiction cannot be modal given that the logic of each world encompassed in modal logic is classical in behaviour, i.e. given that all modal theories are classical. The reason is, once again, that a fictional world may be inconsistent or counterlogical yet nontrivial, in the sense that not everything holds in it; but (as explained in detail in RLR, chapter 1) classical logic trivialises every inconsistent and every counterlogical world.[1]

The logic of fiction differs, of course, from the logic of fictional worlds, much as the logic of modality, which is non-extensional, differs from the logic of modal worlds, of the worlds used in the semantical analysis of modal logic, which logic is essentially classical and extensional, and as relevant logic differs from the logic of the worlds used in its semantical investigation, which worlds are sometimes counterlogical and sometimes inconsistent or incomplete, though relevant logic itself (in main formulations) is none of these things (satisfying the principles of excluded middle and non-contradiction and admitting the rule γ of material detachment). The logic of fictional worlds concerns, in particular, the logical principles that hold for these worlds, especially the closure properties, such as closure under provable logical consequence,[2] which is a requirement on the worlds of relevant and modal logics and is frequently, but erroneously taken for granted in the case of fictional worlds. The logic of fictional worlds is, however, much more anarchical than that of relevant worlds: the logic of a world associated with a work of fiction may be any logic that the author chooses to impose, even none. If an author decides to write an intuitionist or connexivist or nihilist work then he can impose the corresponding logic, and things will occur or be rejected in the world he imagines and describes in accord

[1] Although classically and modally (for normal modal logics) inconsistent and counterlogical worlds coincide, in the wider, and more satisfactory relevant framework they diverge, neither one guaranteeing the other.

[2] That is, if B is a provable logical consequence of set Γ and each element of Γ holds in the world then B also holds in the world.

with the principles of the corresponding logic. Usually, of course, authors do not tax their readers with nonstandard logical, or arithmetic, backdrops (though peculiar geometries are sometimes taken advantage of in science fiction), for obvious reasons – such as lack of author knowledge, reductions in audiences reached when such esoteric tactics are used, etc. Commonly (but not invariably) the background logic is, by default, that of ordinary discourse, and the closure conditions are (as RLR tries to argue) those of relevant logics, e.g. closure under provable entailments, under adjunction, i.e. if A holds in world d and B does also, then the statement $A \;\&\; B$ holds in d, etc.

Given that the logic of a fictional world may be any logic, it follows that *there is no general uniform logic of fiction.* For the intersections of all logics is a null logic, no logic, as each purported logical principle is cancelled out by a logic where it does not hold good. Consider, to illustrate, one of the more promising principles for a logic of fiction, formed by introducing a fictional functor O (Woods' *olim* operator) read, say, "it holds in fiction that", namely the principle $O(A \;\&\; B) \to OA$. Spelled out semantically the principle has it that if $A \;\&\; B$ holds in the world of an arbitrary work N then so does A. But consider now a novel where the principles of connexive logic govern, and where hence $A \;\&\; {\sim}A$ may hold though A does not. The world of such a novel repudiates $O(A \;\&\; B) \to OA$. In claiming that there is no uniform logic of fiction, it is not implied that fiction has no logic, far less that it is illogical. In general, *each work will have its own internal logic*: it is simply that the emerging set of *common* logical principles will be zero. The semantical structure will reflect this situation.

A logic of area x aims to capture, in a certain recognised and not necessarily complete way, a class of true (usually analytic) statements and valid principles concerning x, characteristically those which formulate the logical axioms, or initial truths, of x and the logical principles of inference that apply within or to x. The claim that there is no uniform logic of fiction is *not* based on the harsher empiricist view that there are no truths of, or concerning, fiction to capture in a logic, since all direct statements of fiction, such as that Holmes lived in London, are factually false, while covered statements, such as O(Holmes lived in London), are false (or, on a different view, ill-formed) because unintelligible, because they involve an intensional and opaque functor O of which no sense can be made. In short, fiction is illogical because all its statements are false in one way or another.[1] Such a position is hard to sustain; for example, statements such as "Holmes lived in London, according to the works of Conan Doyle" are perfectly intelligible[2] and, so it certainly seems, true. It is less that empiricism makes it difficult to discern any facts in fiction, and so difficult to have any logic of fiction at all, than that the facts of fiction make life decidedly awkward at least for the harsher empiricist positions. Indeed, so the argument will go (in 9.10 [Vol. 3 $_{-Eds.}$]), the truths of fiction provide yet another set of counterexamples to empiricism; for fiction yields, like theories of nonexistent objects and of intensionality, both truths and knowledge not derived appropriately from experience.

Although there is no uniform logic of fiction there will be an emerging set of logical principles for common works of fiction; it will be, so it will be argued, a kind of relevant logic. It will certainly have to be ultramodal. For simple, and familiar, arguments show *the inadequacy of any modal theory or logic of fiction.* One argument of a syntactical sort – there are, as we have in effect observed, parallel semantical arguments – takes off from the meaning of "modal". A one-place functor C is *modal* if, and only if, whenever $A \equiv B$ (i.e. A is materially equivalent to B) is a provable classical tautology, $C(A)$ iff $C(B)$, i.e. provable tautological equivalents are everywhere intersubstitutable.[3] According to a *modal theory of fiction* each statement A of work N is elliptical, being, when made syntactically explicit, of the form OA where O [strictly $O(N)$, since O depends on N] is a modal functor. The argument uses just one other, rather uncontroversial, assumption, that O is &-distributive, i.e. $O(A \& B)$ iff OA and OB (or at least one half of this principle, namely if $O(A \& B)$ then OA.) If N satisfies $A \& B$ then it surely satisfies A and satisfies B, and conversely. In the modal logic of fiction finally arrived at by Woods (1974, pp. 141–4), &-distribution is an axiom [namely his (A4)]. Now let N be a work which satisfies a contradiction, say $D \& {\sim}D$; that is $O(D \& {\sim}D)$. As Woods explains, the contradiction problem of authored self-contradictions is "economically" solved within the modal framework by O-covering of the contradiction, i.e. by replacing the contradiction $D \& {\sim}D$ of N by $O(D \& {\sim}D)$ (see 1974, p. 139ff.). Then however N is trivial, N satisfies every assertion, i.e. OC for arbitrary C. The argument is as follows: As $D \& {\sim}D \equiv .\ C \& {\sim}C$ is classical tautology, $O(D \& {\sim}D)$ iff $O(C \& {\sim}C)$ by the modality of O. But $O(D \& {\sim}D)$ by choice of N, hence $O(C \& {\sim}C)$ whence by &-distribution

[1] Application of the verification principle, even in some considerably modified forms, will lead to similar results.

[2] Though intensional they even have transparent analogues, as will be seen.

[3] The characterisation extends directly to connectives with several places.

OC. Thus any work such as N which satisfies a contradiction, or more generally some inconsistency, is trivial; and this constitutes a reduction to absurdity of any claims of a modal theory to adequacy.[1]

A modal theory of fiction is bound to be unsatisfactory in other respects as well. For example, it is a feature of any modal logic with operator O that if OA is a thesis for some A then, for any tautology B, OB is also a thesis. Since then every work of fiction satisfies some statement, every work of fiction satisfies all classical tautologies! In Woods' modal theory an even stronger principle is adopted, namely $\Box A \supset OA$, whatever is necessary holds in every work of fiction (axiom (A3) of 1974, p. 141). Now let N' be a work of science fiction where the point of story turns – through exploitation of incompleteness, obtained for example by setting the story in the world of someone's beliefs, or through use of intuitionistic assumptions – on the rejection of the principle of excluded middle, $C \vee \sim C$. In the corresponding logic $O(C \vee \sim C)$ is definitely not wanted, but on any modal logic of fiction the unwanted assertion is automatically forthcoming.

It is a corollary of the inadequacy of modal logics for fiction that no purely classical theory can be adequate either. For any such theory is in effect a modal theory where the modalities collapse: any additional functor admitted, such as O, must meet substitutivity, or extensionality, requirements,[2] such as if $A \equiv B$ then $OA \equiv OB$, so ratified contradictions and tautologies inevitably spread in a quite damaging way. A purely classical theory can no more accommodate fiction than it can accommodate intensionality.

[1] Woods may protest that his "very weak modal" operator O is not modal (in the sense defined), but the protest rings rather hollow given the claims made in 1974, especially p. 143, that we require only the classical truth values "and classical semantical apparatus", that "we do not need to abandon classical negation" etc. etc. For given classical semantical rules for negation and conjunction at each world – rules admissible for even the weakest modal systems – classical contradictions have the same value, namely O(false), at every world, e.g. if $I(D \mathbin{\&} \sim D)(a) = 1$ then $I(D)(a) = 1 = I(\sim D)(a)$, i.e. $I(D)(a) \neq I(D)(a)$ which is impossible, so $I(D \mathbin{\&} \sim D)(a) = 0 = I(C \mathbin{\&} \sim C)(a)$. Thus classical contradictions cannot be discriminated in any modal worlds, and so $O(D \mathbin{\&} \sim D)$ iff $O(C \mathbin{\&} \sim C)$ for any modal operator O.

[2] Wood's quasi-classical satisfaction definition (1974, pp. 133–4) does not meet these requirements, or, so it appears, modal requirements. From the point of view of designing a satisfactory theory this is all to the good, but unfortunately the rather ad hoc stipulations presented remain too close to the modal for much comfort. For example, on half of the O-clause for material implication ensures that where OA and $A \supset .\ B \supset C$ hold valid then so does $O(B \supset C)$ (p. 135, writing "$\supset$" for Woods' "$\rightarrow$", and making the orthodox connections between universal satisfaction and validity). Consider again a work N which ratifies $D \mathbin{\&} \sim D$, i.e. $O(D \mathbin{\&} \sim D)$; then $O(B \supset C)$ for arbitrary B and C. N ratifies *every* material implication, and as a special case every tautology, and, worst of all – upon applying Woods' consistency qualified consequence condition and the fact that $\sim C \supset C$ entails C – every consistent statement.

These problems, and others, discerned in modal and classical logics of fiction can be avoided by resorting to a relevant, or ultramodal, logic of fiction. If the modal plan so far being followed, of taking fictional statements as elliptical and as covered, when set out in uncondensed form, by an operator, is carried out systematically the resulting logic will be a multiply intensional relevant logic (of the type studied in RLR, chapters 7 and 8), with one fictional (or *olim*) functor O_N, or $O(N)$, for each work, or conjunction of works, N. For inclusion of just a *single* functor O to cope with all fiction (the procedure adopted in Woods 1974) is hardly adequate to the range of fiction to be encompassed, especially when the one character, e.g. Orlando, is assigned quite different properties by different authors.

The *common* sentential logic of fiction is, the claim is, a relevant logic which contains, as well as familiar sentential connectives, e.g. from the set $\{\rightarrow, \&, \vee, \sim\}$, the O_N connectives. Let O be a representative connective of this type, i.e. $O = O_N$ for some N. O is a *systemic* connective (in the sense of RLR, chapter 7), that is it conforms to the postulates

R7. Where $A \rightarrow B$ is a theorem, so is $OA \rightarrow OB$

(roughly, fictional worlds are closed under entailment), and

G. $OA \,\&\, OB \rightarrow O(A \,\&\, B)$.

Since $A \,\&\, B \rightarrow A$, by R7, $O(A \,\&\, B) \rightarrow OA$. Thus G can be strengthened to the coentailment $OA \,\&\, OB \leftrightarrow .\ O(A \,\&\, B)$. Several other theorems also follow using the underlying logic of entailment, e.g. $O{\sim}{\sim}A \leftrightarrow OA$, $OA \vee OB \rightarrow .\ O(A \vee B)$ (but not conversely). The logic so far presented is a minimal one for systemic connectives: it admits of substantial strengthening without reinstatement of damaging or especially undesirable theorems. For example, various of the schemes Woods has proposed (1974, p. 141) can be so added, e.g. $OOA \rightarrow OA$ (cf. G8 of RLR), as distinct from $O_M O_N A \rightarrow O_N A$, which is not generally true. Once modal connectives, which are further systemic connectives, are adjoined to the logic other schemes of interest can be investigated, and semantically modelled, e.g. Woods' (A2) $\Diamond(OA \,\&\, {\sim}A)$, i.e. it is logically possible that fictional work N satisfies A though ${\sim}A$. Semantics for all these usual logics, both minimal systems (with various entailmental bases) and enlargements of these systems by further connectives and additional axioms, are readily furnished along explored lines (see again RLR, chapter 7). Connective O_N is modelled using an accessibility relation S_N between worlds. The recursive evaluation rule for O_N is this: $I(O_N A)(a) = 1$ iff for every world b such that $a\,S_N\,b$, $I(A)(b) = 1$, i.e. it takes the form of a modal rule though a *much* wider class of worlds is considered. The modelling condition for optional extra $OOA \rightarrow OA$, for example, is then: if $a\,S\,b$ then for some situation x both $a\,S\,x$ and $x\,S\,b$, i.e. there is an intermediary (fictional) situation, precisely as the "play-within-a-play" phenomena $OOA \rightarrow OA$ acknowledges would lead one to expect.

Just as important as what these logics assert is what they reject, and what they reject the semantics helps disclose. Thus semantical countermodels show that consistency and completeness theses, namely $O{\sim}A \rightarrow {\sim}OA$ and ${\sim}OA \rightarrow O{\sim}A$, are rejected, just as they can be on modal accounts. More interesting is the way in which the shift to the ultramodal escapes modal difficulties. Firstly, in ultramodal logics (of which relevant logics are a subclass) contradictions do not generally spread or trivialise a theory. In particular, $A \equiv B$ and $O_N A$ do not suffice for $O_N B$, so the main case made against modal logics fails with ultramodal logics. Secondly, tautologies no longer hold in every world considered, so many principles as $O_N(B \vee {\sim}B)$ are no longer logically obligatory. Furthermore, a major problem for modal theories, how to amend closure under entailment or logical consequence so as to avert trivialisation of inconsistent stories (a problem explained in Woods 1974, pp. 50–1, with different equally unsatisfactory resolutions suggested on p. 51 and p. 133), is automatically avoided by relevant logics which build in paradox-free accounts of entailment, that is accounts where the consequent of any logical consequence connection must be relevant to the antecedents (see e.g. ABE).

Even an ultramodal theory of fiction based on a multiply intensional relevant logic, despite all its advantages, encounters problems if upgraded to *the* logic of fiction, or acclaimed as a logically comprehensive theory of fiction. As we have noticed, the general semantical framework considered has to be larger than that required simply for relevant logics. For, firstly, works of fiction are not always closed (or intended to be closed) under remote entailment connections. Secondly, a work of fiction can be based upon, or incorporate, a logic (and likewise an arithmetic, a geometry, or a physics) as strange or bizarre or idiosyncratic as the author cares to choose: again, provided the work hangs together appropriately, there are no limits on the organising logic or mathematical theory. The early stories of Borges' (studied in Sturrock 1977) and the stories of the Hoyles give some idea of the (almost unlimited) scope available. In principle, works of fiction can be contrived which, like nonnormal worlds, violate each of the closure conditions of relevant logics; for example a work can be designed where $\sim\sim A$ holds when A does not. Such a wider framework is however readily encompassed within universal semantics, and is not uncongenial to more liberal relevant logics which aim to absorb or accommodate their apparent rivals as far as possible (see the strategy of RLR). But no uniform logic remains: how, in that event, can the semantical theory succeed? The beginning of an answer – which will show that no uniform logic is required – lies with (classical) elliptical theories of fiction: but only a small beginning. For a serious trouble, even with the liberal relevant theories considered, derives from the assumption made – the commonplace philosophical assumption – that fictional discourse is properly handled by covering it with intensional functors.

§3. The main philosophical inheritance: paraphrastic and elliptical theories of fiction.

The main theories of fiction we have inherited are products of the common tradition of empiricism, pragmatism, and more recently classical logic.[1] They are all based on acceptance of the Reference Theory. Accordingly, insofar as they do not try to simply dispense with fiction entirely (a not uncommon philosophical move), they all call for analysis of fictional statements. The favoured method of analysis is linguistic paraphrase.

The method, of paraphrase, was clearly stated by Bentham as part of his more comprehensive theory of fictional objects, for example:

> By the word *paraphrase* may be designated that sort of exposition which may be afforded by transmuting into a proposition, having for its subject some real entity, a proposition which has not for its subject any other than a fictitious entity (Ogden 1932, p. 86).

But Bentham did not apply his joint methods of paraphrasis and archetyptation to fictional objects in the modern sense with which we are here concerned, and which coincide in extension with what Bentham called "fabulous entities". Rather he distinguished *fictions in the logical sense* – which are essential for "the carrying on of human converse"

[1] Thus theories of and deriving from Bentham, Vaihinger, and especially Russell: Bentham's and Vaihinger's theories have had little direct modern impact. There are of course (sketches of) other theories than those of the main inheritance, e.g. what Meinong had to say in "Über Urteilsgefühle: was sie sind und was sie nicht sind" in GA I. The theory eventually arrived at in what follows has, as may well be expected, much in common with Meinong's account.

(p. 18) and which concern fictitious entities – from *fictions in the poetical and political sense* – to which attaches "no coin of necessity" but which serve for amusement or mischievousness and which concern fabulous entities – and applied his methods to logical fiction.[1] Thus Bentham distinguished from

> fictitious entities, which so long as language is in use among human beings, never can be spared, *fabulous* ... for the designation of the other class of unreal entities (Ogden 1932, p. 17).[2]

As examples of fabulous entities, Bentham cited

> Gods of different dynasties; kings such as Brute and Fergus; animals such as dragons and chimaera; countries, such as El Dorado; seas, such as the Straits of Arrain; fountains, such as the fountain of Jouvence (p. xxxvi, note 1).[3]

[1] The theory of items advanced in subsequent chapters tends to agree with Bentham's account of what exist, agrees that many objects which do not exist are indispensible for discourse, but disagrees that these objects are fiction, or entities in any ordinary sense, and disagrees on the need for paraphrase.

Virtually all higher order objects which the theory of items admits as values of variables without further ado, are counted by Bentham as fictitious entities, and as such require analysis. For instance, all the objects falling under Aristotle's categories (or Ten Predicaments), except particulars falling under the first head, substance, are, according to Bentham, fictitious entities (p. 19).

[2] But elsewhere Bentham gives a different characterisation of fabulous entities, inconsistent with that cited in text, e.g. (p. xxxv, extracted from early work):

> Fabulous entities ... are supposed material objects, of which separate existence is capable of becoming a subject of belief ...,

which rules out, for instance, impossible objects. The characterization given by Bentham's nephew, George Bentham, is even narrower and is radically unsatisfactory:

> III. A *fabulous* Entity is one which has been believed in by others, but to the existence of which we attach no belief (Ogden 1932, p. 152)

[3] Vaihinger 1949 makes a parallel distinction, between scientific fictions, which are equated with *fictions* simpliciter, "and the others, the mythological, aesthetic, etc. *figments.* For instance, Pegasus is a figment, a term, a fiction" (p. 81). As in Bentham too, so in Vaihinger, fictions in the sense of fabulous entities or figments, are dismissed as of little or minor importance for a theory of fictions, meaning thereby a theory of scientific fictions.

> ... fictions, such as angels, devils, pixies, spirits, etc., ... are of minor importance for our present theme. At most they concern us only in so far as such a judgement as "matter consists of atoms" or, "the curved line consists of infinitesimals" is to be understood only as a fictive judgement in which no existence is predicated. Otherwise (i.e. if the judgement be not taken to mean that matter is to be regarded *as if* it consisted of atoms), then a correct fiction is changed into an incorrect judgement, in other words into an error. The primary meaning of fiction = mythological entity, is thus distinguished from the scientific fiction (p. 82).

The "primary meaning of fiction" is precisely that which is of present concern. Subsequently the secondary, philosophical, meaning assigned to "fiction" by Bentham, Vaihinger, and many others, will be criticised, for it involves a mistaken theory as regards theoretical objects, namely, the position of fictionalism, that such objects are, and function as, fictions.

Bentham's approach – which is worth noting because it strikes a chord with much modern thinking – appears to have been that discourse concerning fictions in the nonlogical sense, about fabulous objects, is simply dispensible, such objects are not required for what counts, scientific discourse or, above all, for the language of physics (cf. Ogden 1932, p. xix, lxviii).[1] Dispensible how? As not true, according to Bentham.[2] But the case really relies on a blatant application of the Reference Theory:

> Nothing has no properties. A fictitious entity, being as this name imparts – being, by the very supposition – a mere nothing, cannot of itself have any properties: no proposition by which any property is ascribed to it can, therefore, be, in itself and of itself, a true one . . . .[3] (Ogden 1932, p. 86).

This theory has already been examined in detail and found wanting. Most relevant, some statements about fabulous objects are true, beginning with such statements as that they are fabulous and do not exist.[4] Such objects are not nothing (though they are nothing actual), but something; and accordingly they can have and do have, properties. The correct ascription of these properties to fabulous objects yields true propositions. However truth in Bentham's view can only be obtained by return to the referential mode:

---

[1] Similarly in discussing inferential entities, entities which are "not made known to human beings in general, by the testimony of the senses" (p. 8), Bentham remarks (p. 10)

> By the learner as well as the teacher of logic, all these subjects of Ontology may, without much detriment, it is believed, to any other useful art, or any other useful sciences, be left in the places in which they are found;

we need "not trouble ourselves unduly with them" (Ogden's comment, p. liii).

[2] Mathematics, which Bentham regarded as concerned purely with fictitious entities and as basically "a species of short-hand", was similarly dispensible:

> otherwise than insofar as it is applicable to physics, Mathematics (except for amusement, as chess is useful) is neither useful nor so much as true (Bowring 1838-43, vol. xi, p. 72).

Again Bentham's modernity is striking: compare the Quine–Smart position considered in chapter 8 [Vol. 3 —*Eds.*].

[3] Monro (1967, p. 285) concludes, from the assertions he accurately ascribes to Bentham that "a fiction is nothing; and a quality of fiction equally nothing",

> Thus most of our talk is strictly nonsense, though it can be given a meaning by translating it into terms referring to real entities.

Bentham's text does not in fact, seem to put the points this way, in terms of the Reference Theory of Meaning; and given Bentham's repudiation of the traditional doctrine of terms (e.g. Ogden 1932, p. lxvii ff.) he may not have been prepared to put the points in terms of meaning.

[4] On Bentham's account, existence is a quality; but as to how this can be so on his account, he owes us a better explanation. He likewise owes us very many specific paraphrases, if the theory is to achieve much of what is acclaimed.

> Whatsoever of truth is capable of belonging to it cannot belong to it in any other character than that of the representative – of the intended and supposed equivalent and adequate succedaneum – of some proposition having for its subject some real entity (p. 86 continued).

Although Bentham apparently offers no specific paraphrases of claims concerning fabulous objects, his theory does suggest[1] paraphrases of the sort that were subsequently proposed by Carnap (especially in 1949), paraphrases of claims about objects into claims about their names, e.g. of

> Fergus is a fabulous entity

into

> "Fergus" is a fabulous-entity-name.

Such paraphrases have already been considered and found seriously wanting (in chapter 4; the issue is taken up again later in the section).

Vaihinger's account (in 1949), sometimes said to have been anticipated by Bentham, is very different, and (once a complicating modality is removed) rather more promising. The basic logical transformation underlying Vaihinger's philosophy of *as if* consists of the following two elementary stages:

Stage 1. af is transformed to: a must be treated as if it were f. Roughly, af is covered by the functions "It is necessary to treat things as if".

Stage 2. af is transformed to: a must be treated as it would be treated if (it were) f; i.e. "as if" is split into its components "as" and "if".

One of Vaihinger's working examples (p. 92ff.) replaced "Matter consists of atoms" first by "Matter must be treated as if it consisted of atoms", and secondly by "Matter must be treated as it would be if it consisted of atoms". Applied to fictions in the primary sense,

[1] In the constant emphasis on names, and the regular insinuations that fictitious entities are really no more than words. Thus, for instance the famous passage:

> To language, then – to language alone – it is, that fictitious entities owe their existence; their impossible, yet indispensible existence –

to which is appended as a note

> The divisions of entities into real and fictitious is more properly the division of names into *names* of real and *names* of fictitious entities.

this analysis has even less appeal than it has for "scientific fictions". For any figment one cares to consider, it is not *necessary* (in any good sense) to treat it as if . . . . We usually do not proceed *as if* figments existed. But why not delete "necessary to treat things" from "It is necessary to treat things as if". The first stage then simply covers the statement af by the intensional functor "It is as if". The second stage, with a little adjustment, takes "It is as if af" into "If a were to exist then a would be f". For example, "Pegasus is a horse" is transformed to "It is as if Pegasus is a horse" and so is analysed as "If Pegasus were to exist then Pegasus would be a horse". While the intermediary linkages are at best decidedly doubtful,[1] the end connections, which equate an inexistential claim with an intensional conditional, merit consideration. That Pegasus is a horse strictly implies (by a paradox of implications) that if Pegasus exists then Pegasus is a horse, and so presumably yields the subjunctive restatement. But the converse implication does not hold, even materially. It would require a further antecedent to the effect that if Pegasus does not exist then he is a horse: For what "Pegasus is a horse" says materially is: whether or not Pegasus exists, Pegasus is a horse (by $A \leftrightarrow . \ B \supset A \ \vee . \ {\sim}B \supset A$). Such a method of eliminating the inexistential by way of the intensional accordingly fails.

The theory that, intellectually at least, superseded Bentham's theory of fiction, Russell's theory of descriptions and of logical constructions, had much in common with Bentham's theory. It is not just that Bentham's referential and empiricist assumptions went over largely intact into Russell's theory; Bentham's "giving phrase for phrase" in the course of archetyptation is very similar to the method of contextual definition. It is not too inaccurate to say that prevailing theories of fiction (to descend again to the modern nonlogical sense of "fiction") are, for the most part, variations on or elaborations of Russell's theory. Ryle's account, which will be considered in some detail since it still represents a common position, is such a variation.

Russell's theory of descriptions does not, strictly, determine a theory of fiction, but provides a logical framework for such a theory. Such a theory of fictions is, in a quite precise sense, an *application* of the theory of descriptions. Exactly how the theory of descriptions is applied depends however on whether fictional subjects are treated as primary or secondary occurrences, and on how secondary occurrences are syntactically

[1] Given that Pegasus is a horse, it is not as if Pegasus is [were] a horse. "It is as if af" tends to suggest, what af does not, that ${\sim}af$. Conversely, "It is as if af" certainly does not imply af.

Similarly, "It is as if Pegasus is [were] a horse" does not imply "If Pegasus were to exist then Pegasus would be a horse"; nor does the converse implication hold.

produced, i.e. what functor is introduced to turn the occurrence into a secondary one. For on the face of it, fictional subjects such as "Sherlock Holmes" have only one sort of occurrence, a primary occurrence; in sentences like "Sherlock Holmes is a detective" there is no functor to generate scope ambiguities regarding quantificational analysis of descriptive phrases.

The hardline Russellian approach takes such fictional statements at face value, at least as regards the primeness of the occurrence. On this approach names such as "Sherlock Holmes" are first replaced – in one of the usual problem-producing ways – by descriptions, and then the theory of descriptions is applied without further preliminary analysis (at least to noncomplex contexts). The result is that all simple fictional claims (e.g. those of the form af where f is classically admitted as a predicate) are false, since existential upon analysis. For af is transformed first to $((\imath x)xg)f$ with $a = (\imath x)xg$, and then to the existential statement $(\exists x)(xg \;\&\; (\forall y)(yg \supset x = y) \;\&\; xf)$. But in fictional cases the latter is invariably false, since fictional characters do not exist. Separate analyses are, of course required to deal with statements such as "Holmes does not exist (is imaginary, fictional)", etc.

The hardline approach is mistaken for several reasons (most of which have already been discussed); in particular, in intensional contexts not only are there sometimes no descriptive replacements for names, but the theory of descriptions where applied sometimes yields erroneous truth values. But with fictional statements the approach is mistaken even in simple extensional cases, because to assert af is not always to imply, or presuppose, a exists. One who says "Holmes is a detective" does not imply that Holmes exists, he may be well aware that Holmes does not exist. *Simple fictional statements are commonly not existential statements.*

The latter serious problem is avoided by softer secondary approaches, according to which simple fictional statements, such as af, are really, despite appearance, disguised secondary statements, of the form ϕaf where ϕ is a (scoping) functor. Secondary approaches are then elliptical theories of fiction, which combine an elliptical thesis with application, as in the hardline approach, of the theory of descriptions. Specifically the analysis proposed is as follows in the case of af: af, being elliptical for ϕaf, is first expanded to uncondensed form; then ϕaf is transformed to $\phi((\imath x)xg)f$ with $a = (\imath x)xg$; and finally the theory of descriptions is applied, with $(\imath x)xg$ as a secondary occurrence to yield $\phi(\exists x)(xg \ \& \ (\forall y)(yg \supset x = y) \ \& \ xf)$. The result is not an existential statement because the existential quantifier is covered by ϕ. Elliptical approaches differ as to the construct of ϕ. Does it read "It is written that" (a construal that fails for unwritten fiction), "It is told that", "It is said that", or very differently "It is pretended that", or does the functor vary from case to case, e.g. from "Doyle wrote that" to "Billy Graham intoned that" to "The Rolling Stones sang that"? There are telling objections to trying to so combine an elliptical theory with the theory of descriptions. Take almost any sentence of the simple form af in the Holmes stories: then while it is true that Doyle wrote af it is false that Doyle wrote the proposed expansion of af. Consider, for example, the (true) statement that Holmes smoked a pipe: Doyle did not write that there existed a unique object with Holmes' properties who smoked a pipe, nor, very likely, would he have been prepared to write such. For Doyle did not believe that Holmes existed or consider himself to be writing a biography. The failure of the more obvious linguistic-style choices of functors to fit the elliptical approaches may help explain the prevalence of such wild theories as the account of fictions in terms of pretence. For something of the sort is hard to avoid on the softer approach. Given prevailing assumptions, the subject a in af must be eliminated even after expansion to ϕaf (unless like Kripke 1973 we pretend that a exists), as otherwise there would be truths, of an intensional sort, about what does not exist; and for uniformity of elimination a theory of descriptions should be employed, as for other sentence contexts. The functor must be, it is felt, one of, or appropriate to, fictionalisation; and indeed the functor "It is fictionalised that", (or less satisfactorily "Author ... fictionalised that") would give better results than any of the functors that have been proposed for elliptical theories. But theorists would have balked at the verb "fictionalise" (though its rarity is really a great advantage, e.g. against counter cases drawn from established usage); typically they substituted one of the more familiar verbs dictionaries and thesauri couple with "fiction", e.g. "imagine", worse "feign", or worst still "pretend". The initial upshot is appeal to functors of pretence or imagination and corresponding analyses of fictional statements.

However, the pretence theory of fiction, the core thesis of which is that fiction is pretence, and an essential feature of which is that many statements of fiction are elliptical and call for expansion to reveal their real logic, may take more sophisticated forms than simple prefixing by the functor "It is pretended that", "Author a pretended that" or the like. Pretence theories are to be found in Ryle (see 1971, p. 63ff.), where the theory is backed up by a softer secondary approach, and, in a simpler form independent of secondary approaches, in Kripke 1973; pretence theories are also suggested in many other writers, e.g. Smart 1977 [reprinted in this volume $_{—Eds.}$].[1] According to Ryle what an author does is to write sentences which are false but to pretend that they are true. According to Kripke, works of fiction express no propositions, i.e. nothing true or false, by sentences such as "Mr. Pickwick is plump", although they pretend to do so; "in talking of Sherlock Holmes and unicorns we only pretend to express propositions" (1973, p. 8). Ryle even makes the extraordinary claim that Dickens

> fabricated propositions which were as if they were about a man called Mr. Pickwick, who was testy, benevolent, obese, etc. (1971, p. 70)

These claims are simply false to the facts. Most authors are not engaged in an elaborate process of pretence. They tell a story; they do not pretend that it is literally true. Dickens did not pretend that Mr. Pickwick was fat or that he existed. On the contrary, authors (like their audiences) are well aware that fictional characters do not exist; recall, for instance, the notices that it used to be common for authors of fiction to insert at the beginning of their works, which implied, among other things, but not always honestly, that none of the characters ever existed. Moreover authors are usually not trying to take in or otherwise dupe their audiences, or to make out that their assertions are referentially loaded. An author such as Tolkien is not trying to simulate reality, he is not pretending to characterise existing creatures. The pretence account goes very wide of the mark when applied to works such as *Lord of the Rings*. The core thesis of the pretence theory is false: what is right about the thesis that fiction is pretence is that fiction is fiction.

The other part of Ryle's claim, that statements of fiction are uniformly false, will be rejected shortly:[2] but his reason for the claim, that fictional characters do not exist, so works of fiction are about nothing, can be dismissed immediately, since it is yet another

[1] For another example – where the pretence theory is coupled with the dubious thesis that "fictional items are nontemporal" – see Godfrey-Smith 1977, p. 394.

[2] Practically *all* of Ryle's conclusions (1971, p. 81) concerning imaginary objects should be roundly rejected. Contrary to Ryle (the enumeration corresponds to that of Ryle's conclusions):

(1) Being imaginary is an attribute (though not a characterising one).
(2) Mr. Pickwick is an imaginary, specifically fictional, object and hence "Mr. Pickwick" signifies something, namely Mr. Pickwick.
(3) Some of Dickens' propositions are true or false of Mr. Pickwick, but these do not pretend to be true of a Mr. Pickwick.
(4) Propositions from the *Pickwick Papers*, such as "Mr. Pickwick dined at Rochester" are about, and are naturally taken to be about, Mr. Pickwick. They are not about the novel in question, and certainly do not assert that the novel contains such statements.

application of the Ontological Assumption (which is also at the back of the pretence pretence) to discourse that is not, and does not purport to be, referential.

The softer secondary approaches are important examples of elliptical theories of fiction according to which assertions of fiction, apparently about fictional objects, are shorthand for statements – characteristically obtained by introducing covering operators which isolate problematic subject terms – statements not about fictional objects at all. But they do not exhaust elliptical theories; rather the secondary approaches introduce problems, through their elimination of names of fictional characters by way of descriptions (or, to put it in the material mode, of fictional objects by way of complexes of properties) that elliptical theories, which do not require removal of names can avoid. For example, much of the trouble with the attempt to explain the apparent truth of "Holmes smoked a pipe" in a secondary way arose from the replacement of "Holmes" by indefinite description, not, it might well be claimed, by removal of ellipsis.[1] The motivation for elliptical theories is all too evidently referential, but perhaps elliptical theories freed from Russell's theory of descriptions have something else to recommend them, some solid arguments for their

---

(5) Dickens is not pretending to be characterising Mr. Pickwick; he does characterise him. (But Ryle does claim correctly that Dickens' assertions make sense.)

(6) Imagining is *not* always imagining that something is the case; one can imagine much else besides, including objects.

(7) Imagining a thing or person is not always imagining that the object has a complex of characters.

(8) In works of fiction characters are originated ("created" in one sense of the word); these objects reduce neither to mere descriptions nor to complexes of properties.

(9) The phrase "the object or content of an activity of imagining" may be ambiguous, but in one clear sense it can signify an object that does not exist.

(10) The question "What is the status of Mr. Pickwick?" is a not unreasonable one; it can (especially when "status" is clarified) be answered, e.g. Mr. Pickwick is a fictional object, a nonexistent but possible object, he has these properties ... .

The case for theses (1)–(10) and against Ryle can be assembled from elsewhere in the text, e.g. (1) is defended in chapter 1, §17 [Vol. 1 —*Eds.*]. With this material Ryle's arguments, such as they are, are readily demolished. For example, Ryle's defence of (2) depends upon existential-restricted quantifiers, and equating objects with entities. Ryle's arguments for his (3) and (4) (1971, pp. 68–9) turn upon illicitly writing an existence condition into what counts as aboutness. Moore (1959, p. 105ff.) reduces Ryle's conclusion (4) to very small pieces. Ryle's argument for his (6), that one can't strictly imagine imaginary objects, comes down to this (p. 72): imaginary objects do not exist so there can be no correlate to imagining them, since one cannot stand in a relation to what does not exist. But of course one can stand in intensional relations to nonentities, and imagining is an intensional connection.

Similarly rejected are Ryle's main theses concerning imagining and allied mental operations such as picturing and fancying that one can sift out of *The Concept of Mind.* Contrary to Ryle (1949, p. 245 ff.), these operations have objects, but the objects often do not exist. Thus, for example, when imaging occurs images are seen, when tunes are running in my head tunes are heard (contradicting 1949, pp. 247–8).

[1] Again there are trade-offs. If names are not eliminated all the referential difficulties of empty names, beginning with negative existentials, flow in again.

expansion proposals. Not only do they not, but there are other things against such theories.

Elliptical theories come in various forms depending on the sort of operators that occur in the expansion. Intensional expansion, where A is replaced by ϕA with ϕ an intensional functor, such as "It is as if" or "It is pretended that", have already been considered and found unsatisfactory. A further serious problem about the introduction of such functors is that they violate the framework within which elliptical theories are characteristically set; for such functors are highly intensional iterable quantifiable-into operators whose analysis is beyond classical and modal resources (cf. 8.5 [Vol. 3 $_{-Eds.}$]). Thus a more popular alternative, which avoids the referentially problematic intensional covers, is to take the covering operator as involving quotation, as in translations into the formal mode. Assertions of fiction become assertions about words and sentences, about linguistic entities. So, for instance, "Pickwick is a fictional object" is construed as short for "The name 'Pickwick' occurs in some work of fiction", and "Pickwick was plump" is elliptical for "One of Dickens' stories contains sentences implying the sentence 'Pickwick was plump'", or *some such.*[1] What exactly?

Such formal mode elliptical theories have most of the defects (already explained in chapter 4) of theories which would restate theories of objects in the formal mode, in order to avoid violations of the Reference Theory; e.g. the theories (and also most other elliptical theories) are defeated by translation objections, failure-to-generalise objections, and value preservation objections, even modal values not being preserved (consider, e.g. the different effects on consistency of conjoining, sentences like "Dickens never wrote a story", to "Pickwick was plump but Disraeli was not" and to its proposed expansion). Stronger connections such as sameness of content or of meaning, are certainly not preserved in any analysis of a fictional statement that introduces information as to the work in which it occurs or as to the author; for someone may understand, e.g. "Pickwick was plump", without being aware that the source of the assertion was a certain story by Dickens: in short, the analysis conveys different information from what it analyses. It is partly for this reason that elliptical theories cannot do justice to literary criticism and assessment,

[1] An important variant, ostensibly avoiding the serious difficulties of quotation, takes "Pickwick was plump" into a pair of sentences, e.g. "Dickens wrote [differently, imagined] that Pickwick is fat". This theory, what might be called a Davidsonian theory, will be considered in the chapter 8 [Vol. 3 $_{-Eds.}$].

and to such commonplaces as comparisons of fictional characters from different works or of a fictional person with a real one.

More generally, no purely elliptical theory is faithful to ordinary discourse about fictional objects – which has not been shown to be out of order and in need of such drastic reform. For ordinarily we can say truly, at least sometimes, such things as "Holmes was a detective". But what happens to this statement A on the elliptical theory: it is replaced by another A^* (e.g. OA) which is supposed adequately to replace A, and thus be in some sense equivalent to A (or what A was intended to say). What we can say truly with A is said to be properly said with A^*. But now what about the original A, which certainly remains in English even if it disappears in canonicalese? A is not meaningless, it passes all the tests for significance (except perhaps, what does not matter, hit-and-run tests, such as those of empirical verifiability). Nor is A true, as follows from the referential assumptions underlying elliptical theories; and if it were true, elliptical transformation of the sort offered would hardly be required. So it is false, which is what Ryle says and what most other elliptical theorists tend to admit at least when pressed – but which is *not* what we always say ordinarily. And if A is false, then it is not equivalent to A^* which is true. Likewise if A is not truth-valued. So the elliptical theory does not provide a satisfactory paraphrase – unless A is in some way ambiguous. But to insist upon ambiguity of fictional statements such as A is again to diverge from what we ordinarily say. On the other hand, something is right (as will become clearer) about the insistence that fictional statements like A are, in the first place, ambiguous, and, in the second, covertly intensional. (Nothing much is right in the attempt to render intensionality quotationally, e.g. by a formal mode recasting.) But what is right about elliptical theories can be better captured by alternative less damaging theories, such as contextual theories.

The same point applies against partial elliptical theories, i.e. theories according to which not all statements of fiction are elliptical but important classes of such statements are. Kripke's tentative theory of fictional objects appears to be this type. According to this theory, subject-predicate assertions of the form af where a is a fictional object, such as Hamlet, and f a straightforward predicate, such as "soliloquizes", "we have to understand ... as prefixed by the phrase 'In the story, ... '" (Kripke 1973, p. 12). Have to? Hardly; for there are other theories than such an elliptical one; and we can understand af without the prefix. Nor does the prefix help appreciably in this respect; af does not mean "In the story, af", for all the old familiar reasons; e.g. $\sim af$ contradicts af but not "In the story, af"; af entails a exists according to the Reference Theory but "in the story, af" does not, etc. So as a recommendation as to how to escape some of the puzzles fiction generates for referential philosophies (such as Kripke's, see chapter 1, §14 [Vol. 1 —*Eds.*]), the prefixing proposal has its limitations; it reintroduces leading problems of elliptical theories. Kripke proposes to avoid some of the worst of these problems[1] – counterexamples deriving from apparently true statements about fictional objects outside the scope of their source stories, e.g. "Hamlet is a fictional character who wanted to avenge the murder of his father", "There are thousands of fictional characters who have fallen in love", "This literary critic admires Desdemona" – by reserving prefixing for certain usages only, and by adopting what might be dubbed a multiple usage doctrine.[2]

Kripke distinguishes at least these (none too sharply characterised or separated) usages:

1. "A rather special usage, in which one just reports on the story". Here "Hamlet exists" is (alleged to be) true. So presumably is "Hamlet thinks" and "Sherlock Holmes is fit". The philosophically interesting example is Hintikka's "Hamlet thinks, so, Hamlet exists" which Kripke says we must understand as "In the story, Hamlet thinks; so, in the story Hamlet exists". Again we don't have to; indeed this is a most peculiar way to construe the argument.

2. The usage "in which one speaks on the level of reality out and out. Here 'Hamlet exists' is false". So presumably is "Hamlet thinks" and "Sherlock Holmes is fit".

[1] Kripke's claim that "fiction is no problem for *any* theory" of denotation or naming (1973, p. 3) is perhaps surpassed only by Dummett's deliberately provocative claim (already noted in 1.14 [Vol. 1 —*Eds.*]]) that "there is no alternative theory of proper names that can be opposed to Frege's theory" (1973, p. 146, p. 143). Of course both have reasons, bad reasons, for their extravagant claims. Kripke's claim depends on taking for granted a pretence theory of fictions, Dummett's on characterising what a theory of proper names is supposed to do in self-validating Fregean terms. Kripke's argument for his claim is simply that on any theory "it will be part of the pretence of the fiction that the narrator stands in the correct relation to a definite object, and that the appropriate conditions of naming have been met" (1973, p. 3), correctness and appropriateness being determined by the theory. That is, the "no problem" claim depends on the correctness of, what is itself problematic, a pretence theory! This point could (at first sight) be evaded by a fictional theory of fiction, by making it part of (the fiction of) the fiction that ... things are referentially well-behaved. But the trouble Kripke gets into on something as simple as negative existentials concerning fictional objects give the lie to his claim.

Postscript: It is not difficult to unearth claims that vie with Dummett's! One example is Putnam's contention that there are no "longer any *philosophical* problems about Time" (1967b, p. 247).

[2] The doctrine has much in common, both in its referential motivation, and in its results, with Waismann's multi-strata account of language: see Waismann e.g. 1953.

3. The "fictional character" usage, where fictional characters "are abstract objects, but as real as everything else". In this usage,"'Hamlet exists' is true". So presumably are all story-guaranteed ascriptions to those newly located abstract entities, e.g. "Hamlet thinks", "Pickwick is fat".

This is all pretty implausible;[1] such a statement as "Hamlet exists" is not three ways ambiguous. Nothing of course prevents the introduction of new senses for "Hamlet exists", or stops Kripke discerning an abstract object which he can call "Hamlet" if he chooses. That does not show that the object discerned exists, or that it has the ascribed properties, e.g. that it has much to do with Hamlet. In fact Kripke tells us very little about his abstract objects and what properties they have; the only detail given of how the abstraction is effected is the following puzzling remark:

> "Hamlet" is introduced first as a pretended ordinary person: then we move to infer the existence of a pretence – of the fictional character (p. 14).

Enough is revealed however to make it plain that the abstract fictional entities differ from corresponding fictional objects, that the abstract entity *Hamlet* is quite different from Hamlet. For instance, *Hamlet* is abstract (a pretence, etc.), Hamlet is a particular (a prince, etc.); *Hamlet* exists (or at least is supposed to), Hamlet does not (and is not usually supposed to); Shakespeare wrote about (actors study, etc.) Hamlet not *Hamlet*; Kripke introduced us to *Hamlet*, not Hamlet; and so on. But in a very audacious stroke Kripke claims that ordinary language has an ontology of such abstract entities (p. 13, p. 14, p. 15). The claim is false and without foundation; nor does Kripke try to substantiate it. There is ordinary language discourse about Hamlet – but Hamlet does not exist, and so is not part of the corresponding ontology. There is no such discourse about *Hamlet*: it is false that "an ontology of fictional characters [is] ... just a feature of ordinary language" (p. 13). It is not difficult to see why Kripke would think that it was (as he puts it: must be) a feature of ordinary language ontology. The obvious, and natural, way to account for true statements apparently about fictional objects which cannot be

[1] But worse is to follow: Even when the alleged ambiguities are resolved the forms cannot be taken at face value, as otherwise propositions would be expressed, since each of (1)–(3) yield true or false statements. This would contradict Kripke's general thesis that fictional sentences do not express propositions but just pretend to. Kripke tries to solve the problem only in case (2); and there the proposal, a direct application of the Ontological Assumption, is, as we saw in 1.14 [Vol. 1 $_{—Eds.}$], unsatisfactory. A coherent version of Kripke's baroque theory has yet, it seems, to be presented.

set down as internal to their stories, is to allow that they are about what they appear to be about, fictional objects. But then, by the Ontological Assumption, they must exist. No such particulars exist, so they must be abstract objects. Moreover what exists must be the very objects "recognised" by ordinary language. This is essential in reaching (by illegitimately transferring the apparatus of referential logic to what does not exist) supposed counterexamples to the theory (given under head (1)), such as "This literary critic admires Desdemona". Were the theory to supply only *Desdemona* such counterexamples would stand: the critic was not admiring *Desdemona*, such an abstract structure being no substitute for Desdemona.

§4. Redesigning elliptical theories, as contextual theories.

It is often quite unclear, especially in discussions which compare fictional characters, e.g. a as a general with b as a general, what the covering functor elliptical theories lead us to expect is or how it is determined. It is, in fact, unclear in many cases that there is an implicit covering functor, covering apparent truths of fiction or that what is implied thereby holds good, namely that a fictional statement A is ambiguous as between A and $O_N A$ where O_N is the functor supplied by its source N.

Something *like ambiguity* in many statements of fiction – by no means all – is forced by any theory which allows work of fiction N to ratify statements as true, which incorporates a (qualified) *sayso condition of truth* (to adopt again Woods' illuminating terminology and explanation; see 1974, p. 35ff., p. 133ff.). Although something *like* ambiguity is forced, nothing like the ambiguity Kripke discerns is so justified. What has to be taken account of are circumstances like the following, where a fictional object is physically related to an entity: The statement that Holmes lived in London (*hil*, for "Holmes lived in (or inhabited) London" with the predicate infixed, for short) is true, by the author's sayso; but the statement is also, in referential settings, false. For Holmes did not live in London,[1] as empirical scanning would have revealed: a stake-out on Baker St. would have obtained no trace of Holmes, none of the numerous comings and goings said to have occurred there would have been observed. Holmes was not, that is, an historical figure, as we can now convince ourselves in other ways, e.g. there is no record of him in

[1] In terms of the distinctions of 1.21 [Vol. 1 —*Eds.*] – to indicate the direction of travel – "Holmes lived in London" is true, since the relation is reduced; but in a referential context, where the relation is entire, the statement is false.

the British registers for births and deaths, even his address was fictitious as a visit to Baker St. will disclose. Thus London was not where Holmes actually lived: indeed since Holmes did not exist, was a merely fictitious person, Holmes could have lived nowhere actual, it may be said with some feeling.

The consistency (or Brentano) problem[1] for any fictional statement which appears to concern, or relate to, an actual or historical object – e.g. is about Richard III, or relates Holmes to the entity London – arises (so several investigators have observed, e.g. Woods 1974, p. 129, Devine 1974, p. 393) from rival truth standards, or, to put it more fashionably, from rival and conflicting, truth definitions. According to one standard, the sayso condition, *hil* is true, according to another, that of history or, more generally, of empirical fact, *hil* is false. It has been taken as evident that there can be no consistent combining of the results of applying the different standards in the one world (or theory) $\underline{\underline{T}}$, representing the class of truths, without the insertion of some differentiating factor which distinguishes the cases. This factor has been assumed to be ambiguity. Thus Devine (1974, p. 395):[2]

> "Richard III killed the princes" is ambiguous: it may be taken as a statement of historical fact (in which case it had been doubted), or as a statement of what goes on in Shakespeare's play (in which case it cannot be) ... [because] it is true according to another standard of truth, fidelity to Shakespeare's play "Richard III".

Woods makes a similar assumption, and goes on to propose, after some argument that is worth returning to, that the ambiguity of troublesome fictional sentences be resolved by representing each such sentence A by a pair of sentences $\langle A, OA \rangle$. That is, the sentence A of English may mean either A or OA (at least in the logical language where the formal logic of fiction is developed). However there are grounds for unease about each of the shifts made beyond the location of an ambiguity, and especially about the claim to ambiguity.

For "Holmes lived in London" does not appear to be ambiguous in the sense of having more than one sense, and there do not appear to be any structural features which

[1] At least this is a problem if we (correctly) put aside any dialectical grasping of the nettle of more naive theories which accept without qualification both *hil* and $\sim$*hil*.

[2] Devine, despite the title of his paper 1974, offers us no logic of fiction in the modern symbolic sense: there is no formal logic, no syntax, no semantics. Worse, how such a logic would look is left quite obscure. The same criticism applies to many other writers on the "logic" of fiction.

would account for the alleged ambiguity. For consider the parts: "lived in" does not exhibit any relevant ambiguity; and, as Woods remarks (p. 131) "it just does not seem that 'Holmes' is ambiguous", and similarly it seems for the term "London". Nonetheless *every* ambiguity that *could* be ascribed to *hil* (in virtue of its logical form) has been ascribed to it. Thus it has been variously suggested not just that "Holmes" is ambiguous (cf. Kripke above) or that "London" is ambiguous and should be replaced by "the London of the Holmes' stories" or some such duplicate term (cf. Devine), but that "lived in" is ambiguous as between various modes of predication (cf. Castañeda 1979) or that "lived in London" (and likewise "Holmes lived in") is ambiguous. On Parsons' distinctive theory of fiction (1974 and elsewhere), "lived in London" *is* ambiguous; "Holmes lived in London" is ambiguous as between "Holmes [lived in London]" (with "Holmes" as subject and "lived-in-London" as one-place predicate) and "[Holmes lived in] London & Holmes [lived in London]" (which is identified with "Holmes lived in London"), according as the two-place predicate "lived in" is "plugged up" on the right or on both the left and the right to yield one-place predicates. The intuitive base of Parsons' proposal, however, undoubtedly has considerable appeal – in this sort of case,[1] and before one encounters the gory logical details of the cut-and-plug theory of relations – and something rather like it will ultimately be fallen back upon. For there is a rather natural temptation (encountered among the philosophically uncorrupted) to say both that it's true of *Holmes* that he lived in London and that it's not true of *London* that Holmes lived there. But the temptation can be accounted for in other ways than Parsons' theory (as will emerge); and there is unfortunately some temptation among some people to say that inasmuch as it's not true that London was lived in by Holmes, it's not true that Holmes lived in London. What there *isn't* the same temptation to say is that the English sentence "Holmes lived in London" is misleading as to logical form, that it resolves under logical rendition into two different syntactical forms at least one of which does not (really) occur in English. *The statement hil is not ambiguous*: that is a guiding principle that will be adopted and defended in what follows.

If there is no ambiguity of sense or meaning, of the sort disclosed by Castañeda's or Parsons' or Woods' representations, what is the justification for them; for instance, what is the excuse for sometimes replacing A by OA? This seems even less excusable

[1] The case, that is, of physical relations between fictional objects and entities, *not* other cases where the relations are not physical (or extensional) or relate entities and entities or ideal objects and entities or nonentities with nonentities.

when it is realised that the replacement differs in meaning from what it replaces; for example, "Holmes lived in London" is not synonymous with "O(Holmes lived in London)", as elementary translation tests, among others, reveal. If "Holmes lived in London" were translated into another language in the form "O(Holmes lived in London)", or vice versa, the translation would be considered incorrect, a mistranslation. Embedding tests for synonymy, for what they are worth, yield the same results; when a recalls hil he is not recalling $Ohil$, when he asserts or assents to $Ohil$ he is not asserting or assenting to hil, etc. (Rather similarly, though less obviously because of the unfamiliarity of Parson abstracts, when hil is replaced by $h[il]$.) Furthermore $OOhil$ and $Ohil$ are not equivalent (according to Woods 1974, pp. 141–2), as would commonly be expected were the synonymy claim correct; or, in less questionable terms, $Ohil$ and hil are not synonymous because what implies them and what they imply differ. Any elliptical theory (which genuinely shields fictional statements from reality) is going to be open to similar objections from meaning change. Fictional discourse contrasts with literal reportative discourse (as Woods explains, p. 26), but the elliptical theory would have it replaced, according to the empiricist paradigm, by what it is not, discourse reporting facts – in this case about what author wrote or asserted what.

There are still other difficulties with any elliptical theory, some of which transfer to most theories alleging syntactical or semantical ambiguity. For non-negligible portions of typical works of fiction are entirely true (as Urmson has argued, 1976, pp. 153–5): the truths include sociological generalisations, presupposed truths such as that men have such and such physiology, wants and drives, historical and factual information, and background truths of geographical, historical and social setting, often required to make a work comprehensible. (Such truths belong to the source book for the work in question.) Yet such truths require no cover (as an elliptical theory may try to acknowledge by admitting that for such A, $OA \equiv A$). There are accordingly problems as to how to separate such statements not requiring cover from "the essentially fictional statements" of fiction which on elliptical theory do require cover, and as to how to preserve this isolation, since ordinarily statements of the two sorts can be combined, inferentially-linked, and otherwise interrelated (consider, e.g., the trouble Urmson finds himself in, 1976, p. 157, as regards the interpretation, and truth-functional coupling, of truths from fiction with essentially fictional statements, which on his, hardly convincing, account are neither true nor false).

To deny that fictional statement A is synonymous with nonfictional statement OA, to contend that asserting A is not the same as asserting OA, is not however to deny that there are important logical linkages between them. The linkage is this: *the semantical assessments of A and OA often reduce to the same thing.* Although such an admission may suggest to the positivistically inclined that the sameness of meaning of A and OA has now, inconsistently, been conceded – after all the sameness of verification (albeit in a wide sense) has been admitted – examples of the following sort should help bring out the requisite differences: The semantical evaluation of "I am hot", as said by RR, comes down to the evaluation of "RR is hot", since the speaker is RR, but the two sentences do not mean the same. The sentences in their respective contexts have the same semantical content in that, once context is taken up in the course of semantical evaluation, the semantical assessment is the same. For example, $I(Ohil)(\underline{\underline{T}})(s) = I(hil)(\underline{\underline{T}})(c)$, where s is a standard context and c is an appropriate fictional context.

There is no riddle, then, on the contextual theory, as to how the consistency problem for statements such as "Holmes lived in London" is to be resolved without resort to ambiguity in the more ordinary sense: the missing factor, which makes the difference, is context. "Holmes lived in London" is true in one context, c say, and not true in another, a referential context s say. Thus, in the symbolism of the universal semantics of 1.24 [Vol. 1 —*Eds.*], $I(hil)(\underline{\underline{T}})(c) = 1 \neq I(hil)(\underline{\underline{T}})(s)$; there is no inconsistency because c differs from s. The resolution of the contradiction problem is precisely that already proposed in Slog (p. 469, with symbolism adjusted to conform with 1.24):

> ... context-situations [i.e. the pairs formed by taking contexts together with worlds] enable the semantical assessment, but without the usual distortions, of confident truth-claims such as "Sherlock Holmes lived at 221B Baker St." (p_0), said – to supply part of the context c for the assertion p_0 – in answer to the question "Where did Sherlock Holmes live?" This claim is true, in its context, even though it clashes with other factual information we have about Baker St.; yet it is not equivalent to the claim "It is written in Conan Doyle's novels that Sherlock Holmes lives at 221B Baker St." Such claims as p_0 can however be construed as involving a shift of the base situation with respect to which truth assessment is made, from the usual $\underline{\underline{T}}$ to a new situation a, determined by c, i.e. $a_1 = f(\underline{\underline{T}}, c)$ [and from

> c to a new context $s = g(c)$]: roughly a_1 is the situational closure of the world selected by the "it is asserted in Conan Doyle, or Sherlock Holmes', detective novels that". Then $I(p_0)(\underline{\underline{T}})(c) = 1$ because $I(p_0)(a_1)(s) = 1$, i.e. because according to the novels Holmes did live at that address; and this claim is not undermined by objections such as "221 Baker St. was not a residential site" (q_0), because the context of q_0 does not shift the base situation ($\underline{\underline{T}}$). How does one tell when the base is shifted? A non-trivial answer is: the context supplies a base-shifting function, and this figures as part of the modelling. For example, the modelling can, under this extension, associate with each [context-situation (a, θ) a transfer function $\delta = (\theta_b, \kappa)$ where θ_b is] a base-shifting function θ_b from context-situations to situations such that ${\theta_b}^2 = \theta_b$, normally a projection function which makes no shift [and κ is a context-changing function]. Then, where part of the context sets the sentence in another context-situation, $I(A)(a)(\theta) = 1$ iff $I(A)(\theta_b(a, \theta))(\kappa(a, \theta)) = 1$.[1]

The contextual theory alone undercuts the "general lesson" supposed to be learnt from Urmson's account (1976, p. 157), the (alleged) inadequacy of logical theories of fiction:

> There are those who assert that sentences are the bearers of truth or falsity, whereas ... the writer of fiction as such asserts nothing true or false though he employs sentences indistinguishable from those used by those who do. I cannot see how such a sentence, however reformulated in logician's language, could be essentially fictional or non-fictional.

The semantical (meta-)language need only include contextual indicators for requisite distinctions to be made.[2] Parsons' theory does more: it reveals how the logical forms of

[1] A parallel contextual analysis applies also to non-literary fictions; it takes up familiar, more intuitive, assessments such as the following:

> ... when something is said to be a legal fiction, it is sometimes meant that while it is held to be true in courts of law, it is in fact false (Urmson 1976, p. 153).

[2] For a beginning on locating what is wrong with the Strawson–Urmson thesis that "the essentially fictional assertions in works of fiction are neither true nor false", see Blocker 1974. It is worth noting that neither of the analogies on which Urmson's case depends stand up to much examination, and that Urmson has to engage in a certain amount of what looks remarkably like double talk, e.g. (1976, p. 156):

a sentence could show (though in an artificial fashion) whether a sentence was essentially fictional or not.

The contextual theory enables a great many of the commonsense claims that are made regarding fiction and fictions to be made good – many but by no means enough.

§5. Elaborating contextual, and naive, theories to meet objections; and rejection of pure contextual theories.

As Woods has truly said, then, "much of what is distinctive about fictional discourse is ... the distinctive context in which it may be supposed to be embedded" (1974, p. 12). Woods also found a key to the semantical analysis of fictional statements, namely that fictional contexts transfer the base world where truth is assessed from the real world $\underline{\underline{T}}$ to fictional worlds:

> ... mythological statements resemble fictional statements at least insofar as neither kind is thought, by those who recognise them to be mythological or fictional, to describe the real world (1974, p. 30).

But though Woods dallies with the contextual approach (see especially 1974, p. 117–9, p. 131) he eventually rejects it. If the contextual theory of fictions outlined is to stand, it will have to be able to meet such objections as those Woods and others level.

One objection is that the contextual theory still treats fictional claims as ambiguous; it merely relocates the ambiguity, as in the context. If "ambiguity" is stretched in this way, beyond its dictionary senses (such as "double meaning"), if one insists upon calling different contextual resolutions those of ambiguity, then of course the theory is a further "ambiguity" position – any position which resolves the inconsistencies is. So is any statement involving "egocentric" particulars; e.g. "I am hot" is (potentially) infinitely ambiguous! But if due account is taken of what ambiguity *is* like, if different senses are

---

> ... though I can claim philosophically that *Persuasion* begins with no assertion about anything, one can appropriately say that the opening sentence is about Sir Walter Elliot. *No more puzzling* is the fact that if I ... say that Sir Walter Elliot read nothing except the Baronetage, it will be a true statement verifiable by reading *Persuasion*, whereas the first sentence of *Persuasion*, which runs: "Sir Walter Elliot ... never took up any book but the Baronetage", says nothing true or false (emphasis added).

The "philosophical claim" is squarely based, like the Strawson–Urmson thesis, on the Ontological Assumption.

reunited for ambiguity, then the contextual theory is very different from an ambiguity position.

Another objection relies on the shaky distinction between semantics and pragmatics.

> To demand that considerations of structure give way to the pragmatic ones of context and use is tantamount to abandoning a semantics for fictionality (Woods 1974, p. 131).

It is true that calling the contextual theory of fictions outlined a "semantical theory" violates the letter of the received semantics/pragmatics distinction of Morris and Carnap, since contexts (of use) are invoked. Within the confines of the received distinction, the contextual theory proposed, like much else in semantics, really belongs to a part of pragmatics; and if the theory were reclassified, say as a pragmatico-semantical theory, there would not be much point in lodging an objection. But even the received distinction allows semantics to step beyond "considerations of structure". Moreover the received distinction was too narrowly conceived to account for what it was supposed to account for, namely meaning. In order to account for intensional discourse satisfactorily, world relativisation (or an equivalent) of the relations that were taken to be fundamental in semantics, such as truth, satisfaction and designation, has to be admitted, and has been admitted in more liberal quarters. Such world relativisation is, however, much like context relativisation, and can indeed include it (see Slog, pp. 466–8). In fact relativisation with features in common with that required for context is already allowed in received semantics, namely relativisation to a given language. In short, enlargement of what counts as semantics over the received empiricist account is essential if the meaning of much significant discourse is to be accounted for, if the objectives of semantics are to be accomplished, but the enlargement already provides apparatus like or including that employed in contextual analysis. More than this, the semantical apparatus has to include such contextual relativisation if the meaning of tensed expressions and of such particulars as "I", "now", "this" and "that" are to be accounted for satisfactorily. That is, an adequate semantical theory has, to account for significant parts of discourse, to encroach upon what used to be reckoned pragmatics; and the semantics of fiction scarcely encroaches more. The received distinction has, accordingly, to be scrapped.

The scrapping of this distinction is not the only, or worst, trouble that the contextual theory leads us into. The move of taking fictional statements at their face value, as naive views do, and as not analysed away or at least shielded by operators, and thereby quarantined, sets us into big trouble once again with powerful referential interests. For removing the covering is tantamount to allowing that some fictional statements are true and that some fictional objects have genuine properties, and thus to moving back towards more naive theories of fictions, incompatible with assumptions of the Reference Theory and the positions it underpins such as empiricism.

The contextual theory of fiction does enable us, however, to approximate closely naive accounts of fiction and to absorb other theories of fiction, such as the dimensional theory, which cope rather more successfully with the preanalytic data[1] than empiricist moves can do. According to the dimensional theory, which introduces the notion of *linguistic dimension* to solve such problems as the contradiction problem, "Holmes lived in London" is true in one dimension, e.g. the "language of fiction", and false in another, e.g. the "language of reality" (cf. Woods 1974, pp. 117–9, where this popular theory is sketched and dismissed). To take up the theory simply replace "dimension" by "context". Aren't our sayings as dark then as Woods has said those of the dimensional theory are? No, because the notion of context has received a good deal of explanation in the semantical literature, though, needless to say, opponents of its adoption are far from satisfied.[2] What of the problems Woods sets for the dimensional theory? Suppose, to begin on the first, N_1 ratifies p_1 and N_2 ratifies p_2. Then $I(p_1)(\underline{\underline{T}})(c(N_1)) = 1 = I(p_2)(\underline{\underline{T}})(c(N_2))$, where $c(N_1)$ is a requisite context of N_1. But what of logical compounds such as $(p_1 \ \& \ p_2)$, especially where p_1 and p_2 are inconsistent? Suppose N_1 does not ratify p_2, i.e. p_2 is not entailed by anything N_1 ratifies, so $I(p_2)(\underline{\underline{T}})(c(N_1)) \neq 1$. Since $I(p_1 \ \& \ p_2)((\underline{\underline{T}})(c(N_1)) = 1$ iff $I(p_1)(\underline{\underline{T}})(c(N_1)) = 1$ and $I(p_2)(\underline{\underline{T}})(c(N_1)) = 1$, $I(p_1 \ \& \ p_2)(\underline{\underline{T}})(c(N_1)) \neq 1$. If however N_1 does ratify p_2, then $I(p_1 \ \& \ p_2)(\underline{\underline{T}})(c(N_1)) = 1$; that is, an inconsistency is true in the context of $c(N_1)$. But what of combined contexts? Let $c(N_1$ and $N_2)$ be a context generated by N_1 together with N_2; it will ratify whatever either N_1 or N_2 ratifies, so $I(p_1 \ \& \ p_2)(\underline{\underline{T}})(c(N_1 \text{ and } N_2)) = 1$. Such combined contexts, usually obtained by closure of the union of contexts, are important in accounting for literary comparisons, e.g. "Holmes was a detective but Hamlet wasn't", "Holmes was a better strategist than Hamlet", etc.,

[1] It will be complained that many questions have been begged by taking so many fictional statements as true – statements empiricist and reductionist theories would deny. In what follows an attempt will be made to meet some of the complaints that emerge. But there is indeed a problem of data.

[2] For example, Kripke 1973 pooh-poohs use of context dependence and contextual restrictions, but on feeble Quinean grounds, e.g. "how are contexts individuated?".

and in meeting Woods' objection that no room is left for literary criticism, even for such platitudes as "Holmes was a very different sort of strategist from Hamlet". Nor is there anything preventing such combined contexts taking in also sizable slabs of factual material, thus enabling the assessment of such comparisons as "Hamlet was a smarter statesman than Eisenhower". Such comparisons raise problems not intrinsically more difficult than comparisons of historical figures from different times, as for instance in "Jefferson was a smarter statesman than Eisenhower". Temporality provides, as often, a good working model for contextuality. However the way these comparisons are made is not nearly as clear as it might be (and as it is in the integrated theory).

There are, of course, sometimes problems as to which context one is operating in. One can be confused as to what the context is, e.g. as to which work a given character appears in, not knowing what the context is, or be entirely mistaken as to what it is. Such indeed provides part of the evidence for the contextual theory. There is much circumstantial evidence of this sort supporting the contextual account of fiction. For example, there is the experience of not knowing the context, familiar to people who have not read a book, been to a film, or seen a television show, that is under discussion. Such people are left out of discussions by not knowing enough of the context: they cannot play the fictional game properly, a little like someone who does not know a language well. Another piece of evidence comes from the placement and settlement of bets concerning fiction, e.g. as to which street Holmes lived in (cf. Woods' discussion of the bet sensitivity of fiction in 1974). There is also evidence that ordinarily something like transfer is used in avoiding clashes. For instance, J says that Holmes who existed last century lived in Baker St., for instance in answer to a question. K objects that Holmes did not live in London at all and did not in fact exist. J explains that he was talking about a character in stories and that what he said was not intended to hold for the narrow world of fact.

Part of the evidence for the contextual theory of fiction, that is, is the coherent way it enables us to account for much of what we say, and do, as regards fiction, and part of the evidence, the same evidence, is the way the contextual theory enables us to reinstate the ordinary naive theory of fictions. The *ordinary naive theory* (which contrasts with the *absolutely* naive theory) includes, according to Woods, the following theses (see the discussion in 1974, chapter 2, especially pp. 30–1):

(1) Purely fictional items, such as Holmes, do not exist, and never have; they are unreal, do not come into existence, are causally unrelated to actual objects – and similarly for a batch of properties linked with (real) existence.

(2) Notwithstanding (1), what an author ratifies, for example in the way of ordinary features for his characters, holds.

Thus in particular, Sherlock Holmes was male, British, a detective who resided in Baker St., London. He was also both a minded object, capable of action, passion and thought, and a concrete spatio-temporal object, having a body of more or less determinate proportions, a series of sometimes discontinuous geographical locations, a more or less specific temporal history, and so on.

> ... Holmes is a man (or was), and he did live in London, and does not exist and never did. Such ... is the heart of the naive theory of fictionality (Woods 1974, p. 28; cf. also p. 31).

Note that from (1) and (2) a version of Meinong's Independence Principle follows, namely that a fictional item has properties does not imply it exists (or has existed). This suggests that we do, what the contextual theory has thus far signally failed to do, invoke the full theory of items and treat fictional characters as what they are, objects with this-worldly features as the naive theory would have it, not merely with other-worldly features. This is the route to the integrated theory which combines the contextual theory with the theory of objects.

There are several logically important special cases of (2), some of which we have already exploited, which hold on the naive theory:

(2.1) Fictional objects have a wide range of ordinary (and extra-ordinary) properties, e.g. Holmes has or had the property of being a man, i.e. Holmes was a man, Frodo is a hobbit, Gandalf a wizard, and so on.

Exactly which sorts of properties fictional objects have and can have, however, and which (such presumably as existence) are excluded, the naive theory makes none too clear: taking up the matter leads to more sophisticated theories.

(2.2) Identity, difference and numerical statements hold regarding fictional objects, e.g. Holmes is identical with a certain detective who lived in London, Holmes is distinct from Frodo, and Holmes, Lear, and Gandalf are three.

These sorts of truths, in particular, set the ordinary naive theory severely at odds with conventional logical wisdom, e.g. with mainstream theories of quantification, descriptions and proper names.

(2.3) An author may ratify inconsistency and incompleteness, and departures from logical laws or physical laws.

The problem for any reinstatement, or reconstruction, of the ordinary naive view is how all those things can be true, how (1) and all the parts of (2) can hold simultaneously. The contextual account ensures – or claims to ensure – all this, and perhaps consistency as well, by relativising (1) and (2) to different contexts. An author is not an authority on the actual empirical world (on $\underline{\underline{G}}$) and not all he says holds for that, and there what he says (e.g. on radio interviews, and in newspapers, of the world) must take its chance and be tested against what does hold as a matter of fact; but within limits he is an authority on the worlds of his imagination and what he ratifies goes for the worlds of his fiction.

It is a legitimate objection, however, that the contextual theory has so far avoided various important questions to which answers are needed in any logical elaboration of the ordinary theory, questions such as: What properties, other than derivative ones like nonexistence, purely fictional objects do have in respect of standard *nonfictional* contexts? How does fiction relate to reality, by what sort of relations? Unique answers to these questions are hardly to be expected: differently elaborated contextual theories will give different answers.

The ordinary naive theory – at least as so far presented – is deliberately unspecific as to whether authors' ratifications clash with empirical data, e.g. Holmes' living in London with his not living in London, and how, if so, clashes are resolved, if and when they are. It is worth distinguishing various ways of sophisticating the naive theory to cope with this recurring problem. The main options investigated, and not so far completely failed, fall under the following head (*all* options are considered in §7):

(PC) Pure contextual (no clash) theories. According to these, context always serves to prevent clashes: authors' ratifications are only good for fictional contexts, never for what might be called empirical or "everyday" ones (but are really referential ones). Thus these theories can add to theses (1) and (2) the following thesis:

(3) Notwithstanding (1) and (2), fictional truths can be incorporated consistently into the body of ordinary truths (and fictional propositions into the body of ordinary thought).

It may be questioned whether (3) is part of the ordinary theory since it is, on modern conceptions, a meta-statement. However (3) is, it may be contended, part of what is ordinarily believed and assumed, and so is part of the ordinary theory (to the extent that there is an ordinary, but not clearly articulated, theory or set of assumptions). This claim is by no means evident: it is far from as clearcut as empiricists have assumed, that more ordinary views hold the actual world $\underline{\underline{T}}$ to be consistent. More ordinary views are inclined to be far more tolerant than logical empiricism, for example, towards positions of a non-empiricist cast such as, to take a wide but instructive sample, religious, mystical and dialectical positions. The options to pure contextual theories are rather more diverse than might have been expected (see §7) and include both paraconsistent and consistent positions which qualify (3) and allow for some clashes.

Pure contextual theories, even if they can be worked out satisfactorily, are not sufficiently in the spirit of the theory of items to be acceptable. In some respects they belong rather to the *opposition* to object theory; for similar contextual relativisation could be used to dispose – with diminishing plausibility as the method is more widely applied – of most of the data in favour of the theory of objects. For example, "Pegasus is winged" could be said to be true in a suitable context, one concerning objects of mythology.

Nor is the contextually-intensional position that the pure contextual theories lead to, any more than explicitly intensional theories, ultimately satisfactory. (Intensionality results because the base shifting function the context supplies involves world shift.) A serious objection is that fictional objects become mere shells. There is nothing for the intensional properties to grip upon (cf. the transcendental argument of chapter 6), nothing to distinguish fictional objects contingently from one another, etc. Contextual and implicitly intensional positions also fall down badly on comparisons, such as "Holmes is a detective but Harold Wilson is not". Etc.

There are, it seems certain, a variety of statements we make truly concerning fictions which do not seem to be intensional or purely referential, and, secondly, emerging from this, the truth of some of these statements appears to require the truth of other extensional statements about fictions. Among the first class of statements are ones like the following: "Holmes was written about by Doyle", "Green correctly described Holmes", "McX looked like Holmes", "Woods prepared an identikit photograph of Holmes", "Stout was shorter than Holmes". But in order for someone to look like Holmes, or correctly describe Holmes, Holmes must presumably have properties of this and that kind, be a man of such and such a build, gait, etc. Some (quasi-) extensional claims presuppose, or generate, other extensional claims. There are also other good reasons why the contextual intensionality thesis associated with option (PC) has to be modified, namely that there are a variety of extensional claims about Holmes that we have already recorded, e.g. "Holmes did not exist", "Holmes' birth was not recorded on the London register", "Holmes had no funeral", that are not adequately dealt with by referential contexts (cf. again the "problem" of negative existentials).

§6. Integration of contextual and ordinary naive theories within the theory of items.

The way back to positions that are in the spirit of the theory of items from the intricacies to pure contextual theories is not difficult to discern. The way is to combine the ordinary theory of items (as expounded in chapter 1 [Vol. 1 *—Eds.*]), specifically by restricting the correctness of ratifications in principle (2) to *characterising* predicates. So, in particular, a character does not acquire noncharacterising features such as existence because an author ratifies its existence. Thus the ch/nonch distinction renders (2) compatible with (1). Sayso works only up to characterisation, to ch features. In short, (2) is sharpened to

$2A_0$. What an author ratifies in the way of characterising features for his or her characters holds.

Because all characterising features are extensional, difficulties of the absolutely naive theory (of §1) over James Bond's proof that $E = NR$ are automatically avoided. A character cannot be characterised as proving what is impossible, knowing what is false, and so on for other intensional success functors. Naturally the work may ratify that one of its characters did some of these things, and then that the character did so will hold in the world of the work.

To obtain the full intended scope of $2A_0$ a suitable wide construal of "character" has to be taken for granted. For $2A_0$ extends to apply to many objects introduced in a work of fiction that are not ordinarily accounted characters, e.g. natural objects such as mountains, woods, streams, marshes, trees, cliffs, rocks and artefacts such as towns, buildings, rooms furnishings, clothes, coaches, trains, automobiles, ... On the other hand, not all objects mentioned in a work of fiction are "characters", i.e. (in the wide sense) largely characterised in the work; for example, London is not a "character" of the Holmes stories, nor would Mr. Disraeli be if he figured therein. Better than so stretching "character" is to adopt terminology that marks the intended distinction between objects that are characterised in a work and those that are not; and happily suitable terminology is already in use. In 1975 Parsons distinguishes *creative* and *noncreative* uses of names in literature: "The 'creative' use ... is the use of a name in a work of fiction in which that character is created", whereas "noncreative" uses concern characters "whose identity is already established outside the story" (p. 79).[1] This distinction is subsequently (in Parsons 1978) transposed to the more satisfactory material mode distinction between native and immigrant objects: an object *native to* a work is one that is named in a creative use of a name, an *immigrant to* the work by a noncreative use of names. The distinction, that already made, is, as Parsons explains it, "roughly whether the story totally 'creates' the object in question, or whether the object is an already familiar one imported into the story" (1978, III 1). The one object may of course be native to some works, those in which it is characterised, and immigrant to others. The intended scope of principle $2A_0$ is to objects native to the work in question. Thus $2A_0$ may be reformulated as follows, in a way which allows both for plural authors (or no authors) and, by construing "work" widely (as with "opus") for a series of works:

$2A_1$. An object native to a work N has the characterising features the work attributes to it.

Principle $2A_1$ – which will require further minor adjustment – is derivable from a key assumption of the integrating theory of items, a Characterisation Postulate, according to which an object has all its characterising features. To derive $2A_1$ it suffices to identify (extensionally) an object native to a work with an item that has the features attributed to the object in the work: then apply FCP. To illustrate consider again (the) Holmes who is native to a certain series of works of Conan Doyle. He is characterised (now) as a man who was a detective who lived in London in Baker St., etc.; in short as $\imath x x h$.[2] Then, by FCP,

[1] "Creative" in the sense tying with being created in a story: it has nothing to do with having been brought into existence. Parsons adopts the account of creating a character given in Crittenden 1973.

[2] The tensing issue is clarified somewhat when work N is replaced, in the final formulation of 2A, by the source book for the work. For while the work may make present tense statements about its

that object was (necessarily) a detective and lived in London, i.e. $(\imath xxh)d$ & $(\imath xxh)il$. To complete the argument, apply the contingent identity, Holmes = $\imath xxh$. It follows, then, by substitution that Holmes was a detective, lived in London, etc.

Apart from the (*initial*) *Identification Assumption*, that Holmes = $\imath xxh$ – which implies that contingent identities hold true concerning fictional objects, e.g. that it is no mere fictional truth but a matter of fact that Sherlock Holmes is (identical with) the man who was a detective, etc. – just two main assumptions are drawn from the theory of items (both like the first assumption equally questioned by and equally anathema to empiricism), namely FCP and the principle of replacement of extensional identicals in extensional frames.

In a similar way, using the principles, many other commonsense truths, and much high school knowledge, such as that Tom Sawyer ran away from home and that Scrooge encountered several ghosts, may be established (direct confirmation may of course be had by consulting the original works). Such truths are, unlike $(\imath xxh)d$ which is a necessary truth, contingent truths. They are (in contrast with the pure contextual theory) true in the actual world $\underline{\underline{T}}$ in nonfictional contexts. That is, with statements such as (Sherlock Holmes)d no base transfer is required; for, with respect to such statements as (Sherlock Holmes)d, $\underline{\underline{T}}$ and the world of Sherlock Holmes overlap, i.e. such statements hold in both these worlds.

Principle $2A_1$ tells us which characterising features a fictional character has, but gives little or no information as to which features characters do not have. Yet we know quite well that Holmes was not blond and was not eight feet tall and that he did not live in Gin Lane or visit Brasil. How do we know? The initial reasons may appear to be diverse: living in Gin Lane would have been "out of character" and, anyway, don't we know he lived elsewhere?; if he'd have been as tall as that it would surely have been remarked, especially as he would have had considerable difficulties on his numerous travels, and, as with blondness we have counterinformation; and as regards visiting Brasil, well, it does not emerge from anything in the stories. A common denominator of the various characterising features Holmes does *not* have seems obviously to be this: they do not derive from anything, not just in the stories, but in the source book for Holmes. The reasons why it is unsatisfactory to confine consideration to the story are the same as those

characters, a present source book elaborating a past work can make past tense statements about the characters.

given in §1 for adding to what stories say to arrive at the appropriate world; especially the material additions designed to include matters certainly intended but not said in the stories (for derivability will take care of formal closure conditions). But given that a source book is obtained by additions (and perhaps subtractions) like those for a world, and accordingly is appropriately (and so perhaps vacuously) closed, it would be doubling up to say: Among characterising features an object of work N only has features derivable from those ascribed to it by source $S(N)$. As derivability, of whatever requisite sort, has already been allowed for, "derivable from" can be deleted, so yielding

2B. Among characterising features, an object native to work N has only those features its source $S(N)$ ascribes to it.

Since commonly $S(N)$ is closed under entailment, so commonly 2B will not exclude consequences of features characters have. For example, if a normal work ratifies that Hot Pants was placed first, then 2B will not deliver the result that Hot Pants was not placed. Thus 2B is straightforwardly compatible with the common logic of fiction (of §2).

The formulation given of 2B suggests that the formulation of $2A_1$ is insufficiently strong. That is so: the reasons for introducing the source book in the case of 2B similarly show that $2A_1$ should really ascribe to objects all the characterising features their sources attribute to them: else they will be missing features their sources say they have. Call the resulting formulations of $2A_1$, 2A ("the work" in $2A_1$ is replaced by "source $S(N)$"). Then 2A and 2B can be neatly combined in the final form:

2F. An object native to work N has all and only those characterising features which source $S(N)$ attributes to it.

Principle 2F is derivable from the theory of items, given an identification assumption for objects native to works of fiction. For arbitrary N, let d be an object native to N, and let F be the set of characterising predicates $S(N)$ attributes to d. That is, $f \in F$ if $S(N) \vdash df$ (i.e. $df \in S(N)$) and $\text{ch}(f)$.[1] Then d is an object such that for every $\text{ch}f$, $df \equiv f \in F$. What has to be shown is that for every $\text{ch}f$, $df \equiv f \in F$. Consider $d' = \xi x(\text{ch}f)(xf \equiv f \in F)$. By HCP, some object does satisfy the condition C: $(\text{ch}f)(xf \equiv f \in F)$. Hence by $A\xi$, $(\text{ch}f)(d'f \equiv f \in F)$. Further d' is unique up to

[1] Subsequently (in §9) it will be shown, what is important, that this restriction to ch features can be removed. This depends on use of s-features, which are left out of consideration until that stage.

characterising features, i.e. if both d_1 and d_2 satisfy condition C, then $(\text{ch}f)(d_1 f \equiv d_2 f)$. Hence $(\text{ch}f)(df \equiv d'f)$, whence the result.

It remains to further explain source books (the notion of a source (book) for an object was introduced back in 1.25 [Vol. 1 $_{—Eds.}$]). It is advantageous to make use not just of source books *for* objects but also of source books *elaborating* works (or series of works). The approximate model for source books elaborating works is provided by (what is stated in) modern supplemented and annotated editions of the (or a) work of some historical author.

Source books for works of fiction are propositional in character.[1] There need be no written or otherwise recorded account, and in general will not be (that is why the term "source" is sometimes used: to remove any impression that an actual token book is required): source books are like what they are mostly sources for, nonentities. A source book is *compiled* from both *primary* sources, in the case of fictional objects from works of fiction, but in the case of dreams or visions from the experiences, and *secondary* sources, e.g. historical and geographical information, commentaries, discussion (where the object figures as an immigrant object), etc. A source book stands to its primary sources much as a world stands to the same works. The compilation consists of material elaborations, supplied from the secondary sources, together with application of appropriate closure requirements. As with determination of worlds too (especially if a modal account is mistakenly attempted) some deletion may be required, e.g. to filter out inconsistency, as well as extraneous and misleading commentary. The very considerable similarity between source books elaborating works and worlds of works strongly suggests identifying them. The following connection can be argued for: where N is a work, the source book $S(N)$ elaborating N is the propositional representation of world $a(N)$, i.e. $S(N)$ is given by $\{p : I(p, a(N)) = 1\}$. In the limiting case of a work with just one character such a connection results fairly automatically.

In principle, source books *for* objects are excerpted from source books elaborating works to which they are native. Take the (union of the) works to which a given object is native; excerpt the material about or relevant to the object; and then apply essentially

[1] The propositions involved will usually be "tensed", but they do not have to be. Sources can be of a range of sorts, as will emerge. The source book may be past source book for objects presented as being at that time or in the past; they may be present source books for the same objects; they may be "timeless" source books for four-dimensional objects, etc.

the same compilation process as before to obtain the source book for the object. Where source books are obtained only by addition or where the source book $S(d)$ for an object d is compatible with the source book $S(N)$ for the series of works N to which d is native, $S(d) \subseteq S(N)$; and for the theory $S(N)$ can be used in place of $S(d)$ (since the extra information is immaterial and is erased when characterising features of d are considered). The source books for objects may overlap. Consider Frodo and Bilbo. The works or primary sources N are the three volumes of *Lord of the Rings* together with *The Hobbit*. Then S(Frodo) and S(Bilbo) overlap since there are many situations in which both are involved. For the purposes of obtaining characterising features it is enough to consider $S(N)$, since $S(\text{Frodo}) \cup S(\text{Bilbo}) \subseteq S(N)$.

Using source books for objects the native/immigrant distinction – which is somewhat restrictive, since other objects also satisfy a principle like 2F – can be bypassed, and the sayso principle formulated:

2G. A (fictional) object d has just those characterising features its source $S(d)$ ascribes to it.[1]

2G follows from the theory, given an identification assumption. The principle holds much more widely than simply for fictional objects: it holds for imaginary objects, and, with a suitable extension of the notion of source, it holds not merely for nonentities, but for all objects.

The worlds of reality and of works of fiction have, given 2F, limited coincidence. As regards any characterising feature, for any d of N, $I(df)(\underline{\underline{T}}) = 1 = I(df)(a(N))$. However, as regards noncharacterising features and intensional features $\underline{\underline{T}}$ and $a(N)$ are very different. Almost paradigmatically, where d is a main character, $I(dE)(\underline{\underline{T}}) \neq I(dE)(a(N)) = 1$, as even if N does not actually say d exists, d will stand in such physical relations to other objects that are taken to exist that the proposition that d exists will belong to $S(N)$. These agreements and differences will be reflected in agreements and differences concerning truth-value assignments in different contexts. Thus as to characterising features, assignments with respect to fictional and nonfictional contexts will coincide but on noncharacterising and intensional features they will commonly diverge. For example, where c_1 is fictional and c_2 is nonfictional, typically, $I(dE)(\underline{\underline{T}})(c_1) = 1 \neq I(dE)(\underline{\underline{T}})(c_2)$. For, as-

[1] This principle gives the original form of the theory for nonfictional contexts. Formulations $2A_1$–2F are influenced by Parsons' work, especially 1978, borrowing the native/immigrant distinction. Source books can of course be independently explained as in 1.25 [Vol. 1 —Eds.].

suming no other contextual details require taking up, $I(dE)(\underline{\underline{T}})(c_1) = I(dE)(a(N)) = 1$, whereas $I(dE)(\underline{\underline{T}}(c_2) = I(dE)(\underline{\underline{T}}) \neq 1$.

The difference between the pure contextual theory and the integrated theory can be summed up as follows: in the pure theory almost every nonreferential but extensional statement is transferred to the world $a(N)$ of the story for evaluation. What is left of the objects and characters of the story N in world $\underline{\underline{T}}$ is but a shell of certain noncharacterising features – a shell which is not structurally self-supporting or sound. By contrast, in the integrated theory objects of the story have a quite rich make-up in $\underline{\underline{T}}$, for all their characterising features hold. The world $a(N)$ of the story will be as on the pure contextual theory.

§7. Residual difficulties with the qualified naive theory: relational puzzles and fictional paradoxes.

The trouble with the qualified naive theory summed up in 2F is that it still *appears* liable – despite the qualification of (2) to characterising features, the restriction to native objects, and the use of source books – to run us into clashes with empirico-historical facts, through violations of the principle that an author of fiction cannot interfere with or upset empirical fact. It is important to show the appearance is misleading, and that the integrated theory can resolve apparent clashes, in one way or another. The clashes always concern, and are bound to concern, relations between nonentities and entities. (Properties are unproblematic unless they are relationally tied.) There are two main cases to consider: First, there are the *relational puzzles* already considered concerning inferences from relations that do hold, e.g. the apparent clash between the truth that Holmes lived in London, deriving from 2F, and the empirical fact that London had no such Holmes living in it. Secondly, there are *fictional paradoxes* generated by replacements of extensional identicals in truths which relate fictional objects to entities. Neither case is peculiar to fiction; both bear on the general question of relations of nonentities (especially pure ones) to entities. The cases are considered in turn.

1. *Relational puzzles.* A different example will reveal that the puzzle has nothing essentially to do with fictions, but is a puzzle (in the first instance) as to the *relations* between entities and nonentities (and one that has already been alluded to several times, espe-

cially in 1.21 [Vol. 1 $_{—Eds.}$]).[1] It would be a mistake to leap to the conclusion that the puzzles show that there is something wrong with nonentities: they reflect just as much on entities (and many of the problems for the logic of nonentities might be ascribed, not just by a joker, to the erratic behaviour of entities).

Consider a (or the) philosopher who assassinated Russell, and who perhaps has other characterising features as well (they can be rolled in with the predicate "p"). Call the object Hook; so Hook $= \xi x(xp \;\&\; xgr)$ say. Then by FCP, Hook g Russell. But it is fact that it is not the case that Russell was assassinated, i.e. $\sim$(Russell was g).

The resolution (already defended in 1.21) is simply that Hook g Russell does not imply (Russell was g), that the relation is reduced, and so does not sustain intransitive passive conversion. Moreover, to obtain passive conversion in the logic itself – as distinct from in the informal translation into logical form – two additions would have to be made to the second-order logic: it would require, first, the inclusion of passive logical forms (or at least an alternative form to the one it has), and, second, and more important, active-passive conversion axioms. Free λ-categorical logic which can accommodate the forms, still requires the latter axioms. But the latter axioms do not hold for all objects. For example, to vary the vexing stock Holmes–London example, "Holmes blew up London" (after he joined the Goon Show) does not imply "London was blown up".

"The" conversion resolution divides into two somewhat different resolutions, according as it said that the faulted inference in intransitive passive conversion is that to "London was blown up by Holmes" or therefrom to to "London was blown up". Although the first *passive-blocking* resolution was favoured in 1.21 and will be favoured here, both options will be left open. In certain respects the second *demodification-blocking* resolution is more straightforward than the first. It has its problems however, and it is just as readily made fun of: the analogue of the joke "So Holmes is the man who blew up London whom London wasn't blown up by!" is the joke "So London is the city that wasn't blown up that was blown up by Holmes!" Referential jokes like "So Holmes is the man who blew

[1] Hence it reveals another respect in which the purely contextual theory of fiction, whether or not extended to apply to all nonentities, is not in the spirit of the theory of objects. For if it is not extended then fictional objects are treated very differently from other nonentities; while if it is extended, then the theory effectively says that all characterising-type truths about nonentities are at bottom (after taking up context) other wordly statements and not true in nonfictional contexts, which *now* coincide with referential ones.

If it is extended then the theory is like a contextual version of the Castañeda and Rapaport multiple modes of predication theories, dismissed in 12.4 [Vol. $_{—Eds.}$].

up the city that wasn't blown up!" have to be lived with: not a difficult feat compared with what most honest philosophers have to tolerate. Talk of "the object who blew up the city that wasn't blown up" may sound strange to ordinary ears but it is consonant with similar initially "strange" sayings, deriving from the theory of items, such as that someone squared the circle, because a person who squared the circle squared the circle. Of course the someone is an impossibilium; no actual person could have pulled off the feat. In a similar way the mafia Holmes is not an entity.[1]

Such examples as the circle squarer show further that relational puzzles are not confined to relations between nonentities and entities, but also affect relations between nonentities. Let the circle squarer be the person who squared the ideal circle, so the relation is between nonentities. Then conversion would yield the result that the ideal circle was squared. The resolution is as before: such conversion is inadmissible.

While there are alternative resolutions and are objections to the conversion resolution (i.e. resolution through modification of conversion axioms), as will soon become evident, the resolution and the examples will provide a working model for what to say about fictional cases such as the Holmes–London case. Certain options for nullifying the apparent clash of data have already been eliminated or can be eliminated immediately; but is worth being a little systematic in order to try to exhaust the range of options as to what can be said and to reveal just how many options there are that might (just) be pursued. Almost all the options involve either further restricting, in one way or another, the class of characterising predicates, or reducing relations (so they do not support all classical inferences), or, as it often turns out, both. In the classification for nonfictional contexts[2] which follows, predicates are restricted to those that are extensional (at least in the *main* place, i.e. the subject position of traditional grammar):

A. *Relational predicates* of the form gb, where b is a nonentity, *are never characterising.* It would follow that nonentities never stand in characterising-type relations to entities. There are stronger and weaker versions of this position:

B. *Nonentities satisfy no c-extensional predicates*, i.e. extensional predicates of the type previously listed as characterising (to exclude problems over extensional

[1] On some accounts of possibilia, the mafia Holmes is not even a possibilium. That surely tells *against* those accounts of possibilia.

[2] Thus as regards the classification, context plays a more meagre, though by no means negligible, role. Such distinctions as that between referential and nonreferential contexts remain important; so do the usual roles of context in identification of time, place, speaker, etc.

predicates such as "does not exist"). This is the classical position already rejected as radically unsatisfactory.

B*. *Nonentities satisfy some c-extensional predicates.* Among these

C. *They satisfy only one-place predicates.* This previously discarded position involves a hard-to-enforce-or-justify distinction (at least for natural languages) between one- and many-place predicates. The position also largely cripples the theory of objects, not merely in the case of bottom order objects (it prevents us distinguishing in the obvious way between Kingfrance and the present king of China), but especially as regards higher order objects. For it destroys relation theory which is at the heart of mathematics (see chapter 10 [Vol. 3 $_{-Eds.}$]).

C*. *They satisfy some many-place predicates*, as relating to nonentities. This position suffers many of the difficulties of C. But the difficulties can be somewhat mitigated by the strategy of *duplicate replacement*, as illustrated by the replacement of "Homes lived in London" by "Holmes lived in the London of the Holmes' stories", $hil^{\#}$ for short, where the # superscript (read, roughly "duplicate") indicates the duplicate object of the relevant work.[1] The idea, more generally, is that where agb and b exists but a does not, "b" is replaced by "$b^{\#}$": what "agb" was intended to say (or meant?) was $agb^{\#}$. Prima facie, this is false, as respective embedding of "agb" and "$agb^{\#}$" in frames will show, e.g. agb & $\sim agb$ is contradictory but $agb^{\#}$ & $\sim agb$ is not. Such a general strategy has already been jettisoned (in §4) because it forces ambiguities where there appear to be none, e.g. "lived in London" means something different according as it is Holmes or Hampshire we are speaking about. Nor can we infer that they lived in the same city. *The strategy is*, as the difficulties are beginning to reveal, yet another replacement stunt and *open to the crucial objections lodged against such nonuniform replacement moves* in 1.7 [Vol. 1 $_{-Eds.}$]. The strategy also sabotages the theory of items, breaking down some of the data, and enforcing an artificially sharp division between entities and nonentities. For example while the king of Tonga is presumably king of Tonga, Kingfrance is not guaranteed king of France or France$^{\#}$ (however that is characterised?), but Kingfrance$^{\#}$ is king, not of France, but of France$^{\#}$!

[1] When it is necessary to specify the relevant work N, because context fails in the task, $\#(N)$ will be used.

The rejection of the strategy of duplicate replacement has an important corollary, namely that statements of fiction are not about duplicate objects. For example, *hil* is about what it seems to be about, Holmes (primarily) and also London; it is not about London$^{\#}$.

A*. *Some relational predicates of the given form are characterising.* It is evident that sufficiently many predicates of the given form will have to be characterising, including predicates like "lived in London" and "is king of France", else we are back with the positions and difficulties under head A.

D. *There is no problem* with examples like "Holmes lived in London".

E. *There is no clash* with "London was lived in by Holmes". For that is true. What is false is that London was lived in by an existent Holmes. London's population divides (in a way reminiscent of Lewis's distinction of *denotation* and *comprehension*) into an actual population, to which Holmes doesn't belong, and a much larger fictional population, to which Holmes does belong. To explain why no trace is to be found of Holmes in the city records, by the watchers, etc., it is enough to say that Holmes was not a member of London's actual population: the records only pertain to inhabitants who existed. There is no need (here at least) to appeal to reduced relations: conversions can be simply accepted. Reduced relations are essential given the same is to be said of action predicates, such as "blow up", "occupied", "assassinated", etc., which permit intransitive conversion. (The alternative is to try to distinguish such "action" predicates, and to treat them differently, e.g. by adopting the unpalatable course of returning to head A.) For that London was blown up, or occupied in the nineteenth century, *is* false. Given further that full passive conversion is said to be inadmissible, such inferences as "London was blown up by Holmes; so London was blown up" are ruled incorrect. That's alright, it's said: the antecedent shows that the job was the work of a nonentity, so detachment to a statement that removes this qualification is not to be expected. And *that's* why there is no clash. The inference is no better than those to the same conclusion from such premisses as "London was blown up in the military exercise [on the invasion plans, in the film, according to the news, allegedly, ...]". As will become clearer in what follows the demodification blocking resolution encounters serious difficulties. For example, it is in trouble with fictional paradoxes, which appear to compel some qualification on free-wheeling passive conversion. It is in an embarrassing position over truths about London, which depend on what is, or is it what might be, written about London, etc. But many philosophical theories have encountered, or simply ignored, much greater difficulties, and flourished.

E*. *There is a clash; but*

F. *It is only apparent.* This compromise position is obtained by trying to exploit the predicate/sentence negation distinction. Since predicate negation differs from sentence negation, "Holmes did not live in London" differs from "It is not the case that Holmes lived in London". Inconsistency is avoided by saying, none too convincingly perhaps, that all that empirical evidence establishes is "Holmes did not live in London", and that this is not inconsistent with "Holmes did live in London". However unless the theory of items is modified, the position collapses. For as Holmes is possible, $h{\sim}il$ implies $\sim(hil)$. The option can accordingly be discarded.

F*. *It is not merely apparent*, but such clashes are paraconsistently unproblematic. According to dialectical positions clashes do occur, but (in contrast to the absolutely naive theory) clashes are limited because authors do not have an unrestricted licence to ratify whatever they choose. They cannot ratify that their characters have noncharacterising predicates; in particular, they cannot ratify that their characters really exist, i.e. clause (1) of the naive theory cannot be (also) negated. These positions *do* seem to explicate better than any of the options that find ambiguity everywhere, the ordinary naive theory reflected, for example, in the high school teaching and examination of English courses, where no regimentation is required, there no syntactical distinctions are made, e.g. between "fictional" and "empirical" forms, or the like. But to say, as these positions may, that it is simply true both that *hil* (because "*il*" is ch) and $\sim$*hil* (because of the historical records) is not merely rather advanced for these "consistent" times, but does induce contradictions where they are not forced but are misleading and can be avoided by other more satisfactory options. It would be a mistake to dismiss such positions too hastily however; for they do, in contrast to most of the putatively consistent positions, approximate the ordinary naive theory in a way that has little or no artifice. And asked whether Holmes lived in London or not, respondents sometimes do say "Well, he did and he didn't", without any qualification which would indicate ambiguity. But asked to elaborate, respondents typically (in contrast to cases of genuine paradox or inconsistency) resolve the conflict in one way or another, e.g. by contextualisation. With good reason: for *Holmes is not a paradoxical object*, which induces seemingly irreducible inconsistency, but representative of a large class of consistent objects which should not generate inconsistency. Naturally the theory should be able to cope with paradoxical objects, such as the famous village barber who shaves all and only those villagers who do not shave themselves, since a work of fiction may well include paradoxical figures among its characters. But while some large concessions have thus to be made to the paraconsistency point – while indeed a fully adequate theory of fiction will most probably be paraconsistent – the extent of paraconsistency should not be permitted to run wild, and far beyond its proper boundaries. Accordingly option F* can be closed off.

D*. *There is a problem alright: but*

G. *It can be averted in syntactical ways.* The strategies all involve introducing additional notation which separates out elements that are supposed to lead to the clash. (A semantics may accompany, or more likely reflect, the new syntax.)

Thus they all involve finding ambiguity where, so it has been argued, there is none, and making extensive formal replacements where, so it has been claimed, none are required. So eventually they all should be rejected. But it is worth enumerating the options, and remarking upon some of the main options, all of which have been suggested or tried.

G1. The predicate term in sentences like *hil* is ambiguous between the actual London which can be represented "London^{E}", and the fictional London (in question), "London$^{\#}$". The move has already been considered and found wanting under C*.

G2. The copula "lived in" is ambiguous as between two modes of predicates, actuality predication, with copula "lived in^{E}", and fictional predication, "lived in$^{\#}$".

This move, advocated for instance in Castañeda 1979, is sharply criticised in 12.4 [Vol. 3 $_{\textit{—Eds.}}$]. Since it is, like G1 a replacement project (for part of English), it is open to the damaging objections of 1.7 [Vol. 1 $_{\textit{—Eds.}}$]. Both G1 and G2, and a further move

G3. Of finding ambiguity in the subject term (Kripke's move, considered in §3),

are specialisations designed to explain the ambiguity in the sentence *hil*. The parent doctrine, which they all support is

G4. There is an ambiguity in the sentence *hil*, and more generally in all sentences that yield true assertions in both fictional and nonfictional contexts.

The generalisation indicates, what is right, that insofar as there is something to be said for those distinctions, it can be taken up contextually: it is *not* a matter of logical grammar. According to G4, however, the fictional truth or statement, $(hil)^{\#}$, must be given different logical representation from the empirical falsehood $(hil)^{\mathrm{E}}$. (Not so: although different notation *can* of course be introduced, it does not reflect the natural language data, and the requisite point can be much better made by distinguishing, as before, nonreferential and referential contexts, in which the one unambiguous sentence can occur.) As always, it is better to say that there are two different sorts of statements (one true, one false) rather than that there are two kinds of truth. Inevitably then, there is ambiguity. The ambiguity in A, as between A^{E} and $A^{\#}$, must devolve onto sentence parts. There are as many points to locate ambiguity as there are parts of speech, and in complex sentences ambiguity could be located almost anywhere, e.g. in adverbs. But there is a basic form where ambiguity has to be found somewhere, according to the referential thinking that usually underlies option G, namely the relational form agb. That leaves basically 5 proper places at which to account for the ambiguity (excluding agb): a, g, b, ag, gb.[1] It remains to consider the important

[1] Or perhaps 6 if $\langle a, b\rangle$ is counted.

G5. *Relational and adverbial strategies.* Such moves find an ambiguity between *hil* and $h\{il\}$[1] or between $\{hi\}l$ and $h\{il\}$, where $\{il\}$ indicates that i and l *adhere* in some closer way yet to be explained than i and l usually do in relational statements.

Different positions result according to the tightness of the adhesive (ultimately the effect had on logical operations) and/or the leading analogy exploited. Among them are, in approximate order of logical liberality,

(i) concatenation,

(ii) adverbialisation,

(iii) hyphenation, and

(iv) plugging-up á la Parsons.

Some of these devices have already been explained (e.g. (i) and (iii) in 1.21 [Vol. 1 $_{—Eds.}$]) and criticised. Hyphenation and concatenation at least have the advantage of occurring in natural language (though hyphenation is probably being assigned a new, at least a considerably extended, role): they distinguish *hil* from, respectively "Holmes lived-in-London" (*hi-l*) and "Holmes livedinLondon" ($h\widehat{il}$). Adverbialisation, though it has of course natural language analogues, is contrived: *hil* is distinguished from "Holmes lived in-London-mode" or "Holmes lived Londonishly" (*hilly*). Plugging-up[2] distinguishes, like hyphenation which it resembles in the two-place case, $[hi]l$ with the relation plugged-up on the left from $h[il]$ with the relation plugged-up on the right; and both may differ from *hil*. More generally, where f^n is a many-place predicate term and t a subject term, $(ft^i)^{n-1}$ is an $(n-1)$-place predicate term *plugged-up* in the ith place. While the filling-up of places is quite in order classically, the order of occupation does not matter in the way it does with plugging-up.

Each of (i)–(iv) restricts logical operations concerning subjects and other expressions in adhesive positions. (i) blocks virtually all operations; (ii) stops conversion and closes off "London" from all logical operations but permits substitutional quantification on modes (e.g. for some mode m, *himly*); (iii) (presumably), and (iv) allow quantification but preclude (or limit) identity replacement and limit conversion. What logical apparatus like (iv) permits is very much open to determination since there is but little

[1] To take the right-hand side. There is an analogous case for the left-hand side.

[2] Introduced by Parsons 1974, apparently following a suggestion of Chisholm's.

intuitive control. In each case then there is no way to proceed from *hil*, now *replaced* by or properly represented as $h\{il\}$ to $\{hi\}l$, or to *hil* itself.

So the problem is averted. But at a considerable and unnecessary cost. For what does $\{hi\}l$ say? In every case something about London, that London was inhabited Holmesishly, etc. But in this event *the requisite point can be made without resort to special symbolism*: it is that such relations do not convert, that *hil* does not imply "London was lived in by Holmes", li^ph for short. All that is required is English, not such symbolism.[1]

A point about the *logic* of relations has been mistakenly converted into a point about *syntax*. In place of elementary logical restrictions on relations, what is offered is syntactual reformation which can ensure draconian logical restrictions. Positions (i) and (ii) are certainly of this sort; they can be ruled out at once on the grounds that they preclude perfectly legitimate quantificational arguments, e.g. to take a trivial case to "Holmes lived in some place". The situation is rather like the identity issue considered in 1.11 [Vol. 1 —*Eds.*], where complex solutions to puzzles can be avoided by a more straightforward resolution which they effectively guarantee in any case. It is the same with Parsons' theory: for when we climb out of the notation what is said to be true is this:

a. Holmes had the property of living in London; i.e. $h[il]$.

b. London did not have the property of having been lived in by Holmes; $\sim(l[i^ph])$.

c. It is false that Holmes lived in London; $\sim(hil)$.[2]

The key distinction between a and b is (after λ-conversion) simply that between "Holmes lived in London" and "London was lived in by Holmes", in traditional grammatical terms between active and passive. The complexities and restrictions of plugging-up are unnecessary to make the requisite distinction.

Why furthermore, in defiance of the ordinary naive theory, accept c? Parsons identifies *hil* with $h[il]$ & $hi[l]$? (For what *hil* comes to in Parsons' theory is the referential form h^Eil.) Granted one can define a technical predicate $hi^+l =_{\mathrm{Df}} h[il]$ & $[hi]l$, why equate *hil* with hi^+l? The reason for rejecting the equation is, naturally, that *hil* is true,

[1] But perhaps with some help from traditional grammar in selecting a main place, *the* subject in statements, so that it can be said that "*hil*" is primarily about Holmes.

[2] Compare what Parsons says about "Holmes met Gladstone" (1978, III 13), where the case for plugging-up relations is given what little motivation it gets.

not false, but $hi^{+}l$ is false because $[hi]l$ is false. The referential domination assumption which leads to the equation needs its credentials checked. Parsons' theory diverges then from commonsense in rendering "Holmes lived in London" ambiguous between $h[il]$, $[hi]l$ and hil, assigning the statement the truth-value true only for the special predicate $[il]$. The usual objections to such locations of ambiguity where there appear to be none can be brought against the strategy, as well as these evident points. We don't have to learn Parsonese in order to truly ascribe "living in London" to Holmes. Many who will never master plugging-up can do that. If the theory is a rational reconstruction (as most such theories reasonably enough are) then it has moved rather far from its data base, since we use the one predicate in saying hil and (Hampshire)il. There is finally a puzzle as to what the new predicate $[il]$ means, since it doesn't mean il: Parsons' semantics doesn't really explain the matter. The question is especially important for replacements within intensional sentence frames.

In sum, the relational strategies are unnecessary. They are avoided by the simple procedure of using passive forms. They are moreover undesirable in several respects. They lose intuitive control with new symbolism. They multiply up gratuitous ambiguities. They incur the difficulties of replacement moves. They make the wrong assignments to hil.

> G*. *It can be averted without such formal artifice.* For the trouble lies not in the sentential form hil, but in the assumption that the relation expressed is an entire one. The leading idea is that already presented: that the truths supplied by characterisation principles, from source books, about fictional objects such as Holmes take such forms as "Holmes lived in London". But these truths do not entail converted forms, such as "London was lived in by Holmes" for the relations are reduced; and accordingly the truths are not incompatible with information about London that can be gleaned from historical records, such as "It is not the case that London was inhabited by Holmes".

The separation of the active and passive forms agb and $bg^{p}a$ may look artificial to the classically-trained who see no place for and have no notation for passive forms, but it is not. Firstly, it has a long and good pedigree in traditional grammar (nor are the forms on a par in modern grammar). Such formulae as agb are not transformable to

$b\breve{g}a$ (where g is symmetric) without loss, for a is *the* subject of agb but not of the transformed formula. (Nor are they interchangeable in highly intensional frames.) Adopting traditional grammar let us say that agb is *principally about* a while $bg^{\mathrm{p}}a$ is *principally about* b. This is a first step in putting agb and $bg^{\mathrm{p}}a$ to properly separate uses (and why have two forms if this cannot be achieved, one might opportunistically ask?), and leads to a second reason for separating the forms and applying the distinction. The source book for Holmes, compiled from the Conan Doyle series, supplies the ch predicate "lived in London", whence derives the truth "Holmes lived in London". It is not a source for London. The source for London, derived from the historical record, does not deliver the predicate "was lived in by Holmes". Hence, given that this predicate is a characterising one, it is false that London was lived in by Holmes.[1] Therefore the truth of "London was lived in by Holmes" does not follow from that of "Holmes lived in London". (If philosophers were to insist that it does, it would very likely turn out that they were appealing to a referential context, where the conversion is admissible because the relation is not reduced.) The neutral theory of relations not only does not (as was explained in 1.21 [Vol. 1 —*Eds.*] and 5.5) underwrite all instances of active-passive conversion; it would err if it did. Notice further that the separation of forms, though like one of the resolutions of alleged "ambiguity", *avoids* admission of *ambiguity*. For there is not one form which is ambiguous as between two forms, but two forms. Thus is the ambiguity charge escaped.

There is inevitably a hitch to the pleasant separation of forms, and it is this: suppose the source book supplies the predicate "London was lived in by"? There are several options as to what to say:

(i) The source book does not supply such predicates; alternatively, such predicates though supplied are:

(ii) not characterising,

(iii) characterising but misleading as to form,

(iv) characterising and not misleading but simply reflect an inconsistent source book.

(iii) and (iv) are uninviting for reasons already given. It is unsatisfactory to say that such

[1] Naturally this is not what the demodification-blocking resolution says. As to what it does say, there are again various options, most of them problematic.

predicates as a matter of fact do not occur in source material for source books.[1] They might, and we can easily arrange a story in which they do. But a filter *can* be imposed in obtaining source books from source material. So (i) seems to be viable. However if the principle of minimizing deletion, if possible reducing it to zero, is to be seriously retained, then course (ii) is preferable, namely accepting only active forms as characterising. This involves recognising and distinguishing active from passive forms, but again traditional grammar can help out. Nor need it involve any loss of expressive power, e.g. where a primary source is written mainly in the passive. Active closure will simply be adopted as one of the closure principles in compiling the source book.

The differences between the forms can be brought out in symbolic ways that take G back towards G*, for instance by extending what English provides, hyphenation. Then redundancy can be recovered if really desired, by setting *l-ih* equivalent to *hil* and *h-il* equivalent to $li^{\mathrm{p}}h$; so is passive conversion approximated.[2] Yet another way, with a better basis in English is this. The first statement *hil* concerns primarily, Holmes; and, adapting the scoping notation of pp. 197-8 [Vol. 1, §1.14 $_{-Eds.}$] to a new purpose, it could be represented [*h*]*hil*, i.e. "As to Holmes, ..."; whereas the second statement $li^{\mathrm{p}}h$ primarily about London, could be symbolised [*l*]*hil*, "Concerning London, ...". There *is* a case for considering and elaborating on this symbolism (which has its fun aspects and its point, as well as its own problems),[3] even if there is not a case so far for adopting it. For it is important to be able to extend the way the passive changes the main subject to other two-place relations and to many-place relations of high adicity. The operator "[ ]" read, variously, "as to", "concerning", "as regards", "for", "is such that", has the requisite effect. Consider, e.g., "*a* is between *b* and *c*", which has "*a*" as main subject. To emphasize *c*, to make the statement about *c* rather than *a*, apply the operator to *c* and the statement to get [*c*](*a* is between *b* and *c*), read e.g. "*c* is such that *a* is between *b* and it (*c*)". An interesting thing now happens. Much of the common case that is made out for introducing

[1] It is probably the case that the predicate "London was lived in by" does not occur in the Doyle Holmes' stories, and so would not reach the source book. But it would be pointless to conduct a search to find out, since the logical points do not turn on such contingencies as exactly what is not written.

[2] Much as 1.21 [Vol. 1 $_{-Eds.}$], *A-B* is $h(A, B)$ with h the hyphenation operation, usually (but not essentially) restricted to suitably juxtaposed A and B. How hyphenation generalises depends on the conditions imposed on it, e.g. usually hyphenation is associative $(A\text{-}B)\text{-}C \approx A\text{-}(B\text{-}C)$, but ordered hyphenation can be envisaged, whereupon plugging-up would be a special case. It is very open, and very artificial.

[3] One of the problems, that of scoping (1.23 [Vol. 1 $_{-Eds.}$]), vanishes where there are no eliminations to effect, and with it the point to scoping. Here however [ ] takes on a new role.

variables as adjoined to quantificational operators goes over largely intact to showing that [ ] operators should likewise be combined with variables. Consider, for example, the reasons Quine offers (in ML, pp. 66-7) for variablization: compactness and convenience "instead of 'it' with different subscripts". But the *as to* operator leads to subscripted *its* in precisely the way quantifiers can. Apply Quine's argument steps to: number c is either less than or equal to or greater than d, number c h_0 number d, for short. Emphasizing d results in:

7Q. Number d is such that number c h_0 it.

But now similarly emphasize c (inside the "such that"). So results:

8Q. Number d is such that number c is such that it h_0 it,

which is inadequate. What is required is rather:

14Q. Number $d_{(1)}$ is such that number $c_{(2)}$ is such that it$_2$ h_0 it$_1$,

or more elegantly (following Quine's symbolic lead, p. 67):

17Q. (dx) number (cy) number (yh_0x).

Thus emerge *singular quantifiers*, such as $(dx)A(x)$. The notation is an improvement on $[d]A(d)$. For it shows, as does the particle "it", that the operator binds the place within the wff, and so prohibits substitution on the second occurrence of d in $[d]A(d)$.

The formation rule for singular quantifiers which are now tentatively introduced is as follows: where A is a wff not containing subject constant d and x a subject variable, then $(dx)A$ is a wff.[1] Evident axiom schemes for singular quantifiers within the framework of neutral relation theory are these:

1. $(dx_1)((x_1, \ldots, x_n)f) \leftrightarrow (d_1, \ldots, x_n)f$
2. $(dx_i)((x_1, \ldots, x_n)f) \rightarrow (x_1, \ldots, d_i, \ldots, x_n)f$, for $1 \leq i \leq n$.

Classically, where all relations are entire, and so any subject place can be the main place, 2 would be strengthened to a biconditional. And then the point of introducing singular quantifiers would be largely lost, since, at least for all initial wff, they are redundant.

3. $(dx)A \leftrightarrow A$, x not free in A.[2]

[1] A less conservative approach would introduce such quantification for all subjects, whence such wff as $(zx)A(x)$, read for example "z is such that it A".

[2] A distributional scheme,

Some of the earlier notation can be *very* roughly approximated using singular quantifiers, namely passive predicates and main places:

(i) $bg^{\mathrm{p}}a =_{\mathrm{Df}} (bx)(agx)$.

It is not pretended that this definition captures all that the passive achieves. But it does provide some of what is required in the logical theory. Firstly, it brings the exclusion of passive predicates as characterising under the general rubric that predicates involving quantification are not characterising. Secondly it explains, through axiom 2, why a source book commonly includes the active when its primary source includes only a passive form.

(ii) d is the main subject place of $A(a) =_{\mathrm{Df}} (dx)A(x) \leftrightarrow A(d)$.[1]

But a better theory, taking in passive forms and their many-place generalisations, and important parts of traditional grammar, has still to be devised.

2. *Fictional paradoxes and their dissolution.* Fictional paradoxes – which so it turns out are but intensional paradoxes and dissolved in the same way (as in 1.11 [Vol. 1 *—Eds.*]) – take the following form:

$A(a)$	Aeneus defended Troy, a high and windy city.
$a = b$	Troy is a low and airless village in Asia Minor.
$\therefore A(b)$	Aeneus defended a low and airless village in Asia Minor, a high and windy city.[2]

But $\sim A(b)$.

4. $(x)(A \to B) \to . \ (dx)A \to (dx)B$,
may also hold. The semantical theory of singular quantifiers is not yet sufficiently developed for this to be clear.

[1] Singular quantification also enables a start – a very poor start, as results quickly show – to be made upon defining plugging-up within the theory (without hyphenation), thus

(iii) $a[gb] =_{\mathrm{Df}} (ax)(xgb)$; $[ag]b =_{\mathrm{Df}} (bx)(agx)$; $ag^{+}b =_{\mathrm{Df}} a[gb] \ \& \ [ag]b$.

Then $\vdash h[il] \leftrightarrow hil$; $\vdash [hi]l \to hil$; $\vdash ag^{+}b \leftrightarrow [ag]b$. Hence, especially in view of the last equivalence, the approach is inadequate. Nor can it be made good by another theory of singular quantification, e.g. without axiom 2, short of some surprising additions. For although plugging-up has some features in common with binding, e.g. it stops identity substitution, it is not binding. Whereas using singular quantifiers b in $[ag]b$ cannot be quantified – at least on the theory so far elaborated – it can be quantified in $[ag]b$, e.g. to yield $(Py)[ag]y$. In the elaborated theory the latter would become $(Py)(yx)(agx)$.

[2] Both the term "fictional paradoxes" and the example are taken from V. Routley "Lost in Meinong's Jungle" (unpublished lecture notes, 1969), as is other material in this chapter, in particular the thesis that fictional objects have the characterising features determined by their source books.

The logical error is, as before, intersubstituting extensional identicals in intensional places. The only trouble, if it is, is that some of the places where replacements are proposed look extensional. Consider, to get back to Holmes, the following example:

	agb	Holmes lived at 221B Baker St.
	$b = c$	221B Baker St. = Bigshots Brewery.
$\therefore$	agc	Holmes lived at Bigshots Brewery.[1]

The argument involves the following steps when rendered more explicit. By the minor premise, whenever ext f, if (221B Baker St.)f then (Bigshots Brewery)f. So, *assuming* that ext(Holmes lived at ...), if Holmes lived at 221B Baker St. then Holmes lived at Bigshots Brewery. "Holmes lived at" looks extensional, doesn't it? Well does it, when the relation is "plugged-up" with a fictional object? More important, the theory reveals, by a somewhat circuitous but rewarding tour, that it can't be extensional.

Theme 1. The conclusions of fictional paradoxes cannot be merely accepted, e.g. as surprising truths.

Theme 2. As with relational puzzles, "fictional paradoxes" have nothing essentially to do with fictional objects.

It might be thought that the bullet could simply be bitten. Holmes did live at Bigshots Brewery and somehow, despite appearances, his lodgings must have been in the brewery. Contrary to what we thought source books have to be closed under extensional identity! But consider the object, not one of fiction, simply characterised as living at 221B Baker St. and not living in Bigshots Brewery; or to vary the example, the pure object with just these features: living at 13 Bilge Ave., not living at Bilgeview Hotel.

By HCP the latter object, call him Slurb, has just the characterising features of living at 13 Bilge St. and not living at Bilgeview Hotel, and no others. Hence it is not the case that Slurb lives at Bilgeview Hotel. Evidently Slurb is a consistent object. But in fact 13 Bilge Ave. is the Bilgeview Hotel. So if extensional identity could be substituted, Slurb would live at the Bilgeview Hotel, contradicting his not living there and also Slurb's

[1] Adapted from an example discussed in Parsons 1975, p. 84, and ascribed to D. Lewis. In fact Doyle obliged by having in 221B Baker St. a fictitious address, so the Holmes' watcher^{E} or afficionado on Baker St. would not find a 221B. But that the example involves altering the facts a bit doesn't interfere with the point it makes.

consistency.[1] Hence substitution is not admissible, and the bullet cannot be swallowed. Since further the substitution is excluded, "Slurb lived at ..." is opaque, and therefore the place is intensional.

Theme 3. The places held by terms signifying entities in relational characterising predicates are not in general extensional, though the characterising predicates themselves are.

Thus while "lived at Bilgeview Hotel" is extensional, the place held by "Bilgeview Hotel" is not extensional, and hence neither is the predicate "lived at". Though the latter predicate is frequently extensional in both places, for example, but not only, when both places are restricted to entities, it is not invariably extensional.

Theme 4. An important respect in which relations holding of nonentities are sometimes reduced concerns the extensionality of their places.

The general intensionality of minor subject places in characterising predicates is to be expected given the very considerable freedom there is in filling these places, and given the way in which what holds characteristically of a fictional object ties directly, through the source book, with what holds in an alternative world, that is, given the way in which extensional characteristics of fictional objects are, so to say, *intensionally induced.* The general intensionality of the minor places neatly solves a long-standing and vexing problem, namely how is it that the indeterminacy of nonentities does not transfer to entities to which they are related, and so contradict determinacy requirements on existence. The answer is that intensional indeterminacy of entities is unproblematic.

The *dissolution* of the fictional paradoxes is again a virtually minimal one, which *is presupposed by other resolutions.* Only the inferential patterns that lead to the difficulties are qualified (but through the back-up extensional identity theory which is now fitted into reduced relation theory). No syntactical transformations and no replacements are required. Thus the dissolution contrasts with the Parsons–Lewis resolution (in Parsons 1975) which calls for considerable syntactical reorganisation, and also with a duplicate-objects blocking of the argument, by replacing 221B Baker St., for instance, by its duplicate 221B Baker St.$^{\#}$, that is in effect by transferring the intensionality to the objects.

[1] Such replacement failures are not confined to fiction. Replacements of contingent identicals within mathematical frames, e.g. in statements of category theory, will likewise cause problems.

Then of course the minor premiss, *now* 221B Baker St.$^{\#}$ = Bigshots Brewery, is false. But duplicates, though they *can* be introduced (sometimes, but not necessarily, in a way that imports more problems than they are worth), are once again not required: *nor* do they really got to grips with the original problem, which remains unresolved.

§8. The objects of fiction: fictions and their syntax, semantics and problematics.

So far the main focus has been on solving problems concerning the *statements* of fiction. Although these issues are bound to and do bear substantially on the *objects* of fiction – for the reason that true statements of fiction determine the properties of fictions once the objects are discerned – they leave open several vital questions concerning those objects, questions as to their quantificational behaviour, as to their identity, as to their essential nature, etc.

Logically, the main enlargement of the theory called for is the extension of statemental logic to quantificational logic and beyond. The logics – likewise leading features of the semantical theory – are as before (in the relevant case, as in 1.23 [Vol. 1 $_{—Eds.}$]), but there are some additional details, e.g. in extending the common logic of fictions.

1. *Common quantificational and second-order logics of fiction.* The common quantificational logic of fiction results by combining quantified relevant logic (GQ to take a working example [see 1.23(2) $_{—Eds.}$]) with the sentential logic of §2, i.e. the logic adds to GQ the representative functor O subject to the postulates cited in §2 together with the quantificational analogue of $OA\ \&\ OA \rightarrow O(A\ \&\ B)$, namely the Barcan principle: $(Ux)OA \rightarrow O(Ux)A$. The converse $O(Ux)A \rightarrow (Ux)OA$ can be proved, much as $O(A\ \&\ B) \rightarrow .\ OA\ \&\ OA$ is provable, as follows: $(Ux)A \rightarrow A$, by instantiation; $O(Ux)A \rightarrow OA$, by R7; and so, by generalisation and distribution of U, $O(Ux)A \rightarrow (Ux)OA$. Although the Barcan fictional principle is, again like $OA\ \&\ OA \rightarrow O(A\ \&\ B)$, avoidable in more logically esoteric fiction, it holds for common cases. Objections to principles of this type, as to the original Barcan (modal) principle, mostly arise from reading the quantifiers as existentially-loaded and (as before) dissolve given the neutral, existentially non-committed, construals.

The semantical analysis of the common quantification logic of fiction with neutral quantifiers simply enlarges on semantics for relevant logics (of 1.23). It involves, among other things, a class K of normal worlds to which world $\underline{\underline{T}}$ belongs, and a domain D of objects only some of which at most exist. To treat of all of fiction, both usual and logically esoteric, the semantics is embedded in the universal semantics, in particular K is included in the set of all worlds W. The base shifting function of the contextual theory selects in fictional contexts worlds of W, or, in the case of logically common fiction, of K, at which to start on semantical evaluations. In case of nonfictional contexts evaluation commences at $\underline{\underline{T}}$, but when the context is referential the base is further constricted to $\underline{\underline{G}}$ [the referential world which contains only existents —*Eds.*].

The worlds of fiction conforming to the common logic are closed under deducibility, and accordingly pose no difficulty for the thesis that $\underline{\underline{T}} \in K$, that the actual world is appropriately logical. However fictional worlds in $W - K$ which do not conform to the common logic may not be closed under deducibility, but yet supply truths that hold in $\underline{\underline{T}}$, and accordingly *do* appear to cause a problem, *the* $\underline{\underline{T}} \in K$ *problem.* For most purposes outer worlds in $W - K$ present no difficulties because whatever holds in such world is covered by a functor in $\underline{\underline{T}}$. Suppose, e.g., entailment principles break down drastically in d's dream world, so that although A holds B does not though provably A entails B. What holds in $\underline{\underline{T}}$ is not A but aDA, i.e. "a dreamed that A"; and aD is not systemic; the principle, $A \Rightarrow B \twoheadrightarrow aDA \Rightarrow aDB$, is incorrect. What the theory of items does that is different is to allow A in certain cases to hold in $\underline{\underline{T}}$ as well as in d's dream world, e.g. where A is of the form bg with a characterising predicate predicated of a term signifying a dream item. The problem is not as severe as might have been anticipated because of the doctrine of reduced relations (which includes a modified theory of definability). For instance, because of reduced solutions there is little or nothing bg will imply in virtue of g. Certainly bg will imply $bg \vee bh$, but this is immaterial: $\underline{\underline{T}}$ does not have to reflect the logic of d's dream world. Indeed the problem may very well vanish altogether with reduced relations. In case it does not, well the thesis that $\underline{\underline{T}}$ belongs to K can be qualified, without serious damage if it is done carefully, but with some complication of detail.

The common second-order logic of fiction is reached in a similar way to the first-order logic. The postulates of §2 are added to a second-order relevant base together with the Barcan fictional principle formulated for all variables, i.e. $(Uu)OA \rightarrow O(Uu)A$, where u may be a predicate or sentential variable as well as a subject one.

The solid case for introducing neutral quantification becomes especially evident with discourse concerning fictional objects. By contrast, most going logics prohibit any such simple and direct approach as that which neutral logics afford, and immerse us (as we have seen) in a fantasy world that is stranger than much fiction. The basic reason for neutralisation is once again that we often want, and need, to be able to talk objectually about and quantify over items that do not in any way exist, including fictional objects.

For example, we want to be able to infer, by particularisation, from "Da Costa thought about Holmes and Woods thought about Holmes" that there is some one object that both da Costa and Woods thought about, but *not* that there exists an object satisfying this condition, since the object in question, namely Holmes, does not exist. Nonexistential quantifiers are wanted to formalise quite familiar statements and discourse about fictional persons. Nor need fictional subjects reached by quantifiers be always embedded in intensional positions. Consider the evident (but nonetheless disputed) truth "David Copperfield did not exist but might have" from which it follows that something did not exist but might have. Or again, the statement "Sherlock Holmes was written about by Conan Doyle" is true, with the proper name "Sherlock Holmes" as the (apparent) subject, i.e. the statement is about (perhaps among other things) Sherlock Holmes.[1] Since it is about Holmes we can certainly infer that, for some x, x was written about by Conan Doyle, but again we do not infer that there exists such a person. Similar quantified forms are not uncommonly, and quite legitimately, used in statements by people who fail to recall names, consider e.g. the exchange: "Some detective was written about by Conan Doyle. Do you remember his name?" "Sherlock Holmes". "Yes, that's the man".

2. *Avoiding reduced existence commitments and essentialist puzzles*. But all fictional objects exist in some world! (Which is to claim that a weak world-relativised version of the Ontological Assumption can be fallen back upon.) Even if they were to so exist that would hardly be of much help, since the worlds in question are beyond the actual, and accordingly do not exist. (Moreover the problem above concerning apparently extensional frames would remain.) But in fact fictional objects need not exist in any world. A work of fiction may include all sorts of objects that do not exist, e.g. ghosts, witches, gods (even narrators of parts of the story), and in principle there is no reason why even main

[1] The statement concerned is not quotational: Doyle did not write about the name "Sherlock Holmes" but about Holmes. But the Reference Theory requires that the commonsense truth that the statement is about Holmes be rejected.

characters should be taken to exist. A book may introduce leading characters which, so it turns out, or so the author may insist, do not exist (e.g. Virginia Woolf could have re-drafted her final novel so that the shadowy Percival did not exist). Let k be such a figure in work N. Then N ratifies $\sim kE$, i.e. k does not exist, and so $O_N \sim kE$ and $(Px)O_N \sim xE$. Also, given N's logic is usual, $O_N(Px) \sim xE$. Use of existential quantifiers here would be somewhat disastrous for the consistent story.

Mostly we do not make the same presumptions about fictional objects that we make about objects we believe exist (but maybe *don't*). For generally in the case of fiction we are all well aware that the objects do not exist. Nor do we (all of us) "suspend belief" as is so often asserted; we simply do not apply the not uncommon but not invariable contextual assumption that discourse is existentially loaded. Fiction is not "make believe".

A fictional object does not exist, and may not exist in any world, e.g. impossible characters. But could a consistently characterised fictional object exist? Could Holmes have existed? ($\Diamond hE$?) Yes, both on the theory presented and the ordinary naive theory. On the theory, the identity of Holmes with the item with the characterising features of Holmes' source book is an extensional identity, as it should be. For Holmes might have been characterised differently, he might have had a different source book. In fact he might have had $\underline{\underline{G}}$ as a source, and been extensionally complete rather than seriously indeterminate in many respects. Holmes could be different from what he was, much as Zeno Vendler might have been different from what he is (see 1.14 [Vol. 1 $_{\textit{—Eds.}}$]). Some small arguments which purport to show that necessarily Holmes did not exist remain to be dealt with, namely:

(i) Necessarily fictional objects do not exist. Holmes is a fictional object. Therefore necessarily Holmes does not exist. The argument would only be valid were it necessary that Holmes were a fictional object. But it is not necessary. The invalid form may be represented: $\Box(A \supset B)$, $A \therefore \Box B$.

(ii) Necessarily the x which is f is f, where f represents a complete listing of Holmes' characterising predicates. But Holmes is $\imath x x f$. So necessarily Holmes is f. Similarly then, since necessarily the x which is f does not exist, because nothing so incomplete can exist, necessarily Holmes does not exist.
However, the identity replacements made are illicit. $h = \imath x x f$ is an extensional

identity which cannot be used correctly for substitutions within modal contexts. So the arguments fail.

In a similar way other essentialist puzzles may be dissolved. For example, fictional characters may have had features different from the features they in fact have. It is only contingently true that Hamlet was a prince; he might not have been. All *such* essentialist difficulties for a theory of fictions are uniformly avoided through the theory of (extensional) identity.

3. *Transworld identity explained.* The integrated theory presupposes that $(\text{Holmes}, \underline{\underline{T}})$, i.e. Holmes at the real world $\underline{\underline{T}}$, is the same as (Holmes, c), where c is a world of Holmes' stories (e.g. one of the worlds a base-shift function takes evaluations to). That is, it supposes that Holmes can be identified, or traced, across worlds. It is also supposed that quantification across worlds, with Holmes as a value of bound subject variables, is legitimate, as, e.g., in the semantical evaluation of $\Box(x)(x = x) \to \Box(\text{Holmes} = \text{Holmes})$. Such theories have been charged with raising especially severe problems concerning the objects of quantification and their identity conditions, the matter of exact identity conditions for objects at different worlds being dubbed "the problem of transworld identity". But for the most part the "problem" is a further manifestation of the Reference Theory, and arises from thinking that Holmes, for example, must be some sort of subsisting object whose sameness across worlds is to be settled by referential criteria. But the situation is not like that at all.

What has added fuel to the heat of this problem is that even commonly accepted qualified identity principles, such as extensional identity criteria, get left behind, so it appears. For example, one and the same Holmes can be the fictional character Doyle wrote about who did not actually live in London,[1] and, in the world of a Holmes story, the detective who did live in London; that is, Holmes has various properties in one world, that of empirico-historical fact, but not in another, and yet is the same. Examples are easily multiplied. Before too much is made of this phenomenon of change of properties, even of extensional properties, of objects from world to world, it should be remembered that there is a parallel phenomenon with regard to change of properties of objects over time, for example one and the same person is both hirsute and bald, both male and female,

[1] Here "actually" serves as that functor which narrows truth-value assessment to $\underline{\underline{G}}$, and, more generally, constricts holding at a to $c(a)$ (cf. 1.17(7) [Vol. 1 —*Eds.*]). It is not being denied that it is true that Holmes lived in London.

not at the one time naturally, but at different times of their lives. And although more simple-minded versions of Leibnitz's law break down, appropriately qualified versions continue to hold (see chapter 2).[1] There is a parallel problem, with a similar logical form, for objects across worlds, namely when is (x, c), i.e. x at c, the same as (y, d), where c and d are in one case times and in the other worlds? (Temporal sections are like worlds, as semantical analyses of tense logics have emphasised.) In the case of fictional worlds the question as to criteria is very simply answered in a way which has an appealing uniformity with earlier answers delivered by the naive theory: namely sameness is determined (by or) from authors' sayso as for other propositions that can be recorded in source books. To be sure there are limits on what an author can pull off in this respect; he cannot by his sayso make the city of London into a number or even a motor car: there are category and type restrictions he cannot transgress. But the London of an author's work can be very substantially different from the London we have visited, and, for example, rather more like present day São Paulo: his London is much hotter than we remember, the people are very different and speak Portuguese, most of the landmarks we know are gone and we are surprised by the size and efficiency of the downtown bus terminal and the disappearance of the tube system. So much is easily explained through an alternative history, e.g. the world's climate has changed and after the Third World War Brazil assumed world dominance in place of the ruined United States. With a still larger story an author can even put London outside England (in something the way London Bridge is), but limits are now being reached (not to say the point of calling the place "London" at all, if *all* the old features of London begin to disappear, since the associations and "lift off" gained by taking advantage of familiar features of an actual place in a work of fiction are lost). In sum, *what controls sameness across worlds is qualified* author *sayso*, the qualification being that a core of features of the object must be preserved.[2]

There is no need to try to settle every case of identity or difference here, or accordingly to try to explain what is meant by *a core*: these things can be, like many everyday identity puzzles, a matter for cheerful indecision, as Wisdom has explained

[1] There are important questions not decisively settled by duly qualified Leibnitz, e.g. the matter of the identity of a person over time. For even when the criterion is clear, whether dated properties hold may not be.

[2] The answer accordingly has a good deal in common with Kripke's reply to Lewis and others on identity across possible worlds (see Kripke 1971): sayso is stipulation, but Kripke's stipulation has also to be qualified, and there are cases, e.g. of counterfactual worlds, where the matter is not one for stipulation at all.

(1952, chapter 1). There are any number of objects, and entities, for which we cheerfully lack, not clear enough identity principles, but decisive identity criteria – to hit (again) at another dogma of empiricism. To adapt some of the jargon of modern empiricism, *decisive* criteria for object identity are sometimes a Don't Care. It is a little tempting however to try to link the notion of a core with that already introduced, of an alternative history: for example, the core requirement is satisfied in the case of an author's London if there is a path linking his London with the historical London at some time in its history.[1]

4. *Duplicate objects characterised.* Statements about the relations of fictional objects to actual objects cannot be adequately replaced by statements about the relations of the fictional objects to duplicates of the actual objects, e.g. *hil* cannot be satisfactorily paraphrased as $hil^{\#}$. Duplicates do not enable main problems about fictional objects to be resolved, but introduce a new set of difficulties of their own (so it has been argued in §7). Nevertheless duplicate objects can be characterised, and there is occasion to introduce them since they are sometimes spoken about in non-philosophical discourse (sometimes as the result of bad theories no doubt, but not always).

For example, it is true to say that Holmes' London had a 221B Baker St., in symbols $\text{London}^{\#(N)}b$ (whereas on the main option taken it is false that London had a 221B Baker St.). A corollary is that duplicate objects are not extensionally identical with what they duplicate; $\text{London} \neq \text{London}^{\#(N)}$. They may differ, moreover, not just in spatiotemporal location but in central traits: a duplicate object need by no means be *exactly* like or even *very* like what it is duplicate of (thus the term "duplicate" is stretched somewhat in this application, much as Lewis's analogous use of "counterpart" departs from more common usage).

As to identifying duplicates, their leading feature is that they are determined by the worlds in which they occur, for instance $\text{London}^{\#(N)}$ by the world $a(N)$, Shakespeare's England by the England of the worlds of Shakespeare's works, etc. Thus $\text{London}^{\#(N)}$ is an item with just the (characterising) features ascribed to London in $S(N)$. Hence $hi\text{London}^{\#(N)}$ is true since $\text{London}^{\#(N)}$ satisfies the ch predicate *hi*. In elementary cases,

$$I(d^{\#(N)}f)((\underline{\underline{T}}) = 1 \quad \text{iff} \quad I(df)(a(N)) = 1.$$

[1] Here at least we can (tenuously) join forces with causal theories.

More generally the Identification Assumption for duplicates is (anticipating the *final* form of §9) as follows: where d is immigrant to N, F_d is a set of features ascribed to d in $S(N)$, the N-duplicate $d^{\#(N)}$ is identified thus: $d^{\#(N)} = \xi x(f)(xf \equiv f \in F_d)$. Hence *were* d native to N, $d^{\#(N)} = d$; and although d is immigrant to N, $d^{\#(N)}$ is treated as native to N.

§9. Synopsis and clarification of the integrated theory: *s*-predicates and further elaboration.

The integrated theory integrates the contextual theory of fiction with the qualified naive theory of fictional objects. A (purely)[1] *fictional object* is one that is characterised in fiction. Fictional objects are not the only objects that occur in fiction, that works of fiction are about, since for example actual and historical objects also occur. The full characterisation of a fictional object is given in (or by) its *source* book, which is determined from the works of fiction in which the object is characterised (not in which it merely appears incidentally or as introduced from elsewhere). *A fictional object has all and only those characterising features ascribed to it by its source.* This principle, which is a fundamental one in determining what is true in nonfictional contexts of fictional objects, derives from Characterisation Postulates in the following way: Where d is any fictional object and $f_1, \ldots, f_n$ are all the characterising predicates ascribed to d by its source $S(d)$,[2] then $d = \imath x(xf_1 \ \& \ \ldots \ \& \ xf_n)$. More generally, the set of characterising predicates is exhausted in the set $F_d = \{f_i : i \in N\}$ for a suitable indexing set N, and $d = \imath x(f \in F_d)xf$. Hence, by FCP, $df_1 \ \& \cdots \& \ df_n$, and more generally for each $f \in F_d$, df; that is d has all the characterising features ascribed to it by its source. Moreover as regards *characterising* features, these are the *only* characterising features a has, as HCP will show. For by HCP (of 1.21[α]), there is (neutrally) an x such that for every characterising predicate f, xf iff $f \in F_d$. Furthermore x is unique up to characterising features.[3] Let d coincide with this x (or such an x). Then $df \equiv f \in F_d$.

[1] A sometimes-actual object may occur in fiction, and be ascribed features it does or did not have. Such an object is not *purely* fictional, but it *has* a fictional duplicate which differs from it.

[2] Predicate f is ascribed to a by $S(a)$, where $S(a)$ considered as a theory (in the wide sense of RLR) contains af as an element. Put otherwise af is immediately inferable from $S(a)$. (To say af is deducible from $S(a)$ while in order in the usual case, would be wrong for deductively open objects.)

[α] Vol. 1: HCP. $(Px)(Uchf)(xf \equiv A)$, with x not free in A —*Eds.*

[3] Suppose otherwise both x_1 and x_2 satisfied the account, i.e. $(\text{ch}f)(x_1 f \equiv f \in F_a) \ \& \ (\text{ch}f)(x_2 f \equiv f \in F_a)$. Then $(\text{ch}f)(x_1 f \equiv x_2 f)$, i.e. x_1 and x_2 coincide in characterisation.

There are various shortcomings in the theory so far presented which can be overcome by the use of s-predicates which have so far been left out of the (already complicated) picture.[α] For example, just as the theory without s-predicates fails to distinguish the golden mountain from the existing golden mountain (see 1.21), so the theory of fiction fails to distinguish properly distinct characters who are differentiated through noncharacterising features. The difficulty can be evaded by adding the following closure principle to those used in the complication of source books: wherever f is ascribed to an object, ascribe sf also.[1] More serious is a connected inadequacy in the initial Identification Assumption. Holmes is, so it will be objected, the man with *all* the features his source book ascribed to him, not merely those characterising features so ascribed, as the Assumption has. By adopting a different identification assumption, and using s-predicates and KCP′, this telling objection can be met (cake can be had and eaten too).[β] For, in fact, various different fictional objects can be distinguished. Let d be an object and $S = S(d)$ the source for d. The *full object* d is the object that has all the features precisely that S ascribes to d, and the *full extensional object* (fe object) d is the object which correspondingly has precisely the extensional features. The full object has, like the initial object hitherto identified, all characterising features, but it also has all noncharacterising features (including all intensional features) ascribed to it in S in s-forms, e.g. if according to $S(d)$ d exists then dsE, d has presentational existence, etc. Similarly the fe object has as well as all characterising features all remaining extensional features ascribed to it in s-form. Hence the initial object is embedded in the fe object which is in turn embedded in the full object (embedding being defined through properties held: $b < c$ iff $(f)(bf \rightarrow cf)$).

On Parsons' theory 1978 this would imply, what is false, that x_1 and x_2 are identical under a *strong* determinate, namely Leibnitz identity. There is much appeal, however, in the proposition that coincidence in characterisation implies extensional identity, i.e. $(\mathrm{ch}f)(x_1 f \equiv x_2 f) \rightarrow x_1 = x_2$. The reason is that noncharacterising extensional predicates appear to be, by and large, consequential or supervenient: for instance, whether something exists or is incomplete or self-identical depends on what traits it has. But there are also counterconsiderations caused by apparently distinct objects with the same characterising features. So the question has been left open, since such an axiom is nowhere essential. Really a better resolution of so far unclassified extensional predicates into characterising and noncharacterising is needed first.

[α] On s-predicates recall Vol. 1, 1.21(8), (c): "given the predicate f, the operation s yields the predicate $s(f)$ – also written sf – read 'presents itself as f' or 'says of itself that it fs' or 'has supposititious f-ness'." —*Eds.*

[1] That rule generates its own problems, naturally: e.g. for the counterexample writer who tries to characterise a philosopher who satisfies certain predicates but not their matching s-predicates. But since the rule can be avoided, as will become evident, so are the associated problems.

[β] Cf. 1.21(8): KCP′. $(Uf)(z_o sf \equiv A)$, where $z_o = \xi x(g)(xg \equiv A)$ and x is not free in A. —*Eds.*

The *final*, or full, *Identification Assumption* extensionally identifies object d with the full object. Let F_d be as before and let $G_d = \{g_i : i \in M\}$ be the set of all features ascribed to d by S (by suitable set M). According to the Assumption $d = \xi x(Ug)(xg \equiv g \in G_d)$, i.e. d is in fact any object with just the predicates in G_d. Now by KCP′, taking $A(g)$ as $g \in G_d$, $(Ug)(z_0 Sg \equiv g \in G_d)$, where $z_0 \approx \xi x(Ug)(xg \equiv g \in G_d)$. It follows that $d = z_0$. It further follows, as $\mathrm{ext}(sg)$ because $\mathrm{ch}(sg)$, that $(Ug)(z_0 Sg \equiv g \in G_d)$ is extensional (by compounding principle (Ext P) of p. 301 [Vol. 1]).[α] Hence by replacement

d$_1$. $(g)(dsg \equiv g \in G_d)$.

What has to be shown is

d$_2$. $(\mathrm{ch}f)(df \equiv f \in F_d)$.

Suppose $f \in F_d$ and $\mathrm{ch}(f)$. Then $f \approx sf$ (see p. 352 [Vol. 1][β]), and as $F_d \subseteq G_d$, $f \in G_d$. Hence by d$_1$, dsf, and so df. Suppose conversely df for f characterising. Then $f \approx sf$, so dsf, whence by d$_1$, $f \in G_d$. But $\mathrm{ch}f$ so $f \in F_d$.

Therefore, by d$_2$, the full object d *has* precisely the characterising features ascribed in $S(d)$. Accordingly d *does not differ in characterising features from the initial object* (assuming now s-features are counted in), say $d' = \xi x(\mathrm{ch}f)(xf \equiv f \in F_d)$; for by HCP,

d$_3$. $(\mathrm{ch}f)(d'f \equiv f \in F_d)$,

whence $(\mathrm{ch}f)(d'f \equiv df)$. Thus *as far as characterising features go* – perhaps as far as extensional features go – *it does not matter whether the initial or full object is taken* in the identification assumption. But it surely matters intensionally.

Consider now an elementary statement of the form af for some fictional object a and some (extensional) predicate f; e.g. af is that worn-out statement *hil* with subject "Holmes". To evaluate af in context c, there are, on the integrated theory, three main cases:

α. c is *fictional*. Then af is evaluated in the way explained in the pure contextual theory, i.e. the base is shifted to the relevant world and af is evaluated there.

[α] (Ext P): A functor is extensional if it is defined in extensional terms only. —*Eds.*

[β] Operator s is said to be subject to the condition: if $\mathrm{ch}(f)$ then $sf \approx f$. —*Eds.*

α^*. c is *nonfictional.*

β. c is *referential.* Then af is evaluated at $\underline{\underline{G}}$. So, in particular, where g is any characterising predicate, ag is assigned false (i.e. $I(ag)(\underline{\underline{G}}) = 0$), because qu($a$) lacks a referent, a being a fictional object. This is the classical assignment, where ag is evaluated just like $a^{\mathrm{E}}g$.

β^*. c is *nonreferential.* Then af is evaluated at $\underline{\underline{T}}$. Where f is a characterising predicate $I(af)(\underline{\underline{T}}) = 1$ according as f belongs to F_a, i.e. according as f is ascribed to a by its source.[1]

Thus the example "Holmes lived in London" is true in cases α and β^* and false in case β. Naturally the cases are not exhaustive, for "Holmes" has other roles,[2] e.g. in a suitable context "Holmes" could signify a certain Braidwood gander in which event the statement would in fact be referential and false.

Cases α–β^* provide an important part of the base case for an inductive truth definition (of the sort included in the universal semantics); they are the crucial cases for a theory of fiction. Indeed given a language in which all noncharacterising predicates (such as E and $=$) are defined and all intensional predicates can be regarded as being composed of a sentence functor in combination with an extensional predicate, almost enough is already given to apply (in a notional way) the universal semantics and so derive definitions of central notions from the theories of truth and meaning (as given in 1.24 [Vol. 1 —*Eds.*]).

§10. The extent of fiction, imagination and the like.

Fiction in both philosophical and older popular senses includes much that is not fiction. But what it also includes is amenable to treatment within the theory of items (of course): much of what deserves to be accounted fiction or fictionlike or associated with fiction – imaginative or artistic products – operates logically and semantically according to the general principles already sketched for fiction in the narrower modern sense.

[1] The astute reader will have observed that this neutral classification of contexts corresponds in a *very* rough way to Kripke's platonistic classification of usages (set out in §3).

[2] So do the other terms. In particular, "London" may signify Holmes' London, in which event there is a further (quasi-analytic) case to consider under β^*.

1. *"Fictions" in the philosophical sense.* The fictions of literature and the media, of novels and some poetry, of science fiction and comics, of much of television and film, of drama and opera, to take main examples, are only some among the types of fictions that figure in philosophical discussion, especially older discussions (e.g. Bentham's). Much of philosophy, especially reductive empiricism, has been concerned with other "fictions", logical and scientific fictions and legal and political fictions, including, for example, universals such as Matter, Time, the State, the Law, or again such objects as average items, nations, multinationals, and so on. Detailed classifications of such objects may be found in Bentham 1932 and in Vaihinger 1949. Most of these objects, which are abstractions (including abstractly defined items), have markedly different logics and properties from those of fictional objects, and should be classified separately, and given the different treatment they deserve (see chapters 5, 8 and 9 [Vol. 3 $_{-Eds.}$]). Since it is, it turns out, decidedly misleading to account such objects "fictions", the philosophical sense is best abandoned.

In addition, there are many objects not considered of much importance by Bentham and Vaihinger that resemble the fiction of literature and the media much more closely than the "fictions" they concentrate upon, the fictions of myth and legend and, differently, what may be called the fictions of imagination, the objects of dreams and daydreaming, of private or shared fantasy and imagining.

Despite the significant differences between these very various types of objects – some of which we will come to – they have important features in common. One feature they all have is nonexistence; they are certainly not the solid material preceptible preferably-kickable individuals empiricists favour, and in many philosophers' eyes are to be shunned and never spoken about, or, if not so avoidable, reduced to items that do exist. But, as the later Meinong might almost have put it, none of these objects exist in any way at all, yet they are in general irreducible to objects that do exist, and require for their integrity no such reduction. Like other objects that do not exist they have extensional features, they have all their characterising features, as guaranteed by appropriate Characterisation Principles. But they differ significantly in their sources, and sometimes in the way they are apprehended.

2. *Imaginary objects, their features and their variety: initial theory.* Basic to a comprehensive theory of fiction is a direct analogue of Meinong's unrestricted assumption

principle, the *unrestricted imagination*, or freedom of imagination, *thesis*: there are no restrictions on what is imaginable (even the unimaginable is imaginable). One can imagine anything one likes, of any sort, abstract or particular, bottom or higher order, no matter how bizarre or whether inconsistent, incomplete, paradoxical or absurd. Indeed one of the few special principles of the logic of imagination is the thesis $(x)\Diamond Ix$, for any item it is possible to imagine it.[1] Implicitly restricted quantifiers are of no avail here: items violating any restriction can be imagined. Only a logic of all items, a theory of items, is adequate to encompass the range of items concerned.

Restrictions come, not with what is imagined, but with the features had. Just as an assumed item may not possess all the properties it is assumed to have, so an imagined or imaginary item may not have all the features it is imagined or otherwise supposed to have. What features are had by imaginary objects are, as with that subclass of imaginary objects, purely fictional objects, in fair measure determined by their *core* features, characterising features supplied by their sources (as in principle 2F). As with fictional objects, there are limits on what can be so ascribed; in particular severe limits are imposed by empirico-historical facts. Other features had, ontic, intensional and so on, are typically by no means independent of the core characterising features had, though they are also externally constrained, e.g. by $\underline{\underline{T}}$, by the havers of intensional attitudes, etc.

The major difference between various sorts of imaginary objects – as with types of fiction – lies in their *sources*, especially in their *primary* sources. An important difference that may be included under difference in source, resides in the way the sorts of objects are apprehended or encountered. For an important class of imaginary objects, distinct from fictional objects, presentation is experiential. Of course images are had, objects imagined, and so forth, in the course of absorbing a story, but the mode of presentation is different. Such objects as those of visions and dreams are *perceptually presented*, and their primary source consists of such presentation. Thus the primary source for dream objects is dreams, "series of pictures or events, presented perceptually to the sleeping person" (OED) or creature (since other animals dream); for daydream objects, daydreams or series of daydreams, which resemble dreams except that the subject is awake; for objects of visions, "supernatural or prophetic apparitions"; etc. Other features that are distinctive

[1] A special case of imagination is that where the objects are propositional in kind, e.g. given in the form, that A. In this event the statements can be reexpressed using such sentence functors as xI "x imagines that" and I "it is imagined that". Many of the points made in 8.12 [Vol. 3 $_{—Eds.}$] as regards belief then apply, *mutatis mutandis*, except that the logic of imagination is nonsystemic.

or idiosyncratic for objects of perceptually presented sorts also derive from their source, e.g. the fragmented nature of dreams, their unmemorabillty.

Given the primary source N the rest can be elaborated more or less as before. For instance, where N is a dream or a series of dreams, $S(N)$ is the source compiled from N, and $a(N)$ is the dream world of N ("dream world" is a term that can be transposed directly from ordinary discourse to worlds semantics). In "McX dreamed that Holmes was tailing him" the transfer in semantical evaluation is to the world of McX's dream. Where d is an object, unlike Holmes native just to N, $S(d)$ is included in $S(N)$ and can for most purposes be equated with $S(N)$. But there are certain notable features, e.g. the closure conditions in determining $a(N)$ are usually exceedingly weak. Correspondingly, there seems to be no common logic of dreams as there is, it has been argued, for fiction. Some dreams will have their internal logic, but again, as with fiction, there can be no logic of dreams.

Previous problems will reappear too in a somewhat new setting, relational puzzles, dream paradoxes, questions as to how to identify extensionally dream objects, etc. With imaginary objects such as dream objects it certainly seems preferable to identify the object with the fully described item that has *all* features described in the source, i.e. the *full* object. The requisite Characterisation Postulate will then act as a filter to deliver only the characterising features ascribed as held, others will be merely presented, e.g. the object which is green will be presented as existing, presented as valuable, etc.

3. *Works of the fine arts and crafts, and their objects*. Works of the fine arts are almost invariably works of the imagination (cf. OED, under "art"). Since works of the imagination are frequently about what does not exist, works of the fine arts are frequently about what does not exist. Even where the works are occupied with what does or did exist (or maybe will exist), as in portraiture the (former) entity portrayed is imaginatively conceived, a duplicate is envisioned, and if the work is realised or actualised, what is produced is a duplicate, an object which resembles the original, the model, but also differs in ways not merely due to the medium in which it is produced, but reflecting the way the model is imagined.[1] If the work is not actualised or not completed or not perfectly realised, the intended object is again one that does not exist. The point about nonexistence generalises: much of what count as the arts, much of what is accounted

[1] The objects of portraits have much in common logically with those of legends, considered below.

culture, is concerned with objects of the imagination, with what does not exist. It is no mere accident that a predominantly commercial material civilization like our own is, despite its great monetary wealth, so culturally impoverished.

The fine arts are something of a mixed bag philosophically: they include, according to OED (under "fine"), "those [arts] appealing to sense of beauty, as poetry, music, and especially painting, sculpture, architecture". In principle they include tapestry, weaving, landscape gardening, and indeed *any* of the *crafts* that aim to produce particular and distinctive objects of beauty. Nor is beauty of the essence; the aim may be to produce something of resplendent ugliness or repulsiveness. What is more fundamental is that they are works of skill and imagination. Many of these art and craft forms and activities involve (the production of) works of the imagination which depict objects that, like the objects of fiction, do not exist. A water jug or basket in the shape of a mythical beast or fish depicts an object in much the way that a piece of sculpture or a painting depicts an object (or several objects): an object that does not exist by way of something, some medium, perhaps of a rather different kind, that does exist.[1] The latter is the primary source; secondary sources include details of cultural and ideological setting, historical background, special features of the medium concerned, etc. Museum presentations illustrate both well: for typically both the prime source (the painting, the piece of "primitive" art, etc.) will be on display, and accompanying notes will supply some of the secondary information. Together these supply the material for the source (book) for objects native to the prime source and confined to it. The source books for other imaginary objects depicted differ only in elaboration, e.g. consideration of all the works in case of objects native to several sources, and of material from its native sources for immigrant objects. Such an imaginary object will have, like a fictional object, all the characterising features ascribed to it by its source, and it will be in fact the object with all the features its source attributes it. There will be contexts too, like fictional contexts, where we speak of the objects taking it for granted that they have all the features they are ascribed; in such cases semantical evaluation is ultimately made in the worlds of the objects spoken about. In short, *the general theory transfers largely intact to the objects depicted in imaginative works of art and craft.* The works of the "especially" fine arts that depict objects do, however, have the feature that commonly a direct representation of the objects depicted is obtained, so

[1] As with fiction, an adequate theory of the matter cannot be restricted to primary sources that exist, or have existed, or will exist, though they don't yet.

the objects, which can be *seen*[1] (the main pictorial arts being visually oriented), are often a good deal more determinate than the objects of purely literary fiction which readers must to some extent picture for themselves.

Much music, some poetry and some modern art falls outside this setting in that no imaginary objects are depicted. Poetry, which when it portrays fictional objects belongs strictly with literary fiction, may be nonfictional. Music is often not pictorial, and a modern painting may be entirely formal, simply concerned with the arrangement of shapes. With architecture too, especially modern functional architecture, no objects may be depicted in the final product: with architecture which is symbolic or includes sculpture or paintings it is different.

Works of art generally involve both a production stage and a product: the work is in the making. So far we have been largely concerned with the product and with objects depicted in the product. But the process is also of considerable noneist interest (as it was of Aristotelian interest). For the artist is engaged with designs and models and sketches and images of what does not (yet) exist. The artist envisages, and as necessary changes, a product which does not exist (though materials may exist which go into the final arrangement); and, like a builder with whom she may work, is involved in later stages in the actualisation of the object imagined (cf. 1.25 [Vol. 1 $_{—Eds.}$]).

4. *Types of media and literary fiction.* There are distinctive features of various literary and media types, mainly connected with their sources, that have not been elucidated. The primary sources of the various types differ significantly in form, e.g., comics, like the films in which they may be presented and like nondocumentary film generally, have a visual component consisting of a sequence of illustrations, balloons or photographic stills. Indeed the *form* of primary sources is the main feature distinguishing the types. Differences in content or theme are never essential in distinguishing types, though such differences may yield *sub*classifications, e.g. situation drama, the western, the social novel.[2] Many of the differences in source of the types are obvious, e.g. an opera has as a primary source a score which includes both a libretto and an often elaborate musical score, and it may contain

[1] Compare the usage in which the average Australian TV viewer who has never seen a lyrebird (in the flesh) sees lyrebirds; and see further 8.10 [Vol. 3 $_{—Eds.}$].

[2] Somewhat as a scientific theory may succeed, by considering simplified models with idealised nonentities, in explaining what happens in $\underline{\underline{G}}$, so a social novel may make or register its social or moral points regarding $\underline{\underline{T}}$ by looking at suitably arranged situations involving fictional characters where the points can be brought out better or more sharply.

stage and other production directions. Art forms such as opera, masques, plays, ballet and the like that have, unlike novels (unless they are filmed, televised, etc.), both primary sources and performances; they have a similarity set $\{N_i : i \in I\}$ of intermediate sources for I a set indexing the performances and the primary source $N_0 = N$. For each N_i with $i \neq 0$, N_i answers to N_0 and interprets it. The set $\{N_i : i \in I\}$ extends to a cluster of similar worlds $\{a(N_i) : i \in I\}$ with each N_i $(i \neq 0)$ overlapping $\underline{\underline{T}}$ since the performances occur.[1] Likewise as regards the objects of such art forms there will be similarity sets; for object d of N a set $\{d_i : i \in N\}$ of d together with interpretations by performers, and a cluster $\{S(d_i) : i \in I\}$ of source books for d and its interpretations.[2] Some of the conditions governing similarity and interpretation are not too difficult to formulate. N_i, for $i \neq 0$, must answer to N_0 in including or conforming to it. For example, where N_0 is a musical score and N_i a musical performance of N_0 – to take in yet another art form – then N_i cannot strictly (or at least only to a very limited extent) change the notes or their sequence as shown in N_0. But N_0 will be characteristically incomplete – very incomplete with earlier musical scores, which allow for variations – and leave scope for interpretation within the given scheme by performers. Interpretation and performances are governed not only by N_0 but by what $\underline{\underline{T}}$ permits.

Some literary forms are more heavily constrained by $\underline{\underline{T}}$ than those primarily considered, e.g. legends, sagas, and myths. Legends comprise stories in the tradition of archetypal forms presenting the lives of saints or heroes. They thus have a main character whose life is based on the life of someone who did exist (in Greek "legends" the latter condition may not be satisfied).[3] Let d be such a character and d^* the former entity on which d is based. Then almost invariably $d \neq d^*$, since they differ on characterising features. Then typically d never exists: d is an immigrant, a duplicate of d^*. To say

[1] At least tokens of the type do. Difficulties as to types recede on any theory of objects, nonexistent types representing no problem *per se*.

A work of fiction itself, being characteristically a type determined by certain tokens, is not a particular and does not exist. Thus modern works of fiction contrast with works of art such as paintings or pieces of sculpture or historic buildings or gardens which are particular, though they may have copies.

[2] A rather similar complication of theory may be called for in the case of a work of fiction N which has rival interpretations. Then N may have several elaborations $S_i(N)$ for i in an indexing set J of interpretations. If deletion is permitted (as it may have to be in such cases, e.g. where a literary critic manages to discount certain passages), then an *elaboration* is not simply an extension governed by material and closure conditions. $\{S_i(N) : i \in J\}$ is *like* a set of models for a theory N, and will ideally include an *intended* interpretation $S_0(N)$ (which may not, however, be known).

[3] Epics, which have a given literary form, need not satisfy the condition either. However, the basis relation is not exacting; e.g. as regards heroic facts, it matters not that the historical documents of the time make no mention of dragons.

that d is based on d^* is to say that the source book $S(d)$ for d is a variation on the part of the chronicle of the life of d^*. Legends are special cases of sagas, which (freed from their original Scandinavian setting) may be tied not to an individual d^* but to an historic family or group.

What is distinctive about myths is that they (traditionally) serve or are intended to explain natural phenomena or events or things, i.e. roughly aspects of world $\underline{\underline{G}}$. This is achieved characteristically in traditional stories of narrative type with supernatural characters, who help in accounting for things. Thus where N is a myth, $a(N)$ overlaps $\underline{\underline{G}}$ – at least that is an adherent's or insider's view. The non-adherent's or total outsider's view on myths is more or less summed up in the OED definition of "myth" as "purely fictitious narrative usu. involving supernatural persons and embodying popular ideas on natural phenomena". There are significant differences between insiders' and outsiders' treatment of a given myth,[1] differences reflected in their respective (inferential) behaviour, differences we shall see again as regards scientific theories. For myths provide a bridge between pure fiction and theory: myths embody primitive or rudimentary theories.

5. *Fictional objects versus theoretical objects, and the mistake of fictionalism.* A theoretical object is, crudely, one of the "characters" of a (scientific) theory, of a theory that, in live cases, purports to elucidate empirical reality. Objects of a dead or failed theory behave very like a fictional object, like the objects of myth. Most obviously, theoretical objects belong to some theory while fictional objects are supplied by a work of fiction or the like, not a very hard and fast boundary. To sharpen the contrast, let us restrict theoretical objects to the objects of live theories.

Then theoretical objects are, according to insiders, always deductively closed, fictional objects are sometimes not; theoretical objects stand in entire relations to their own kind and to entities, fictional objects do not. So in particular, with statements concerning theoretical objects, passive conversion is always admissible and frames are commonly extensional and so admit replacement of identicals: many of the contextual intensional aspects of genuine fiction are removed. Insiders operate in these logical ways with the theory and its objects: outsiders are more cautious. The key distinguishing features derive from the intended role of theoretical objects in giving a description and explanation of the

[1] The inside/outside distinction is, like the worlds picture, appreciated in literary criticism. Thus, for example, Tolkien can say of the *Beowulf* poet: "he could view from without, but could feel ... from within the old dogma".

empirico-historical world (i.e. $\underline{\underline{G}}$). So too do other distinctive constraints on theoretical objects: that the theories to which they belong are subject to empirical falsification, etc. In marked contrast, the source books of fictional objects are not so subject. Holmes does not have to answer to historical facts. With respect to Holmes we are all (well, almost all) outsiders.

The basic mistake of fictionalism, the thesis of which is that theories are fictions, is that it ignores these crucial differences.[1]

6. *The incompleteness and "fictionality" of the theory of fictions advanced.* The integrated theory of fictions outlined is like most philosophical theories, rather incomplete. It admits of, and needs, extension; for, as so far outlined, it leaves several important areas open. To take a rather different example from those that will have been evident to the careful reader: though the theory admits of extension, like orthodox semantics it only deals with sentences one at a time, not with whole chunks of discourse at once (the effect of larger chunks can however be obtained, firstly through logical compounding, e.g. by extensional and intensional conjunction, and secondly using context). Of course it is commonly supposed that the theory extends directly from one sentence to several, by the rule $I(A \,.\, B) = I(A \,\&\, B)$, i.e. full stops are evaluated like conjunctions. Davidson's work (e.g. 1968–69) has the accidental benefit of indicating that this assumption – long under suspicion, since for example a book as a whole is not rendered false just because one of its statements is – is mistaken, and not mistaken only where some of the sentences are of non-indicative moods. Consider "Galileo said that . The earth is round". In this case the intended evaluation rule is not I(Galileo said that . The earth is round) = I(Galileo said that and the earth is round), but, what it had better be if Davidson's account is to begin to serve, I(Galileo said that . The earth is round) = I(Galileo said that the earth is round).

The theory may also be alleged to be fictional in the philosophical sense. If to include classes of objects which do not exist and whose nonexistence is irreducible under analysis, is to engage in work of fiction, the theory presented is a work of fiction. For in this enlarged – but eminently questionable and rejectable – sense, almost all the apparatus is fictional: the worlds of fiction are "fictions", logicians' "make-believe" so to say, the

[1] The differences are likewise not observed in Bentham and Vaihinger. A brief discussion of fictionalism is to be found in Harré 1972, p. 80ff.

objects of fictions are commonly fictions, and stand in what may be said to be "fictitious" relations. And like more ordinary fictions sayso determines much of the character of such logicians' fictions, for instance, what has what properties, whether given objects are the same; e.g. we can stipulate (transworld-wise) that a city London with different properties from the actual London of 1977 is the same as the actual one. Once again, neither the purely stipulative nor the purely given accounts of such fictional objects is correct: what is right is a mixture of these accounts. In all these respects, and others, the integrated contextual theory proposed is, in the wide and erroneous philosophical sense of "fictional", a fictional theory of fiction and fictions.

BIBLIOGRAPHY

Works referred to in the text

ABE Anderson, A. R. and N. D. Belnap, Jr., *Entailment*, vol. 1, Princeton University Press, Princeton, N. J., 1975.

BP Routley, R. and V., *Beyond the Possible*, in preparation.

DCL Routley, R. and Meyer, R. K., "Dialectical logic, classical logic, and the consistency of the world", *Studies in Soviet Thought* **16** (1976) 1–25.

DLSM Routley, R., "Dialectical logic, semantics and metamathematics", *Erkenntnis* **14** (1979) 301–331.

DS Routley, R., "Domainless semantics for free, quantification and significance logics", *Logique et analyse* **14** (1971) 603–626.

EI Routley, R., "Existence and identity in quantified modal logics", *Notre Dame Journal of Formal Logic* **10** (1969) 113–149.

ENP1 Routley, R. and V., "Human chauvinism and environmental ethics", in *Environmental Philosophy*, edited by D. Mannison, M. McRobbie and R. Routley, Research School of Social Sciences, Australian National University, Canberra, 1980, 96–189.

ENP2 Routley, V. and R., "Social theories, self management, and environmental problems", in *Environmental Philosophy*, edited by D. Mannison, M. McRobbie and R. Routley, Research School of Social Sciences, Australian National University, Canberra, 1980, 217–332.

EP Meinong, A., *On Emotional Presentation*, trans. M. L. Schubert-Kelsi, Northwestern University Press, 1972.

ER Meyer, R. K. and Routley, R., "Extensional reduction I", *Monist* **60** (1977) 355–369.

Erfgl. Meinong, A., *Über die Erfuhrungsgrundlagen ausseres Wissens*, Springer, Berlin, 1906.

FLP Quine, W. V., *From a Logical Point of View*, Second edition, revised, Harvard University Press, 1961.

GA Meinong, A., *Gesamtausgabe*, edited by R. M. Chisholm, R. Haller and R. Kindinger, Akademische Druck- und Verlagsanstalt, Graz, 1969–1973.

GB Geach, P. and Black, M., eds., *Translations from the Philosophical Writings of Gottlob Frege*, Blackwell, Oxford, 1952.

ILT Strawson, P. F., *Introduction to Logical Theory*, Methuen, London, 1952.

LSL Carnap, R., *The Logical Syntax of Language*, Routledge and Kegan Paul, London, 1949.

ML Quine, W. V., *Mathematical Logic*, Revised, Harvard University Press, Cambridge, Mass., 1951.

MN Carnap, R., *Meaning and Necessity*, Enlarged, University of Chicago Press, Chicago, 1956.

MNG Grossmann, R., *Meinong*, Routledge and Kegan Paul, London, 1974.

Mög Meinong, A., *Über Möglichkeit und Wahrscheinlichkeit*, Leipzig, 1915.

MP Russell, B., *Introduction to Mathematical Philosophy*, Allen and Unwin, London, 1919. Chapter 16 is reprinted in R. Ammerman, ed., *Classics of Analytic Philosophy*, New York, 1965. Page references are to the reprint.

MTD Routley, R., "Meaning as semantical superstructure: a universal theory of meaning, truth and denotation?", *Philosophica* **19** (1977) 33–68.

NE Routley, R., "Non-existence does not exist", *Notre Dame Journal of Formal Logic* **11** (1970) 289–320.

NNL Sylvan, R., "Nihilisms and nihilist logics", *Discussion Papers in Environmental Philosophy* **4** (1983).

OD Russell, B., "On denoting", *Mind* **14** (1905) 479–493. Reprinted in Marsh 1956, pp. 41–56. All page references are to the reprint.

OED *Oxford English Dictionary*, Clarendon Press, Oxford, 1971.

OP Hawkins, D. and Pauling, L., eds., *Orthomolecular Psychiatry: Treatment of Schizophrenia*, W. H. Freeman, San Francisco, 1973.

OR Strawson, P. F., "On referring", *Mind* **59** (1950) 320–344.

OT Russell, B., "On order in time", *Proceedings, Cambridge Philosophical Society* **32** (1936) 216–228. Reprinted in Marsh 1956, pp. 347–363.

P Russell, B., *The Principles of Mathematics*, Second, Allen and Unwin, London, 1948.

PB Putnam, H. and Benacerraf, P., eds., *Philosophy of Mathematics*, Prentice-Hall, Englewood Cliffs, N. J., 1964.

PC *Pears Cyclopedia*, 80th, Pelham Books, U. K., 1971.

PLA Russell, B., "The philosophy of logical atomism", *Monist* **28** (1918) 495–527 and **29** (1919) 33–63, 190–222 and 344–380. Reprinted in Marsh 1956, pp. 177–281. Page references are to the reprint.

PLO Morscher, E., Czermak, J. and Weingartner, P., eds., *Problems in Logic and Ontology*, Akademische Druck- und Verlagsanstalt, Graz, 1977.

PM Whitehead, A. N. and Russell, B., *Principia Mathematica*, Second edition, Cambridge University Press, Cambridge, 1950. (Editions 1 and 2 are distinguished where necessary as PM1 and PM2).

PT O'Donnell, R., Stevens, P. and Lennie, I., eds., *Paper Tigers: an Introduction to the Critique of Social Theory*, Pilot, Department of General Philosophy, University of Sydney, 1978.

RLR Routley, R., Meyer, R. K. et al., *Relevant Logics and Their Rivals*, RSSS, Australian National University, Canberra, 1980.

SE Routley, R., "Some things do not exist", *Notre Dame Journal of Formal Logic* **7** (1966) 251–276.

Slog Goddard, L. and Routley, R., *The Logic of Significance and Context*, Scottish Academic Press, Edinburgh, 1973.

SMM Routley, R., "The semantical metamorphosis of metaphysics", *Australasian Journal of Philosophy* **54** (1976) 187–205.

ST Routley, R., "On a significance theory", *Australasian Journal of Philosophy* **44** (1966) 172–209.

Stell Meinong, A., *Über die Stellung der Gegenstandstheorie im System der Wissenschaften*, R. Voitlander Verlag, Leipzig, 1907.

TO Meinong, A., "The theory of objects", in *Realism and the Background of Phenomenology*, edited by R. M. Chisholm, Free Press, Glencoe, Ill., 1960, 76–117. Translation of "Über Gegenstandstheorie" in *Untersuchungen zur Gegenstandstheorie und Psychologie*, Barth, Leipzig, 1904.

UA Meinong, A., *Über Annahmen*, Second, Barth, Leipzig, 1910.

UL Routley, R., "Ultralogic as universal?", *Relevance Logic Newsletter* **2** (1977), 50–90 and 138–175. [Included as an appendix to the original volume; reprinted here in Vol. 4, part I. —*Eds.*].

US Routley, R., "Universal semantics?", *Journal of Philosophical Logic* **4** (1975) 327–356.

WN Routley, R., "What numbers are", *Logique et analyse* **8** (1965) 196–208.

WO Quine, W. V., *Word and Object*, MIT Press, Cambridge, Mass., 1960.

Ackermann, R. J., *Belief and Knowledge*, Macmillan, London, 1972.

Ajdukiewicz, K., "Die syntaktische Konnexität", *Studia Philosophica* **2** (1935) 1–27. Translated as "Syntactic connexion" in S. McCall, ed., *Polish Logic*, Oxford, 1967, pp. 207–231. Page references are to the translation.

Anderson, J., *Studies in Empirical Philosophy*, Angus and Robertson, Sydney, 1962.

Andrews, P., *A Transfinite Type Theory with Type Variables*, North-Holland, Amsterdam, 1965.

Anscombe, G. E. M., *An Introduction to Wittgenstein's Tractatus*, Hutchinson, London, 1958.

Anson's, *Law of Contract*, edited by A. G. Guest, 24th, Clarendon, Oxford, 1975.

Armstrong, D. M., *Perception and the Physical World*, Routledge and Kegan Paul, London, 1961.

Armstrong, D. M., *A Materialist Theory of the Mind*, Routledge and Kegan Paul, London, 1968.

Armstrong, D. M., *Belief, Truth and Knowledge*, Cambridge University Press, Cambridge, 1973.

Armstrong, D. M., "Naturalism, materialism, and first philosophy", in *Ist Systematische Philosophie Möglich?*, (Stuttgarter Hegel-Kongress 1975), edited by D. Henrich, Bonn, Grundmann, 1977, 411–425.

Armstrong, D. M., *Nominalism and Realism: Universals and Scientific Realism, volume I*, Cambridge University Press, Cambridge, 1978.

Armstrong, D. M., *A Theory of Universals: Universals and Scientific Realism, volume II*, Cambridge University Press, Cambridge, 1978.

Arruda, A. I., "A survey of paraconsistent logic", in *Proceedings of the Fourth Latin-American Logic Conference*, Amsterdam, North-Holland, 1979.

Asser, G., "Theorie der logischer Auswahlfunktionen", *Zeitschrift für mathematische Logik und Grundlagen der Mathematik* **3** (1957) 30–68.

Austin, J. L., *Sense and Sensibilia*, Clarendon Press, Oxford, 1962.

Ayer, A. J., *Language, Truth and Logic*, Gollancz, London, 1946.

Ayer, A. J., *The Problem of Knowledge*, Macmillan, London, 1956.

Baldwin, J. M., ed., *Dictionary of Philosophy and Psychology*, Macmillan, New York, 1901–1905.

Barber, K. F., *Meinong's Hume Studies: Translation and Commentary*, University Microfilms, Ann Arbor, 1966. Ph.D., University of Iowa, 1966.

Barcan, R. C., "A functional calculus of first order based on strict implication", *Journal of Symbolic Logic* **11** (1946) 1–16.

Barcan-Marcus, R., "Extensionality", *Mind* **69** (1960) 55–62.

Barker, J. and Grigg, G., *A Field Guide to Australian Frogs*, Rigby, Melbourne, 1977.

Belnap, N. D. and Dunn, J., "Entailment and the Disjunctive Syllogism", **1** (1981), edited by G. Floistad and G. H. von Wright 337–366.

Bentham, J., see C. K. Ogden, *Bentham's Theory of Fictions*, Kegan Paul, Trench, Trubner, London, 1932.

Bergmann, G., *Meaning and Existence*, University of Wisconsin Press, Madison, 1959.

Bergmann, G., *Logic and Reality*, University of Wisconsin Press, Madison, 1964.

Bergmann, G., *Realism. A critique of Brentano and Meinong*, University of Wisconsin Press, Madison, 1967.

Berkeley, G., "The Principles of Human Knowledge", in *The Works of George Berkeley, Bishop of Cloyne*, edited by A. A. Luce and T. E. Jessop, vol. 2, Thomas Nelson, London, 1949.

Bernays, P., "On platonism in mathematics", in PB, 1964, 274–286.

Black, M., *Language and Philosophy*, Cornell University Press, Ithaca, 1949.

Blackburn, S., ed., *Meaning, Reference and Necessity: New Studies in Semantics*, Cambridge University Press, Cambridge, 1975.

Blake, W., *A Descriptive catalogue*, 1809.

Blanshard, B., *The Nature of Thought*, Allen and Unwin, London, 1939.

Blocker, H. G., "The truth about fictional entities", *Philosophical Quarterly* **24** (1974) 27–36.

Bowring, J., ed., *The Works of Jeremy Bentham*, 11 Volumes, Edinburgh, 1838-43.

Bradford, D. E., "A bibliography on the topic of existence", *Philosophy Research Archives* **2**.1137 (1976) 20–58.

Bradley, F. H., *Appearance and Reality*, Clarendon Press, Oxford, 1930.

Brady, R. and Routley, R., "Don't care was made to care", *Australasian Journal of Philosophy* **51** (1973) 211–225.

Brady, R. and Routley, R., "The non-triviality of extensional dialectical set theory", in Priest, Routley and Norman, 1989, 415–436.

Bressan, A., "On the semantics for the language ML", and discussion, in PLO.

Bressan, A., "Metodo di assiomatizzazione in senso stretto della meccanica classica. Applicazione di esso ad alcuni problemi di assiomatizzazione non ancora completamente risolti", *Rendiconti di Seminario matematico, Università di Padova* **32** (1962) 55–212.

Bressan, A., *A General Interpreted Modal Calculus*, Yale University Press, New Haven, 1972.

Bressan, A., *Relativistic Theories of Materials*, Springer, Berlin, 1978.

Bressan, A., "On Physical Possibility", in *Italian Studies in the Philosophy of Science*, edited by M. L. Dalla Chiara, Springer, Dordrecht, 1980, 197–214.

Broad, C. D., *Religion, Philosophy and Psychical Research*, Routledge and Kegan Paul, London, 1953.

Brouwer, L. E. J., *Collected Works*, North-Holland, Amsterdam, 1975–76.

Bunge, M., *Treatise on Basic Philosophy,* Volume 3, *Ontology I. The Furniture of the World*, Reidel, Dordrecht, 1977.

Bunting, I. A., "Perception and the existence criterion", *Philosophical Studies (Ireland)* **20** (1972) 77–89.

Cahn, S., "Fatalistic arguments", *Journal of Philosophy* **61** (1964) 295–304.

Capra, F., *The Tao of Physics*, Wildwood House, London, 1975.

Cartwright, R. L., "Negative existentials", in *Philosophy and Ordinary Language*, edited by C. E. Caton, University of Illinois Press, Urbana, 1963, 55–66.

Castañeda, H. N., "Thinking and the structure of the world", *Philosophia* **4** (1974) 3–40.

Castañeda, H. N., "Identity and sameness", *Philosophia* **5** (1975) 121–150.

Castañeda, H. N., "Rejoinder to Professor Michael Loux", *Critica* **8** (1976) 109–116.

Castañeda, H. N., "Fiction and reality: their relations and connections", *Poetics* **8** (1979) 31–62.

Caton, C., "Overview", in *Semantics*, edited by D. D. Steinberg and L. A. Jakobovits, Cambridge University Press, Cambridge, 1971.

Chalmers, A., *What Is This Thing Called Science?*, University of Queensland Press, St Lucia, 1976.

Chalmers, A. and Suchting, W., "Empiricism", in PT, 29–49.

Chihara, C. S., "Truth, meaning and paradox", *Nous* **10** (1976) 305–311.

Chisholm, R. M., "Problems of identity", in Munitz 1971, 3–30.

Chisholm, R. M., *Realism and the Background of Phenomenology*, Free Press, Glencoe, Ill., 1960.

Chisholm, R. M., "Meinong, Alexius", in *The Encyclopedia of Philosophy*, edited by P. Edwards, vol. 5, Macmillan, London, 1967, 261–263.

Chisholm, R. M., "Review of Kindinger 1965", *Philosophical Review* **77** (1968) 372–375.

Chisholm, R. M., "Beyond being and nonbeing", *Philosophical Studies* **24** (1972) 245–257. Also in Haller 1972.

Chisholm, R. M., "Homeless objects", *Revue Internationale de Philosophie* **27** (1973) 207–223.

Chisholm, R. M., *Person and Object*, Allen and Unwin, London, 1976.

Church, A., "A formulation of the simple theory of types", *Journal of Symbolic Logic* **5** (1940) 56–68.

Church, A., "On Carnap's analysis of statements of assertion and belief", *Analysis* **10** (1950) 97–99.

Church, A., *Introduction to Mathematical Logic*, Princeton University Press, Princeton, N. J., 1956.

Cresswell, M. J., *Logics and Languages*, Methuen, London, 1973.

Cresswell, M. J., "Critical study of Montague 1974", *Philosophia* **6** (1976) 193–207.

Crittenden, C., "Ontology and the theory of descriptions", *Philosophy and Phenomenological Research* **31** (1970–71) 85–96.

Crittenden, C., "Thinking about non-being", *Inquiry* **16** (1973) 290–312.

Curry, H. B., *Foundations of Mathematical Logic*, McGraw-Hill, New York, 1963.

Davidson, D., "Truth and meaning", *Synthese* **17** (1967) 304–323.

Davidson, D., "On saying that", *Synthese* **19** (1968–69) 130–146. Also in *Words and Objections*, edited by D. Davidson and J. Hintikka, Reidel, Dordrecht, 1969, pp. 158–174.

Davidson, D., "Thought and talk", in *Mind and Language*, edited by S. Guttenplan, Clarendon Press, Oxford, 1975, 7–23.

Davis, J. W., Hockney, D. J. and Wilson, W. K., eds., *Philosophical Logic*, Reidel, Dordrecht, 1969.

Dedekind, R., *Was Sind und Was Sollen die Zahlen?*, Sixth edition, Vieweg, Braunschweig, 1930.

Devine, P. E., "The logic of fiction", *Philosophical Studies* **26** (1974) 389–399.

Donnell, F. H., "A note about presupposition", *Mind* **81** (1972) 124–125.

Donnellan, K., "Proper names and identifying descriptions", in *Semantics of Natural Language*, edited by D. Davidson and G. Harman, Reidel, Dordrecht, 1972. Originally published in *Synthese* **20** (1970).

Donnellan, K., "Speaking of nothing", *Philosophical Review* **83** (1974) 3–31.

Dretske, F. I., *Seeing and Knowing*, Routledge and Kegan Paul, London, 1969.

Dummett, M., *Frege: Philosophy of Language*, Duckworth, London, 1973.

Dyche, R. E., "Meinong's Theory of Objects: a Reinterpretation and Defense", PhD thesis, Rutgers University, The State University of New Jersey, New Brunswick, 1976.

Edwards, P., "Bertrand Russell's doubts about induction", in *Logic and Language*, edited by A. Flew, First series, Blackwell, Oxford, 1960, 55–79.

Feyerabend, P., *Against Method*, New Left Books, London, 1975.

Feys, R., *Modal Logics*, edited by J. Dopp, Nauwelaerts, Louvain, 1965.

Findlay, J. N., "Use, usage and meaning", *Aristotelian Society Supplementary Volume* **35** (1961) 231–242.

Findlay, J. N., *Meinong's Theory of Objects and Values*, Second Edition, Clarendon Press, Oxford, 1963.

Findlay, J. N., "Meinong the phenomenologist", *Revue Internationale de Philosophie* **57** (1973) 161–177.

Flew, A., *Introduction to Western Philosophy*, Thames and Hudson, London, 1971.

Forster, E. M., *Aspects of the Novel*, E. Arnold, London, 1949.

van Fraassen, B., "Singular terms, truth-value gaps, and free logic", *Journal of Philosophy* **63** (1966) 481–495.

van Fraassen, B., *An Introduction to the Philosophy of Time and Space*, Random House, New York, 1970.

van Fraassen, B., *Formal Semantics and Logic*, Macmillan, New York, 1971.

Fraenkel, A. A., Bar-Hillel, Y. and Levy, A., *Foundations of Set Theory*, North-Holland, Amsterdam, 1973.

Freeman, K., *Ancilla to the Pre-Socratic Philosophers*, Blackwell, Oxford, 1947.

Frege, G., "Über Sinn und Bedeutung", in *Zeitschrift für Philosophie und philosophische Kritik*, 1892. Translated as "On sense and reference" in GB.

Frege, G., *The Foundations of Arithmetic*, translated by J. L. Austin, Blackwell, Oxford, 1950.

Gale, R. M., ed., *The Philosophy of Time*, Doubleday, New York, 1967.

Geach, P. T., *Reference and Generality*, Cornell University Press, Ithaca, 1962.

Geach, P. T., *Logic Matters*, Blackwell, Oxford, 1972.

Goddard, L. and Routley, R., "Use, mention and quotation", *Australasian Journal of Philosophy* **44** (1966) 1–49.

Gödel, K., "Über formal unentscheidbare Sätze des Principia Mathematica und verwandter Systeme I", *Monatshefte für Mathematik and Physik* **38** (1931) 173–198.

Godfrey-Smith, W., "Time and Existence", Doctoral Dissertation, King's College, Cambridge, 1976.

Godfrey-Smith, W., "Beginning to exist and ceasing to exist", *Philosophical Studies* **32** (1977) 393–402.

Godfrey-Smith, W., "The generality of predictions", *American Philosophical Quarterly* **15** (1978) 15–25.

Goodman, N., "A world of individuals", in *The Problem of Universals*, Notre Dame University Press, 1956. Reprinted in N. Goodman, *Problems and Projects*, Bobbs-Merrill, Indianapolis, 1972.

Goodman, N., *The Structure of Appearance*, Third edition, Reidel, Dordrecht, 1977.

Goodman, N. and Leonard, H. S., "The calculus of individuals and its uses", *Journal of Symbolic Logic* **5** (1940) 45–55.

Goodman, P., *Community of Scholars and Compulsory Miseducation*, Random House, New York, 1966.

Gram, M. S., "Ontology and the theory of descriptions", in Klemke 1970, 1970, 118–143.

Grave, S. A., *The Scottish Philosophy of Common Sense*, Clarendon Press, Oxford, 1960.

Grave, S. A., "Common Sense", in *Encyclopedia of Philosophy*, edited by P. Edwards, vol. 2, Macmillan, New York, 1967, 155–160.

Gregory, R. L., *Eye and Brain: the Psychology of Seeing*, Weidenfeld and Nicolson, London, 1966.

Grene, M., *Introduction to Existentialism*, University of Chicago Press, Chicago, 1962.

Grice, H. P., "Utterer's meaning, sentence meaning, and word meaning", *Foundations of Language* **4** (1968) 225–242.

Griffin, N., *Relative Identity*, Oxford University Press, London, 1977.

Griffin, N., "Russell's 'horrible travesty' of Meinong", *Russell: the Journal of the Bertrand Russell Archives* **25–28** (1978).

Griffin, N., "The independence of *Sosein* from *Sein*", *Grazer Philosophische Studien* **9** (1979).

Griffin, N. and Harton, M., "Sceptical Arguments", *Philosophical Quarterly* **31** (1981) 17–30.

Grossmann, R., "Meinong's doctrine of the *Aussersein* of the pure object", *Noûs* **8** (1974) 67–82.

Grünbaum, A., *Philosophical Problems of Space and Time*, Routledge and Kegan Paul, London, 1964.

Haack, S., *Deviant Logic: Some Philosophical Issues*, Cambridge University Press, London, 1974.

Hailperin, T., "Quantification theory and empty individual-domains", *Journal of Symbolic Logic* **18** (1953) 197–200.

Haller, R., ed., *Jenseits von Sein and Nichtsein*, Akademische Druck- und Verlagsanstalt, Graz, 1972.

Hansson, B. and Gärdenfors, P., "A guide to intensional semantics", in *Modality, Morality and Other Problems of Sense and Nonsense*, Gleerup, Lund, 1973, 151–167.

Harré, R., *The Philosophies of Science*, Oxford University Press, London, 1972.

Hempel, C. G., "Problems and changes in the empiricist criterion of meaning", *Revue Internationale de Philosophie* **4** (1950) 41–63.

Hempel, C. G., *Philosophy of Natural Science*, Prentice-Hall, Englewood Cliffs, New Jersey, 1966.

Henkin, L., "The completeness of the first-order functional calculus", *Journal of Symbolic Logic* **14** (1949) 159–166.

Henkin, L., "Completeness in the theory of types", *Journal of Symbolic Logic* **15** (1950) 81–91. Reprinted in *The Philosophy of Mathematics*, edited by J. Hintikka, Oxford University Press, London, pp. 51–63.

Hilbert, D. and Bernays, P., *Grundlagen der Mathematik*, Springer, Berlin, 1934–39.

Hinckfuss, I., "Necessary existential dependence", *Australasian Journal of Philosophy* **54** (1976) 123–132.

Hintikka, J., "Towards a theory of definite descriptions", *Analysis* **19** (1958–59) 79–85.

Hintikka, J., "Existential presuppositions and existential commitments", *Journal of Philosophy* **56** (1959) 125–137.

Hinton, J. M., "Quantification, Meinongianism, and the ontological argument", *Philosophical Quarterly* **22** (1972) 97–109.

Hirst, R. J., "Realism", in *Encyclopedia of Philosophy*, edited by P. Edwards, vol. 7, Macmillan, New York, 1967, 77–83.

Hobbes, T., *Leviathan*, Clarendon Press, Oxford, 1909.

Hollis, M. and Nell, E. J., *Rational Economic Man*, Cambridge University Press, London, 1975.

Honoré, A. M., "Reference to the non-existent", *Philosophy* **46** (1971) 302–308.

Hospers, J., *An Introduction to Philosophical Analysis*, Routledge and Kegan Paul, London, 1956.

Howell, R., "Fictional objects: how they are and how they aren't", *Poetics* **8** (1979) 129–177.

Hull, D. L. and Snyder, D. P., "Contemporary logic and evolutionary taxonomy: a reply to Gregg", *Systematic Zoology* **18** (1969) 347–354.

Hume, D., *A Treatise of Human Nature* and *On the Academical or Sceptical Philosophy*, in *Hume's Philosophical Works*, edited by T. H. Green and T. H. Grose, 4 volumes, London, 1874–75.

Huxley, A., *The Doors of Perception and Heaven and Hell*, Penguin, 1959.

Jeffrey, R., *The Logic of Decision*, McGraw-Hill, New York, 1965.

Johnson, W. E., *Logic*, Cambridge University Press, Cambridge, 1921–1924.

Kalish, D. and Montague, R., *Logic: Techniques of Formal Reasoning*, Harcourt, Brace and World, New York, 1964.

Kant, I., *Critique of Pure Reason*, Unabridged edition, translated by N. Kemp Smith, Macmillan, London, 1929. Abridged edition, Macmillan, London, 1934.

Kaplan, B., *The Inner World of Mental Illness*, Harper and Row, New York, 1964.

Kaplan, D., "Quantifying in", *Synthese* **19** (1968–69) 178–214. Reprinted in *Words and Objections*, edited by D. Davidson and J. Hintikka, Reidel, Dordrecht, 1969, pp. 206–242.

Kemeney, J. G., "Models of logical systems", *Journal of Symbolic Logic* **13** (1948) 16–30.

Kemeney, J. G., "A new approach to semantics – parts I and II", *Journal of Symbolic Logic* **21** (1956) 1–27, 149–161.

Kenny, A., *Descartes*, Random House, New York, 1968.

Kimball, J. P., "Predicates definable over transformational derivations by intersection with regular languages", *Information and Control* **11** (1967) 177–195.

Kimball, J. P., *The Formal Theory of Grammar*, Prentice-Hall, Englewood Cliffs, N. J., 1973.

Kindinger, R., ed., *Philosophenbriefe aus der Wiesenschaftlichen Korrespondenz von Alexius Meinong*, Akademische Druck- und Verlagsanstslt, Graz, 1965.

Kiteley, M., "Is existence a predicate?", *Mind* **73** (1964) 364–373.

Kleene, S. C., *Introduction to Metamathematics*, North-Holland, Amsterdam, 1952.

Klemke, E. D., ed., *Essays on Bertrand Russell*, University of Illinois Press, Urbana, 1970.

Kneale, W., "Is Existence a Predicate?", *J. Symbolic Logic* **2** (1937) 60–61.

Kneale, W., *Probability and Induction*, Clarendon Press, Oxford, 1949.

Kneale, W. and M., *The Development of Logic*, Clarendon Press, Oxford, 1962.

Kneebone, G. T., *Mathematical Logic and the Foundations of Mathematics*, Van Nostrand, Princeton, 1963.

Koestler, A. and Smythies, J. R., eds., *Beyond Reductionism: New Perspectives in the Life Sciences*, Hutchinson, London, 1969.

Kripke, S., "Semantical considerations on modal logic", *Acta Philosophica Fennica* **16** (1963) 83–94. (Proceedings of a Colloquium on Modal and Many-Valued Logics).

Kripke, S., "Identity and necessity", in Munitz 1971, 1971, 135–164.

Kripke, S., "Naming and necessity", in *Semantics of Natural Languages*, edited by D. Davidson and G. Harman, Reidel, Dordrecht, 1972.

Kripke, S., *Empty Reference*, Shearman Lectures, University of London, 1973. Notes taken by I. Hacking, J. Heale and T. Baldwin.

Kuhn, T. S., *The Structure of Scientific Revolutions*, Second enlarged edition, University of Chicago Press, Chicago, 1970.

Lakatos, I., "Proofs and refutations", *British Journal for the Philosophy of Science* **14** (1963–64), 1–25, 120–139, 221–245 and 297–342. Reprinted as *Proofs and Refutations, the Logic of Mathematical Discovery*, Cambridge University Press, London, 1966.

Lakatos, I., "A renaissance of empiricism in the recent philosophy of mathematics", in *Mathematics, Science and Epistemology, Philosophical Papers of Imré Lakatos*, edited by J. Worrell and G. Curry, vol. 2, Cambridge University Press, Cambridge, 1978.

Lambert, K., "Existential Import Revisited", *Notre Dame Journal of Formal Logic* **4** (1963) 288–292.

Lambert, K., "Being and being so", in Haller 1972, 1972, 37–46.

Lambert, K., "The theory of objects: a review discussion of J. N. Findlay's *Meinong's Theory of Objects and Values*", *Inquiry* **16** (1973) 221–224.

Lambert, K., "Impossible objects", *Inquiry* **17** (1974) 303–314.

Lambert, K., "On 'The durability of impossible objects'", *Inquiry* **19** (1976) 251–253.

Lambert, K. and van Fraassen, B., "On Free Description Theory", *Zeitschrift fur Mathem. Logik und Grundlagen der Mathematik* **13** (1967) 225–240.

Lambert, K. and van Fraassen, B., *Derivation and Counterexample: an Introduction to Philosophical Logic*, Dickenson, Encino, Calif., 1972.

Leblanc, H., *Truth-Value Semantics*, North-Holland, Amsterdam, 1976.

Leblanc, H. and Hailperin, T., "Non-designating singular terms", *Philosophical Review* **68** (1959) 239–243.

Leblanc, H. and Thomason, R., "Completeness theorems for some presupposition-free logics", *Fundamenta Mathematica* **62** (1968) 123–164.

Lejewski, C., "Logic and existence", *British Journal for the Philosophy of Science* **5** (1954–55) 104–119.

Lenoci, M., "Bibliografia degli studi su Alexius Meinong", *Rivista di Filosofia Neo-scolastica* **62** (1970) 437–473.

Leonard, H., "The logic of existence", *Philosophical Studies* **7** (1956) 49–64.

Lewis, C. I., "Facts, systems, and the unity of the world", *Journal of Philosophy* **20** (1923) 141–151.

Lewis, D., "General semantics", *Synthese* **22** (1970–71) 18–67. Reprinted in *Semantics of Natural Language*, edited by D. Davidson and G. Harman, Reidel, Dordrecht, 1972, pp. 169–218.

Lewis, D., *Counterfactuals*, Blackwell, Oxford, 1973.

Lewy, C., *Meaning and Modality*, Cambridge University Press, Cambridge, 1976.

Linsky, L., *Referring*, Routledge and Kegan Paul, London, 1967.

Linsky, L., ed., *Reference and Modality*, Oxford University Press, London, 1971.

Linsky, L., *Names and Descriptions*, University of Chicago Press, Chicago, 1977.

Locke, J., *An Essay Concerning Human Understanding*, edited by P. Nidditch, Clarendon Press, Oxford, 1975.

Loparić, A., "Une étude sémantique de quelques calculs propositionnels", *Comptes Rendus de l'Academie des Sciences de Paris* **284 A** (1977) 835–838.

Lukes, S., *Individualism*, Blackwell, Oxford, 1973.

Lycan, W., "The trouble with possible worlds", in *The Possible and the Actual*, edited by M. Loux, Cornell University Press, Ithaca, 1979, 274–316.

Malcolm, N., "Anselm's ontological arguments", *Philosophical Review* **69** (1960) 41–62.

Marcuse, H., *One-Dimensional Man: Studies in the Ideology of Advanced Industrial Society*, Routledge and Kegan Paul, London, 1964.

Margolis, J., *Knowledge and Existence: an Introduction to Philosophical Problems*, Oxford University Press, New York, 1973.

Marsh, R., ed., *Logic and Knowledge*, Allen and Unwin, London, 1956.

Martin, B., *The Bias of Science*, Society for Social Responsibility in Science, Canberra, 1979.

Martin, R. M., "A homogenous system for formal logic", *Journal of Symbolic Logic* **3** (1943) 1–23.

Massola, A., *Bunjil's Cave: Myths, Legends and Superstitions of the Aborigines of South-East Australia*, Lansdowne Press, Melbourne, 1968.

Matilal, B. K., *Epistemology, Logic and Grammar in Indian Philosophical Analysis*, Mouton, Paris, 1971.

McKinsey, J. C. C., "A solution of the decision problem for the Lewis systems S2 and S4, with an application to topology", *Journal of Symbolic Logic* **6** (1941) 117–134.

Meinong, A., ed., *Untersuchungen zur Gegenstandstheorie und Psychologie*, Barth, Leipzig, 1904.

Mendelson, E., *Introduction to Mathematical Logic*, Van Nostrand, Princeton, 1964.

Miliband, R., *The State in Capitalist Society*, Quartet Books, London, 1973.

Mill, J. S., *A System of Logic*, Longmans, Green, London, 1947.

Mish'alani, J., "Thought and object", *Philosophical Review* **71** (1962) 185–201.

Monro, D. H., "Bentham, Jeremy", in *The Encyclopedia of Philosophy*, edited by P. Edwards, vol. 1, Macmillan, New York, 1967, 280–285.

Montague, R., "Universal Grammar", *Theoria* **36** (1970a) 373–98.

Montague, R., "English as a formal language", in *Linguaggi nella società e nella tecnica*, edited by Bruno et al., Edizioni di Comunità, Milan, 1970b, 189–224.

Montague, R., *Formal Philosophy*, edited by R. H. Thomason, Yale University Press, New Haven, 1974.

Moore, G. E., "Is existence a predicate?", in *his* Philosophical Papers.

Moore, G. E., *Principia Ethica*, Cambridge University Press, Cambridge, 1903.

Moore, G. E., *Some Main Problems of Philosophy*, Allen and Unwin, London, 1953.

Moore, G. E., *Philosophical Papers*, Allen and Unwin, London, 1959.

Mortensen, C., "A theorem on verisimilitude", *Bulletin of the Section of Logic, Polish Academy of Sciences* **7** (1978) 34–43.

Munitz, M. K., *The Mystery of Existence*, Appleton-Century-Crofts, New York, 1965.

Munitz, M. K., ed., *Identity and Individuation*, New York University Press, New York, 1971.

Munitz, M. K., *Existence and Logic*, New York University Press, New York, 1974.

Naess, A., "An Empirical Study of the Expressions 'true', 'perfectly certain', and 'extremely probable'", *Avhandlinger utgitt av Det Norske Videnskaps-Akademi i Oslo. II. Hist.-Filos. Klasse* 4 (1953).

Nagel, E., *The Structure of Science*, Routledge and Kegan Paul, London, 1961.

Nakhnikian, G. and Salmon, W., "'Exists' as a predicate", *Philosophical Review* **66** (1957) 535–542.

Nerlich, G. C., "Presupposition and entailment", *American Philosophical Quarterly* **2** (1965) 33–42.

Niven, B. S., "The formal definition of the environment of an animal", *Australian Journal of Zoology* **5** (1980).

Noll, W., "Foundations of continuum mechanics and thermodynamics", *Ardi Bol. Mech. Anal.* **52**.62 (1973).

Ogden, C. K., *Bentham's Theory of Fictions*, Kegan Paul, Trench, Trubner, London, 1932.

Pap, A., *Semantics and Necessary Truth*, Yale University Press, New Haven, 1958.

Pap, A., *An Introduction to the Philosophy of Science*, Free Press, New York, 1962.

Parsons, T., "A prolegomenon to Meinongian semantics", *Journal of Philosophy* **71** (1974) 561–580.

Parsons, T., "A Meinongian analysis of fictional objects", *Grazer Philosophische Studien* **1** (1975) 73–86.

Parsons, T., "Nuclear and extranuclear properties, Meinong, and Leibniz", *Nous* **12** (1978) 137–151.

Parsons, T., *Nonexistent Objects*, typescript 1978, Yale University Press, forthcoming.

Passmore, J. A., *A Hundred Years of Philosophy*, Duckworth, London, 1957.

Passmore, J. A., *Philosophical Reasoning*, Second, Duckworth, London, 1970.

Peacocke, C. A. B., "Proper names, reference and rigid designation", in Blackburn 1975, 1975, 109–132.

Peters Jr., P. S. and Ritchie, R. W., "On the generative power of transformational grammars", *Information Sciences* **6** (1973) 49–83.

Plutarch, "Adversus Colotem", in *volume 14 of* Plutarch's Moralia, Heinemann, London, 1967, 190–315.

Pollock, J. L., "On logicism", in Klemke 1970, 388–395.

Popper, K., *The Logic of Scientific Discovery*, Hutchinson, London, 1959.

Popper, K., *Conjectures and Refutations*, Routledge and Kegan Paul, London, 1969.

Popper, K., *Objective Knowledge*, Oxford University Press, Oxford, 1972.

Priest, G., *The truth about mathematical objects*, typescript, 1978.

Priest, G., *The logical paradoxes: a unified account*, typescript, 1978a.

Priest, G., "The logic of paradoxes", *Journal of Philosophical Logic* **8** (1979) 219–241.

Priest, G., "Classical logic: a degenerate research programme in crisis", Paper presented at the 1979 Australasian Association of Philosophy Conference, Melbourne, 1979a.

Priest, G., Routley, R. and Norman, J., eds., *Paraconsistent Logic; Essays on the Inconsistent*, Philosophia Verlag, Munich, 1989.

Prior, A. N., *Time and Modality*, Clarendon Press, Oxford, 1957.

Prior, A. N., "A family of paradoxes", *Notre Dame Journal of Formal Logic* **2** (1961) 16–32.

Prior, A. N., *Formal Logic*, Second, Clarendon Press, Oxford, 1962.

Prior, A. N., "Is the concept of referential opacity really necessary?", *Acta Philosophica Fennica* **16** (1963), (Proceedings of a Colloquium on Modal and Many-Valued Logics).

Prior, A. N., *Past, Present and Future*, Clarendon Press, Oxford, 1967.

Putnam, H., "The thesis that mathematics is logic", in Schoenman 1967, 273–303.

Putnam, H., "Time and physical geometry", *Journal of Philosophy* **64** (1967b) 240–247.

Quine, W. V., "The problem of interpreting modal logic", *Journal of Symbolic Logic* **12** (1947) 43–48.

Quine, W. V., "Three grades of modal involvement", in *Proceedings of the XIth International Congress of Philosophy*, vol. 14, Brussels, 1953, 65–81.

Quine, W. V., *Methods of Logic*, Revised, Holt, Rinehart and Winston, New York, 1959.

Quine, W. V., *Philosophy of Logic*, Prentice-Hall, Englewood Cliffs, N. J., 1970.

Quine, W. V., "On the individuation of attributes", in *The Logical Enterprise*, edited by A. R. Anderson, R. Marcus and R. M. Martin, Yale University Press, New Haven, 1975, 3–12.

Quine, W. V. and Goodman, N., "Steps toward a constructive nominalism", *Journal of Symbolic Logic* **12** (1947) 105–122.

Ramsey, P. F., *The Foundations of Mathematics and other Logical Essays*, Kegan Paul, London, 1931.

Rapaport, W. J., "Meinongian theories and a Russellian paradox", *Nous* **12** (1978) 153–180.

Rashdall, H., *The Theory of Good and Evil,* Second edition, Oxford University Press, Oxford, 1948.

Redmon, R. B., "Exists", *Mind* **82** (1973) 56–72.

Reichenbach, H., *The Philosophy of Space and Time,* Dover, New York, 1958.

Reid, T., *The works of Thomas Reid,* with notes and supplementary dissertations by Sir William Hamilton, Eighth edition, vol. I, J. Thin, Edinburgh, 1895.

Rennie, M. K., *Some Uses of Type Theory in the Analysis of Language,* Dept. of Philosophy, Research School of Social Sciences, Australian National University, 1974.

Rescher, N., "Truth and necessity in temporal perspective", in Gale 1967, 183–220.

Rescher, N., "On the logic of existence and denotation", *Philosophical Review* **68** (1959) 157–180.

Rescher, N., *Topics in Philosophical Logic,* Reidel, Dordrecht, 1968.

Rietdijk, C. W., "A rigorous proof of determinism derived from the special theory of relativity", *Philosophy of Science* **33** (1966) 341–344.

Robinson, A., *Non-Standard Analysis,* North-Holland, Amsterdam, 1970.

Rorty, A., "Realism and reference", *Monist* **59** (1976) 321–340.

Rosser, J. B., *Logic for Mathematicians,* McGraw-Hill, New York, 1946.

Roszak, T., *Where the Wasteland Ends: Politics and Transcendence in Post-Industrial Society,* Anchor, New York, 1973.

Routley, R., "What numbers are", *Logique et Analyse* **8** (1965) 196–208.

Routley, R., "Exploring Meinong's jungle I. Items and descriptions", typescript, Sylvan Papers, Fryer Library, University of Queensland, 1967.

Routley, R., "A simple, natural deduction system", *Logique et Analyse* **12** (1969) 129–152.

Routley, R., "Existence and identity in quantified modal logic", *Notre Dame Journal of Formal Logic* **10** (1969a) 113–149, (cited as EI).

Routley, R., "Domainless semantics for free, quantification and significance of logics", *Logique et Analyse* **14** (1971) 603–626.

Routley, R., "Another 'fatal' objection to Meinongian objects", *Philosophical Studies* **25** (1974) 131–135.

Routley, R., "A rival account of logical consequence", *Reports on Mathematical Logic* **3** (1974a) 41–51.

Routley, R., "The durability of impossible objects: a reply to Lambert", *Inquiry* **19** (1976) 247–257.

Routley, R., "Problems and solutions in the semantics of quantified relevant logics I", in *Proceedings of the 4th Latin-American Symposium of Mathematical Logic*, edited by R. Chauquí et al., Amsterdam, Nort-Holland, 1979a, 305–340.

Routley, R., "Constant domain semantics for quantified non-normal modal logics and for certain quantified quasi-entailment logics", *Reports on Mathematical Logic* **10** (1979b) 103–121.

Routley, R., "The theory of objects as common-sense", *Grazer Philosophische Studien* **9** (1979c) 1–16.

Routley, R., Review of R. Grossmann's *Meinong*, (forthcoming).

Routley, R., "The choice of logical foundations: nonclassical choices and the ultralogical choice", *Studia Logica* **39**.1 (1980) 77–98.

Routley, R. and Loparić, A., "Semantical analyses of Arruda-da Costa P systems and adjacent non-replacement systems", *Studia Logica* **37** (1978) 301–320.

Routley, R. and Macrae, V., "On the identity of sensations and physiological occurrences", *American Philosophical Quarterly* **3** (1966) 87–110.

Routley, R. and Meyer, R. K., "Every sentential logic has a two-valued worlds semantics", *Logique et Analyse* **19** (1976) 174–194.

Routley, R., Meyer, R. K. and Goddard, L., "Choice and descriptions in enriched intensional languages – I", *Journal of Philosophical Logic* **3** (1974) 291–316. This paper, together with parts II and III and a Postscript also appears in PLO.

Routley[2] 1972, Routley, R. and V., "The semantics of first degree entailment", *Nous* **6** (1972) 335–59.

Routley[2] 1973, Routley, R. and V., "Rehabilitating Meinong's theory of objects", *Revue Internationale de Philosophie* **27** (1973) 224–254.

Routley[2] 1975, Routley, R. and V., "The role of inconsistent and incomplete theories in the logic of belief", *Communication and Cognition* **8** (1975) 185–235.

Routley[2] 1975a, Routley, R. and V., "Ideal objects on a Meinongian theory of universals", in *Proceedings of the XVth World Congress of Philosophy*, (Varna, 17–22 September 1973), vol. 5, 1975, 581–589.

Russell, B., *A Critical Exposition of the Philosophy of Leibniz*, Cambridge University Press, 1900.

Russell, B., "Meinong's theory of complexes and assumptions", *Mind* **13** (1904) 204–219, 204–219.

Russell, B., "Review of Meinong, *Uhtersuchungen zur Gegenstandstheorie and Pyschologie*", *Mind* **14** (1905) 530–538.

Russell, B., *The Problems of Philosophy*, Williams and Norgate, London, 1912.

Russell, B., *Introduction to Mathematical Philosophy*, Allen and Unwin, London, 1919, (cited as MP).

Russell, B., *The Principles of Mathematics*, Second edition, Allen and Unwin, London, 1937, (cited as P).

Russell, B., "Reply to Criticisms", in *The Philosophy of Bertrand Russell*, edited by P. A. Schilpp, Open Court, La Salle, Ill., 1944.

Russell, B., *A History of Western Philosophy*, Allen and Unwin, London, 1946.

Russell, B., *My Philosophical Development*, Allen and Unwin, London, 1959.

Ryle, G., "Review of Findlay, *Meinong's Theory of Objects*", *Oxford Magazine* **52** (1933–34) 118–120.

Ryle, G., *The Concept of Mind*, Hutchinson, London, 1949.

Ryle, G., *Dilemmas*, Cambridge University Press, Cambridge, 1954.

Ryle, G., *Collected Papers*, Hutchinson, London, 1971.

Ryle, G., "Intentionality-theory and the nature of thinking", in Haller 1972, 1972, 7–14. Also published in *Revue Internationale de Philosophie* **27** (1973) 255–264.

Scheffler, I., *The Anatomy of Inquiry*, Routledge and Kegan Paul, London, 1964.

Schilpp, P. A., ed., *The Philosophy of Bertrand Russell*, Library of Living Philosophers, Evanston, Ill., 1946.

Schilpp, P. A., ed., *The Philosophy of Rudolf Carnap*, Open Court, LaSalle, Ill., 1963.

Schoenman, R., ed., *Bertrand Russell, Philosopher of the Century*, Allen and Unwin, 1967.

Scott, D., "Existence and description in formal logic", in Schoenman 1967, 1967.

Searle, J. R., "Proper names", *Mind* **67** (1958) 166–173.

Searle, J. R., *Speech Acts*, Cambridge University Press, Cambridge, 1969.

Shapere, D., ed., *Philosophical Problems of Natural Science*, Macmillan, New York, 1965.

Slade, C. J., "The myth of mistake in the English law of contract", *Law Quarterly Review* **70** (1954) 385–408.

Smart, J. J. C., *Philosophy and Scientific Realism*, Routledge and Kegan Paul, London, 1963.

Smart, J. J. C., "Science as an approximation to truth", in *Papers Presented to the Annual Conference of the Australasian Association for History and Philosophy of Science*, Melbourne, 1976.

Smart, J. J. C., "A critique of Meinongian semantics", Paper presented at the Australasian Association of Logic Conference, Melbourne, 1977. [Reprinted in this volume —*Eds.*].

Smart, J. J. C., "Is Occam's Razor a physical thing?", *Philosophy* **53** (1978) 382–385.

Smart, J. J. C., "The reality of the future", typescript, Canberra, 1979.

Smiley, T. J., "Sense without denotation", *Analysis* **20** (1959–60) 125–135.

Smullyan, A. F., "Modality and description", *Journal of Symbolic Logic* **13** (1948) 31–37.

Smythies, J. R., *Analysis of Perception*, Routledge and Kegan Paul, London, 1956.

Spengler, O., *The Decline of the West: Form and Actuality*, vol. 1, translated by C. F. Atkinson, Allen and Unwin, London, 1926.

Stein, H., "On Einstein-Minkowski space-time", *Journal of Philosophy* **65** (1968) 5–23.

Strawson, P. F., *Individuals*, Methuen, London, 1959.

Strawson, P. F., "Identifying reference and truth-values", *Theoria* **30** (1964).

Sturrock, J., *Paper Tigers: the Ideal Fictions of Jorge Luis Borges*, Clarendon Press, Oxford, 1977.

Tarski, A., "The semantical conception of truth and the foundations of semantics", *Philosophy and Phenomenological Research* **4** (1944) 341–376.

Tarski, A., *Logic, Semantics and Metamathematics*, translated by J. H. Woodger, Clarendon Press, Oxford, 1956.

Thom, R., "Role and limits of mathematization in applied sciences", Paper presented at the 6th International Congress of Logic, Methodology and Philosophy of Science, Hanover, Aug. 1979.

Thomason, R., "Indirect discourse is not quotational", *Monist* **60** (1977) 340–354.

Tichý, P., "An Approach to Intentional Analysis", *Nous* **5** (1971) 273–297.

Tooley, M., "Some objections to Meinongian semantics", seminar paper, Philosophy Department, Research School of Social Sciences, Australian National University, 1978.

Urmson, J. O., "Fiction", *American Philosophical Quarterly* **13** (1976) 153–157.

Vaihinger, H ., *The Philosophy of "As If"*, translated by C. K. Ogden, Routledge and Kegan Paul, London, 1949.

Vendler, Z., "Each and every, any and all", *Mind* **71** (1962) 145–160. Reprinted in his *Linguistics in Philosophy*, Cornell University Press, Ithaca, 1967. Page references are to the reprint.

Vendler, Z., "A Note to the Paralogisms", in *Contemporary Aspects of Philosophy*, edited by G. Ryle, 1976, 111–121.

Vendler, Z., "Reference and introduction", Paper presented at the Australasian Association of Philosophy Conference, Canberra, 1978. Published in *Philosophia* **10** (1981) 253–273.

von Wright, G., *An Essay in Modal Logic*, North-Holland, Amsterdam, 1951.

von Wright, G., *Time, Change and Contradiction*, Cambridge University Press, Cambridge, 1969.

Waismann, F., "Language strata", in *Logic and Language, Second series*, edited by A. Flew, Blackwell, Oxford, 1953, 102–121.

Wang, H., "Truth definitions and consistency proofs", *Transactions of the American Mathematical Society* **73** (1952) 243–275.

Wang, H., *A Survey of Mathematical Logic*, Science Press, Peking and North-Holland, Amsterdam, 1963.

White, M., "The analytic and the synthetic: an untenable dualism", in *John Dewey: Philosophy of Science and Freedom*, edited by S. Hook, Dial Press, New York, 1950, 316–330.

Williams, B., "Comment on Kenny's Descartes' ontological argument", in *Fact and Existence*, edited by J. Margolis, University of Toronto Press, Toronto, 1969, 55–56.

Wisdom, J., *Interpretation and Analysis*, 1931.

Wisdom, J., *Other Minds*, Blackwell, Oxford, 1952.

Wisdom, J., *Philosophy and Psycho-Analysis*, Blackwell, Oxford, 1953.

Wittgenstein, L., *Tractatus Logico-Philosophicus*, Routledge and Kegan Paul, London, 1947.

Wittgenstein, L., *Philosophical Investigations*, Blackwell, Oxford, 1953.

Woods, J., *The Logic of Fiction*, Mouton, The Hague, 1974.

Woods, J., "Critical notice of S. Haack, *Deviant Logic*", *Canadian Journal of Philosophy* **7** (1977) 651–666.

Zemansky, M. W., *Heat and Thermodynamics*, McGraw-Hill, New York, 1943.

Correction to: Noneist Explorations I: The Sylvan Jungle — Volume 2

Richard Routley and Val Routley

Correction to:
Dominic Hyde (ed.), *Noneist Explorations I: The Sylvan Jungle — Volume 2*, Synthese Library 415, https://doi.org/10.1007/978-3-030-26309-6

We regret that the editor's name 'Dominic Hyde' was mistakenly appeared as a co-author in all the chapters on Springerlink and third-party websites. The book and chapters have been updated with changes.

The updated original version of the book can be found at
https://doi.org/10.1007/978-3-030-26309-6
https://doi.org/10.1007/978-3-030-26309-6_1
https://doi.org/10.1007/978-3-030-26309-6_2
https://doi.org/10.1007/978-3-030-26309-6_3
https://doi.org/10.1007/978-3-030-26309-6_4
https://doi.org/10.1007/978-3-030-26309-6_5
https://doi.org/10.1007/978-3-030-26309-6_6

D. Hyde (ed.), *Noneist Explorations I*, Synthese Library 415,
https://doi.org/10.1007/978-3-030-26309-6_7

SUPPLEMENTARY ESSAYS

A critique of Meinongian semantics

Jack Smart[α]

In this paper I intend to try to say what I think may be wrong with Richard Routley's semantics, in so far as it is based largely on ideas derived from Meinong. My understanding of Routley's position comes mainly from the illuminating paper Routley and Routley 1973 and also from the so-far unpublished paper by the same authors, "Exploring Meinong's Jungle".[β] Certainly it transpires that a Meinongian semantics is harder to refute than is commonly believed. In the same number of the journal in which Routley and Routley 1973 appears there is an immediately adjacent paper by Gilbert Ryle, in which Ryle rashly asserts: "Let us frankly concede from the start that *Gegenstandstheorie* itself is dead, buried and not going to be resurrected. Nobody is going to argue again that, for example, 'there are objects concerning which it is the case that there are no such objects.'" (Ryle 1973, p. 255). Meinong has been wrongly understood as holding that such objects as the golden mountain and the round square *exist* (or perhaps *subsist*) in a sort of Platonic heaven, whereas it is the case that in talking of "objects" Meinong wished to have no ontological commitments: the golden mountain and the round square do not exist at all, and *a fortiori* they do not exist in a Platonic heaven. Meinong asserts as true "The golden mountain is golden" and "The round square is round", but denies that "The golden mountain exists" and "The round square exists". Contrary to the Russellian analysis of definite descriptions, for Meinong a phrase of the form "The so-and-so" carries no ontological commitment. Nor does his term "object".

The Routleys in Routley and Routley 1973 put forward a number of Meinongian theses.

[α] Paper presented at the Australasian Association of Logic Conference, Melbourne, 1977. —*Eds.*

[β] Subsequently published as chapter 1 of *Exploring Meinong's Jungle and Beyond*; reprinted here: Volume 1, chapter 1. —*Eds.*

D. Hyde (ed.), *Noneist Explorations I*, Synthese Library 415,
https://doi.org/10.1007/978-3-030-26309-6

(1) Rejection of the *Ontological Assumption* that one can not make true statements about what does not exist.[1] Contrary to Quine's "To be is to be the value of a variable" non-entities can be values of variables. (But the Routley's predicate "is a value of" is intensional and so is not the same as Quine's "is a value of". Nor is it the "Ryle" or substitutional sense either.[2])

(2) *The Assumption Postulate.*[3] Objects have the properties which characterise them, whether or not these objects exist. ((2) is a reformulation of (1).) Thus it would appear that a Meinongian set theorist could talk a lot about sets without committing himself to saying that sets exist. It also follows that there can, according to Meinong, be contingent extensional identities between non-entities, as is asserted by the sentence "What I am thinking about = the golden mountain".[4]

(3) *Distinction between wide and narrow negation.*[5] According to Meinong's theory we can talk about impossible objects, ones which are both F and not-F. Here the "not" in "not-F" is a narrow negation sign. To say that the round square is both round and not-round is not to say (using wide negation) that it is both the case and not the case that the round square is round. The law of non-contradiction holds only for wide or "external" negation, not for the narrow or "internal" negation.

Now I want to say something about the notions of "reference" and "aboutness". The Routleys distinguish these two. Thus they are happy to say that a name "a" refers to an object a, if and only if a exists. I shall here use "exists" tenselessly, which I think goes against what the Routleys say in the second half of their "Exploring Meinong's Jungle". I don't want, at present, to buy into a discussion of what (to me) seem to be their strange ideas about time and tense: I don't think they are relevant to the present discussion. (If I can refute a Meinongianism which goes half way to meet me by not being tendentious about time, *a fortiori* I can refute a stronger Meinongianism.) Thus I shall interpret the Routleys as allowing that "Socrates" refers to Socrates because Socrates exists. (If they don't like my use of the tenseless present here, they can replace "Socrates" by "Malcolm Fraser".) However they would not say that "Mr. Pickwick" refers to Mr. Pickwick, because Mr. Pickwick does not exist: Mr. Pickwick is a non-entity. They would say that

[1] See Routley and Routley 1973, pp. 227–8.
[2] See Quine 1966, p. 180.
[3] See Routley and Routley 1973, pp. 228–9.
[4] *Ibid*, p. 229.
[5] *Ibid*, p. 231.

"Mr. Pickwick" is nevertheless *about* Mr. Pickwick. They might also say that "The set of prime numbers" is about the set of prime numbers but does not refer to it. According to them (I think) sets are non-entities. Nevertheless set theory is "about" sets, though it does not *refer* to sets. If you will forgive me for using a tired old Russellian simile, it does seem to me that such an avoidance of Platonism has all the advantages of theft over honest toil.

The Routleys reject the view that truth is a function of reference. They hold that "a" and "b" can be about the same object, but that even so "Fa" can be true and "Fb" false. Thus someone may believe something of the morning star but not believe it of the evening star, since he may not believe that the morning star is the evening star.

If one could accept the Routleys' notion of "being about" it would provide a smooth way of dealing with referentially opaque contexts. Thus Russell had great difficulty of preventing such sentences as "Meinong believes that the round square is round" coming out false (because $\sim(\exists x)$ round square x) whereas our intuition makes it true. Of course classical logicians do have ways of dealing with this sort of sentence. I think that the best way is Davidson's paratactic analysis. "Meinong believes that. The round square is round." Here the second sentence is not asserted, but exhibited so that its content can be taken to be what is referred to by the demonstrative "that".

It is clear that the Routleys' phrase "'N' is about ..." is an intensional, or what Quine would call a "referentially opaque" context. Similarly the phrase "Richard talks about ..." is referentially opaque. "Richard talks about Pegasus" does not imply that Pegasus exists, in the way that "Richard kicks Pegasus" does. It is this referential opacity or intensionality of "talks about" which gives Meinongian semantics its apparent advantage. For example it seems to allow the Meinongian to say that mathematicians talk about sets and other mathematical objects without the Meinongian committing himself to saying that these objects exist. It also enables a Meinongian to allow science to "talk about" ideal entities (such as dimensionless particles in Newtonian theory) without being committed to their existence. Such idealised theories present a problem to the philosopher of science. I think that they can be dealt with within the framework of both an extensional logic and a realist theory of science, but to justify this much "honest toil" is needed. The opacity of "talks about" also enables the Meinongian to give an account of discourse about fictional objects, such as Mr. Pickwick, though I regard the implications with respect to fiction as of lesser theoretical importance. It would not matter so much if we were simply to say that "Mr. Pickwick wore trousers" was simply *false*. It would be shocking to say that mathematics is false because mathematics, unlike fiction, is an essential part of science. However the Meinongian semantics does allow us to say that "Mr. Pickwick wore trousers" is true and that "Mr. Pickwick wore a kilt" is false. This fits in with our intuitions. However I myself would prefer to treat fiction simply as a pretence use of language with a classical semantics, rather than as a serious use of language with a Meinongian semantics.

Now I want to say that semantics must use an extensional predicate like "satisfies", or must use " '*N*' is about ..." in a referentially transparent sense. I am aware of course that the Routleys would regard this assertion of mine as merely dogmatic. However let me for the moment carry on regardless, and state the classical point of view. "Pegasus" can't be about Pegasus, because there is in the universe no Pegasus for the name "Pegasus" to be about. So when the Meinongian *says* that "Pegasus" is about Pegasus, I interpret him as meaning that discourse containing the term "Pegasus" is about something else, namely the property of being a winged horse. If we can for the moment put aside Quinean doubts about postulating such things as properties, then we can say that talking about Pegasus is talking about the property of being a winged horse, and talking about unicorns is talking about the property of being a one-horned quadruped. This *extensional* use of "talk about" allows us to bring out the ontic commitments hidden by the Routley's *intensional* use of the word. In the case of set theory the ontic commitments are rather embarrassing: it is hardly an advance to be "nominalist" about sets while covertly being Platonistic about properties. (Whitehead and Russell got into this sort of bother in PM.[1])

Terence Parsons has worked out this sort of account of Meinongian semantics in some details in Parsons 1971. I shall give a brief account of parts of his paper.

Malcolm Fraser has a lot of "ordinary" properties such as being over 6 ft. tall. A Meinongian would also ascribe to him a lot of rather odd properties, such as *existing* and *being thought about by Richard.* Parsons calls the former set of properties "nuclear" properties, and the latter ones he calls "extra-nuclear". Objects are "represented" by sets of nuclear properties. In the case of a Meinongian object i its correlated set of properties is $i^c = \{p\colon i \text{ has } p\}$. If i is an existent object, like Fraser, the set i^c will be *complete*:

[1] See W. V. Quine's objections to this feature of PM, for example in Quine 1958, p. 121, f.n.

i.e. every property or its contradictory property is in the set. In the case of Pegasus or Sherlock Holmes the set will not be complete, since it is indeterminate whether Pegasus ever sneezed on his birthday or whether Sherlock Holmes ever saw W. G. Grace hit a six (unless some Holmes scholar can prove me wrong!)

Now to say that Sherlock Holmes smoked a pipe is just to say that the property *smokes a pipe* is a member of the set of properties $\{p$: Sherlock Holmes is said or implied by Conan Doyle's stories to have $p\}$. If we wanted to say (falsely) that Sherlock Holmes existed we would say that $(\exists i)(i^c = 0)$, where $0 = \{p$: Sherlock Holmes is said or implied by Conan Doyle's stories to have $p\}$.

Let me illustrate Parsons' ideas by showing one way in which he deals with the notorious Meinongian sentence "The existent golden mountain does not exist". We can distinguish two senses of "exists" or "is existent". The first is the one which expresses the universal nuclear property of individuals: y exists $=_{\text{df}} (\exists x)(x = y)$.[1] Then there is the extra-nuclear "exists": in this sense to say that an object exists is to deny that it is a non-entity. Then when the Meinongian says that "the existent golden mountain does not exist" he is using "existent" in the nuclear sense and "exist" in the extra-nuclear sense. In "Although the existent golden mountain is existent, the existent golden mountain does not exist", we have three cases of the nuclear "is existent" and one of the extra-nuclear "exists".

Parsons goes on to sketch out a semantics based on this representation of an object as a set of properties. It probably makes good sense (*modulo* the acceptance of properties), though I have one or two qualms. Can one really replace talk of objects by talk of properties? In the case of incomplete objects like Pegasus, and neglecting properties which are logical consequences of other properties, the set $\{p$: Pegasus has $p\}$ is finite. I do not think that such an expression as $\{p$: Sherlock Holmes has $p\}$ can itself be part of Parsons' extensional semantics, because "... has p" is an intensional context. Similarly, for $\{p$: Sherlock Holmes is said or implied by Conan Doyle to have $p\}$. Perhaps these expressions should be reformulated in some such way as $\{p$: p corresponds to a predicate applied to the name "Sherlock Holmes" or to a "Sherlock Holmes" description by Conan Doyle$\}$, or something of the sort. However in what follows I shall, like Parsons, continue to speak loosely. But in the case of actually existing objects, like W. G. Grace,

[1] See Parsons 1971, p. 574.

the set $\{p\colon \text{W. G. Grace has } p\}$ is complete and presumably infinite. We can specify this infinite set only by talking about (in the referentially transparent sense) W. G. Grace himself. Now what about Homer, supposing that we are doubtful whether or not Homer exists? Is the set of properties we are talking about complete or incomplete, infinite or finite? I sense a serious objection to Parsons' theory here.

Nevertheless, something which is either the same as Terence Parsons' theory or else is something along the same lines is surely needed if we are to give an extensional semantics for Meinongian discourse. If there are no unicorns, then in the referentially transparent sense we must say that Routley talks about unicornhood, not that he talks about unicorns.

Now we come to the crunch. I say that the Meinongian is committed to an ontology of properties. This is because *I* hold that semantics has to use expressions like "is satisfied by" and "talks about" and "denotes" in a referentially transparent way. Semantic predicates should be like "kick" or "is less than", not like "desires" (which in a sense is not a predicate, though "desires-true" in a sentence "Smith desires-true the proposition ..." *is* a predicate). The reason I have for saying this is that it is the business of semantics to say how language hooks on to the world, and a referentially opaque expression doesn't fill this bill.

I think that Parsons has probably put up a pretty convincing case for saying not only that Meinongian logic is consistent, but also that it has a sensible interpretation, given an ontology of properties. Of course a Meinongian language could be consistent without having a sensible interpretation. If one were concerned not with interpretation but with mere modelling, one could no doubt give an arithmetical model. Thus we might regard the set of prime numbers as the set of actual individuals, and the set of squares of prime numbers as the set of non-entities. One has to be a nominalist to deny that integers exist, and so the "non-entities" in the model would exist all right. What I can't understand is the idea of an interpretation or a model which contains among its elements non-entities themselves. I want to say that non-entities would have to exist in order to be constituents of the model.

You can see, therefore, why I want to say that a proper semantics must use only of referentially transparent senses of "about", "is satisfied by" and the like. They must be like "kicks" or "is less than": for A to kick B, B must exist to be kicked by A, and for a number x to be less than a number y then y must exist for it to be greater than x.

How am I to convince the Meinongian about this? He will simply deny my contention, and will use a Meinongian metalanguage. Indeed if he is like Richard Routley his metalanguage will be the same as the object language, because of Routley's use of paraconsistent logic to defuse the semantic paradoxes. But whether or not the metalanguage is identical with the object language, the Meinongian will use a Meinongian metalanguage. He will just deny that semantic predicates must be like "kicks" and "is less than" and unlike, say, "desires". I can get no purchase on him if I want to convince him that an extensional metalanguage is needed for semantics, so as to hook language on to the world. This leads to an interesting situation. I can argue on my own terms that Meinongian logic commits us to an ontology of properties, but an essential premiss of my argument would be denied by the Meinongian.

In the discussion I would welcome two sorts of things. (1) I would like any Meinongians present to say whether I have misrepresented the Routleys' position in any important way. (2) I would like any anti-Meinongian present to help me find some purchase on the slippery skin of the Meinongian theory, either by helping to show that the Meinongian does *not* after all get out of trouble by using a Meinongian metalanguage, or else by helping to show the Meinongian that it is not legitimate to use a Meinongian metalanguage, or at least that somewhere up the hierarchy of metalanguage metametalanguage and so on, one must eventually have recourse to a non-Meinongian language.

Non-entities surely can't be *explanatory* in the way in which entities can. The Routleys in their unpublished paper "Exploring Meinong's Jungle" talk of the discourse about non-entities (sets) as a "logical juice extractor". If this phrase implies instrumentalism, then I think that this picture is unsatisfactory. We should, following Quine, believe in as many objects as are needed for scientific explanation, including the ones which are postulated by that much mathematics which is needed for science. Surely assertions "about" non-entities could not constitute an explanation at all.

The Routleys say "The richness and diversity of objects far outruns that of the Universe, and reflects rather the richness of the Universe of discourse, of what we talk and think about" (Routley and Routley 1973, p. 243). I can form no idea of a universe of discourse which is not part of the actual universe. What on earth *is* a universe of discourse other than a set of actual things? The Routleys will say "It is what we talk about, and we can talk about non-entities". Back to square one: I'm using "talk about" transparently and they are using it opaquely.

There is a disturbing *impasse* here which needs to be resolved.

References

Parsons, Terence, "A prolegomenon to Meinongian semantics", *Journal of Philosophy* **71** (1971) 561–580.

Quine, W.V.O., *Mathematical Logic*, Harvard University Press, 1958.

Quine, W.V.O., "Reply to Professor Marcus", in *The Ways of Paradox and Other Essays*, Harvard University Press, Cambridge, Mass., 1966, 175–182.

Routley, Richard and Routley, Val, "Rehabilitating Meinong's theory of objects", *Revue Internationale de Philosophie* **27** (1973) 224–54.

Ryle, Gilbert, "Intentionality-theory and the nature of thinking", *Revue Internationale de Philosophie* **27** (1973) 255–65.

Routley's theory of fictions

Maria Elisabeth Reicher

Routley devotes chapter 7 ("The Problems of Fiction and Fictions") of *Exploring Meinong's Jungle and Beyond* to fictions and fictional objects, though the topic occurs at several other places in the book as well. In what follows, I shall focus on chapter 7, starting, in the first section, with a (sympathetic, or at least neutral) outline of the theory that is developed there; in the second section, I shall critically discuss several issues raised by this theory.

Fictions and fictional objects in the framework of noneism: an outline of Routley's theory

In this section, I shall report on Routley's views on fiction and fictions in a rather neutral way, postponing critical remarks to the next section. To avoid a multitude of cumbersome declarations of reservation in the text, I'd like to announce right now, at the beginning, that I deeply disagree with some of Routley's basic noneist assumptions and, consequently, also with fundamental tenets of his theory of fictions. However, it is one of the tasks of this article to introduce the reader to Routley's theory, and I try to do so as concisely as possible. I just want to alert the reader to the fact that statements like, e.g., "There are fictional worlds but they do not exist and have no kind of being whatsoever" are meant as mere descriptions of Routley's views and do not imply agreement on my part, even if I refrain from commenting on them here.

Initially, Routley explicates the notions of fiction and works of fiction as follows:

> A piece, or work, of fiction typically has an author or set of authors (perhaps unknown or anonymous) and is intended to reach some audience: thus it is a *communication*, or message, in the wide sense. More generally, (a piece of) *fiction* is authored discourse (whether written or spoken) or communication which consists of imagined or invented statements or narrative, which conveys a *story* as contrasted with factual or reportative discourse. A work of fiction will generally be about fictional items or creatures (i.e. *characters*), that is the story is about *fictions* in another important sense of that term. ($EMJB_2$ this volume, p. 228f.[1])

As Routley notes, works of fiction are not necessarily *literary* works; there are non-literary works of fiction as well, e.g. film works: "Since a film may be silent and contain no script, discourse in the usual sense, though characteristic of fiction, is not essential for nonliterary fiction: potential communication or *portrayal of an imaginary situation* through some medium is". (p. 229)

There are, however, not only fictional *works*, but also fictional *worlds* and their objects:

> A work of fiction, e.g. a novel or film or series of such, N can be regarded as depicting a (dynamic) world, the world of the story. The world $a(N)$ is given (in a canonical form) by the class of statements that hold according to N; it has as domain $d(N)$ the items of N, e.g. the characters and things which it is about. (p. 230)

Routley's favourite examples for fictional characters are the standard ones: Sherlock Holmes, Mr. Pickwick, Hamlet. Routley's theory of fictions, however, is quite distinctive. It is embedded in and an integral part of a much more general ontological theory called *noneism*, which is developed and defended in *Exploring Meinong's Jungle and Beyond*. Fictional worlds and their objects (characters and things) are *nonentities*. The notion of a nonentity is one of the central notions of noneism, along with the notion of an item. The term "item"

> is introduced as an ontologically neutral term: *it is intended to carry no ontological, existential, or referential* commitment whatsoever. In

[1] All page references in the essay are to this volume, unless stated otherwise.

> particular then, talk of items carries no commitment to, and should be sharply distinguished from, the *subsistence* of items; for "subsists" means, in the relevant senses, "exists, in some weak or low grade way". ($EMJB_2$, Vol. 1, p. 7)

Items divide into particulars and non-particulars, where non-particulars include "all abstractions such as universals of one kind or another, attributes, classes, propositions, objectives, states of affairs, etc.". (*Ibid.* p. 9) All nonparticulars are nonexistent. By contrast, some particulars exist, but not all do. Those particulars that exist (at some time) are called *entities*, those particulars that do not exist (at any time) are called *nonentities*. (*Ibid.*) Nonentities are further divided "into *possibilia*, those which are logically possible, and *impossibilia*, those which are logically impossible". (*Ibid.*)

In contrast to Alexius Meinong, on whose *theory of objects* noneism is based to a large degree, Routley rejects the assumption of different "modes" or "grades" of being (like subsistence, which is often interpreted as a weaker mode of being) as an absurdity:

> The necessary conditions on existence are neglected in the ever-popular kinds-of-existence doctrines, under which nonexistent objects are alleged to have existence of some other kind. Thus objects drawn from fiction exist in fiction, mental objects exist in the mind, theoretical objects exist as part of their theory, etc. The doctrine admits of ready send-up, for kinds are easily multiplied up; e.g. paradise and holes have geographical existence, God may not exist but he has religious existence, the superego and all sorts of factors and effects exist in psychology, absolute values exist in objective morals, canned peaches exist as grocery supplies, and perfect spuds and indigo-flowered plumwoods have agricultural, or is it horticultural, existence. (p. 104)

There is just one kind of being, namely existence. The domain of objects that have being coincides with the domain of existent objects. Items that do not exist have no being whatsoever. ($EMJB_2$, Vol. 1, p. 5) To sum up: nonentities, and thus fictional objects, are particular (i.e., nonuniversal, nonabstract) items that do not exist and have no kind of being whatsoever at any time.

Since the theory of fictions is an integral part of noneism, in order to outline the former, it is unavoidable to mention at least some of the basic principles of the latter. Here are some fundamental tenets of noneism:

> (I) [*The significance thesis*] Very many sentences the subjects of which do not refer to entities[,] e.g. "the round square does not exist", "Primecharlie (the first even prime greater than two) is prime", are significant. Furthermore the significance of sentences whose subjects are about (or purport to be about) singular items is independent of the existence, or possibility, of the items they are about. (*Ibid.* p. 18)
>
> (II) [*The content thesis*] Many different sorts of statements about non-existent items, including many of those yielded by single subject-predicate sentences, are truth-valued, i.e. have truth-values true or false. (*Ibid.* p. 19)
>
> (III) [*The basic independence thesis*] That an item has properties need not, and commonly does not, imply, or (pre)suppose, that it exists or has being. Thus statements ascribing features to nonentities may be used, and are used, without involving *any* existential or ontological commitment. (*Ibid.* p. 31)

(III) is to be distinguished from the following two (stronger) principles:

> (i) the Advanced Independence Thesis (AIT), according to which nonentities (can and commonly do) have a more or less determinate nature [...], and
> (ii) the Characterisation Postulate (CP), according to which nonentities have their characterising properties [...]. (*Ibid.* p. 31f; see also p. 58f.)

Noneism includes both AIT and CP. CP raises the question of what, exactly, the "characterising properties" of nonentities are. To start with, consider standard examples of Meinongian nonentities: among the characterising properties of the golden mountain are the properties of being golden and being a mountain; among the characterising properties of the round square are the properties of being round and of being a square (but also the property of being nonsquare, see p. 176). Thus, the golden mountain does not exist and never did exist and does not have any other kind of being, but it has the property of being a mountain, i.e., it is a mountain (in the usual sense of the word), and thus the sentence "The golden mountain is a mountain" is a true sentence about the golden mountain, just as "Mt. Everest is a mountain" is a true sentence about Mt. Everest. Thus, the golden mountain is not an abstract object (a type or class of properties or the like) but as concrete and particular as every existent mountain.

To apply the above-mentioned principles to a fictional object: Sherlock Holmes does not and never did exist and does not have any other kind of being, yet "he" has a number of properties, such as being a detective, smoking the pipe etc., and thus sentences like "Sherlock Holmes is a detective" (where "is a detective" is used in the usual sense) are true of him. Thus, Sherlock Holmes is not an abstract object but a man of flesh and blood, a concrete particular, though a nonexistent one.

> [T]he abstract entity *Hamlet* is quite different from Hamlet. For instance, *Hamlet* is abstract (a pretence, etc.), Hamlet is a particular (a prince, etc.); *Hamlet* exists (or at least is supposed to), Hamlet does not (and is not usually supposed to); Shakespeare wrote about (actors study, etc.) Hamlet, not *Hamlet*; Kripke introduced us to *Hamlet*, not Hamlet; and so on. (p. 256)

Yet there are differences between entities and nonentities. For one thing, entities are always fully determined, that is, for every predicate F: if x is an entity, then either Fx or non-Fx. For each existing mountain (problems of vague boundaries of geographic entities left aside), either it is higher than 3000 meters or it is not higher than 3000 meters (i.e., it is either exactly 3000 meters high or less than 3000 meters high). Not so the golden mountain: it seems that neither "The golden mountain is more than 3000 meters high" nor "The golden mountain is not more than 3000 meters high" is true. Similarly, it seems that neither "Holmes has a mole on his back" nor "Holmes does not have a mole on his back" are true (given that the Holmes stories do not tell us anything about this matter). Fictional worlds and fictional objects are usually (perhaps always) incompletely determined. (pp. 109–111, p. 236f.)

Another important difference between entities and nonentities is that entities cannot have contradictory properties, but nonentities can do so. ($EMJB_2$, Vol. 1, p. 9)

This, however, does not make noneism an inconsistent theory. Inconsistency occurs when logical principles whose validity is restricted to possible objects are applied – erroneously – to impossible objects. This does not show that something is wrong with the assumption of impossible objects, but rather that something is wrong with the logic that allows the application of these principles unrestrictedly to entities and nonentities alike:

> Objections to impossible objects are almost invariably derived by taking over principles for possible objects and entities and showing that they lead to trouble – inconsistency or the like – when applied to impossible objects. They depend, that is, upon assuming that the logic of impossible objects is the same as, or very similar to, that of objects which are not impossible. But [...] the logic is not the same; and moreover only a little reflection indicates that it would not be expected to be the same. (p. 144)

Fictional worlds and objects may have contradictory characterising properties and thus may be, in this sense, impossible objects. But what, exactly, are a fictional object's characterising properties? It suggests itself that a fictional object's characterising properties are determined by the fictional work or works in which the object occurs, such that, for instance, Sherlock Holmes has exactly those characterising properties that are ascribed to him in the Sherlock-Holmes stories by Arthur Conan Doyle.

Yet, Routley rejects this picture as too simple: First, he notes that "what holds according to [a fictional work] N does not coincide with what is presented by, e.g. written in, N". (p. 230) Rather, the set of those sentences that are true according to N is bigger than the set of sentences that N consists of (given that N is a literary work). The sentences that are true according to N are those that N consists of plus additional sentences. (p. 230f.)

> The material additions include factual or quasi-factual material, e.g. descriptive geographical or historical material, alluded to in setting the scenes of N, that may be made use of in N, but which is not explicitly spelt out in N. The material – which consists primarily of relatively common knowledge of parts of the current world and its history – may include, usually in very sketchy form, such geographical information as the arrangements of certain streets of London, of the

> upper reaches of the river Thames, of a minor rail network and its stations, of the climate of a certain region; such historical details as the prime minister of a certain government, parts of a genealogical tree of an "important" family; such scientific information as that normal physical forces, like gravity, operate on characters and that characters are in main psychological and sociological respects like sometime-actual humans of such and such a place and time; knowledge of other works of fiction or the like in the tradition of which the work is a part; information as to social customs; etc. (p. 231f.)

Secondly, Routley makes use of his technical notion of a *source book* (or *source*, for short) in order to answer the question of fictional objects' characterising properties:

> A (fictional) object d has just those characterising features its source $S(d)$ ascribes to it. (p. 275)

More elaborately stated:

> A (purely) *fictional object* is one that is characterised in fiction. Fictional objects are not the only objects that occur in fiction, that works of fiction are about, since for example actual and historical objects also occur. The full characterisation of a fictional object is given in (or by) its *source* book, which is determined from the works of fiction in which the object is characterised (not in which it merely appears incidentally or as introduced from elsewhere). *A fictional object has all and only those characterising features ascribed to it by its source.* This principle, which is a fundamental one in determining what is true in nonfictional contexts of fictional objects, derives from Characterisation Postulates in the following way: Where d is any fictional object and $f_1, \ldots, f_n$ are all the characterising predicates ascribed to d by its source $S(d)$, then $d = \imath x(xf_1 \,\&\, \ldots \,\&\, xf_n)$. (p. 300)

Originally, the notion of a source is introduced not for fictional objects only, but for objects, or at least, nonentities, in general:

> Much as a character in a work of fiction is characterised in the book, so more generally an object may be said to be characterised in a source

> book for it, or its characterisation given by a *source book*. For each object there is a source book which tells its "story" what it is like, what it does, and so forth. The source book is the *source* for characterising details of the object: it is not the source, of course, for what we think or feel about the object, for all intensional relations to the object. Any actually *specified* nonentity does have an actual source – whether it consists of what is written in books or documents, verbal or film material, or is in the form of plans, blueprints, specifications or the like. All these sources will be accounted source books or part of source books in the technical sense. ($EMJB_2$, Vol. 1, p. 464)

A fictional object's source book includes the work in which the object is "native", but it is not identical to this work:

> It is advantageous to make use not just of source books *for* objects but also of source books *elaborating* works (or series of works). The approximate model for source books elaborating works is provided by (what is stated in) modern supplemented and annotated editions of the (or a) work of some historical author.
> Source books for works of fiction are propositional in character. There need be no written or otherwise recorded account, and in general will not be [. . .]. A source book is *compiled* both from *primary* sources, in the case of fictional objects from works of fiction, [. . .], and *secondary* sources, e.g. historical and geographical information, commentaries, discussion (where the object figures as an immigrant object), etc. [. . .]
> The compilation consists of material elaborations, supplied from the secondary sources, together with application of appropriate closure requirements. (p. 274)

An important qualification is still left to be made: As has been stated above already, "[a] (fictional) object d has just those characterising features its source $S(d)$ ascribes to it". (p. 275) However, not all features that a source ascribes to d are *characterising* features of d; some of the features a source ascribes to d are noncharacterising features of d.

The distinction between a nonentity's characterising and noncharacterising features is an important component not only of Routley's theory of fiction but of noneism in general. Its function is to avoid paradoxes (or "clashes with empirical reality"): According to noneism, there is (in an ontologically uncommitting sense of "there is") an item that is characterised as "the golden mountain", i.e., an item that is both golden and a mountain. This assumption does not lead to a clash with empirical reality, since "there is the golden mountain" does not imply "The golden mountain exists". Consider, however, the *existent* golden mountain. If to exist were a characterising feature, then, according to the Characterisation Postulate (CP), the existent golden mountain would not only have the features of being golden and being a mountain but also the feature of being existent. This assumption, however, obviously clashes with empirical reality, for as a matter of fact, no thing exists that is both golden and a mountain. Routley avoids this clash by stipulating that existence is not a characterising feature.

The point applies, of course, in an analogous fashion, to fictional characters: Surely, according to Sherlock Holmes' source book, Holmes exists. Yet, as a matter of fact, Holmes is a fictional character and as such does not and never did exist. But since existence is not a characterising feature, the source book cannot bring it about that Holmes is an existent object, no matter what is told in it; and thus, the clash with empirical reality is avoided.

> As repeatedly seen, nonentities do not have, by any means, all extensional features their descriptions or (more generally) source books present them as having. Nonentities cannot settle their own ontological status, their own identity, nor can they by their mere sayso [sic!] impose relations and conditions on actual things and places. (p. 192)

There is no hard and fast criterion for the distinction between characterising and non-characterising features, as Routley himself admits. The distinction is explicated by means of examples. Characterising features are features like being red (or non-red), being a mountain, being a detective; non-characterising features are, for instance: all features concerning an item's ontological or modal status, like being existent, being possible, being real, being imaginary, being fictional, being created (*EMJB*$_2$, Vol. 1, pp. 60f., 331f.; this volume p. 251, fn.); intensional features, like being thought of by someone, being imagined by someone in such-and-such a way (*EMJB*$_2$, Vol. 1, p. 464); being red or green (this volume, pp. 116–119); being determinate (*ibid.*, p. 174); being such that it is not the case that p (*ibid.*, p. 177); being identical with something (*ibid.*, p. 192).

For Routley, the most disturbing problem of fiction concerns fictional characters' relations to entities (i.e., existing objects):

> What has to be taken account of are circumstances like the following, where a fictional object is physically related to an entity: The statement that Holmes lived in London (*hil*, for "Holmes lived in (or inhabited) London" with the predicate infixed, for short) is true, by the author's sayso; but the statement is also, in referential settings, false. For Holmes did not live in London, as empirical scanning would have revealed: a stake-out on Baker St. would have obtained no trace of Holmes, none of the numerous comings and goings said to have occurred there would have been observed. Holmes was not, that is, an historical figure, as we can now convince ourselves in other ways, e.g. there is no record of him in the British registers for births and deaths, even his address was fictitious as a visit to Baker St. will disclose. Thus London was not where Holmes actually lived: indeed since Holmes did not exist, was a merely fictitious person, Holmes could have lived nowhere actual, it may be said with some feeling. (p. 257f.)

One might expect, in light of what has been said so far, that Routley resolves this problem by stipulating that relations to entities do not belong to a fictional object's characterising features. Indeed, Routley considers this option – but ultimately he rejects it. (p. 189f.) Rather, Routley attempts to resolve the apparent clash with reality by recourse to contexts of utterances: "'Holmes lived in London' is true in one context, *c* say, and not true in another, a referential context *s* say". (p. 261)

For Routley, it is a desideratum for a theory of fiction that both of the following two theses hold:

> (1) Purely fictional items, such as Holmes, do not exist, and never have; they are unreal, do not come into existence, are causally unrelated to actual objects – and similarly for a batch of properties linked with (real) existence.
> (2) Notwithstanding (1), what an author ratifies, for example in the way of ordinary features for his characters, holds. (p. 267)

These two theses belong to what Routley calls "the ordinary naive view". Relational features, such as living in London, are supposed to belong to a fictional character's "ordinary features". (p. 267) It is one of the major advantages of the "contextual theory" of fictions that it allows us to hold to (1) and (2) simultaneously:

> The problem for any reinstatement, or reconstruction, of the ordinary naive view is how all these things can be true, how (1) and all the parts of (2) can hold simultaneously. The contextual account ensures – or claims to ensure – all this, and perhaps consistency as well, by relativising (1) and (2) to different contexts. An author is not an authority on the actual empirical world (on $\underline{\underline{G}}$) and not all he says holds for that, and there what he says (e.g. on radio interviews, and in newspapers, of the world) must take its chance and be tested against what does hold as a matter of fact; but within limits he is an authority on the worlds of his imagination and what he ratifies goes for the worlds of his fiction. (p. 268)

So far, in this section, I have presented a rough scaffolding of Routley's theory of fiction and fictional characters in *Exploring Meinong's Jungle and Beyond.* There are several ramifications and refinements of this theory. Since some of them shall be discussed in the following section, I have left them out in my first exposition of the theory, in order to avoid unnecessary repetition.

Some critical remarks on Routley's theory of fictions and fictional objects

Fictions and scientific theories

Right from the start of *Exploring Meinong's Jungle and Beyond*, and throughout large portions of this book, Routley draws strong parallels between fictional objects (in the sense explicated in the previous section) and objects of mistaken scientific theories (i.e., objects that have been erroneously assumed), often mentioning them in one breath. Consider, for instance, the following passages:

> Logically these objects [i.e., fictional objects, like Pegasus, Sherlock Holmes etc.] have a good deal in common with the objects of mistaken scientific theories. ($EMJB_2$, Vol. 1, p. 40)
>
> Just consider "what is" questions concerning fictional objects or the object of false theories, e.g. "What is a hobbit?", "What is phlogiston?" [...]. (p. 67)
>
> Of course it is true that Mr. Pickwick wore gaiters, that Sherlock Holmes lived in London, and that phlogiston is a heat substance; and it is true that James Bond stood in, or at least passed through, various doorways; [...]. (p. 78, note 1)
>
> Semantical difficulties concerning certain literary phenomena, those of fiction and fictions in particular, are not however so easily set aside or separated from general semantical puzzles – puzzles firstly over talk about nonexistent things and about intensional discourse, especially the language of invention and of imagination and the language of scientific discovery, conjecture, and speculation; and puzzles secondly about scientific theories, especially false theories, which after all predominate among theories. As to the first, the logical theory, logic and semantics, of fiction and fictions is a special, and very instructive, case of any general logical theory of nonexistence and intensionality. As to the second, scientific theories, especially mistaken ones, closely resemble fiction – much more closely than has usually been allowed by accounts other than fictionalism [...]. (p. 226)

Incidentally, according to Routley, nonentities occur even in (allegedly) correct scientific theories, namely as useful abstractions and idealisations, such as ideal gases, ideal fluids, perfectly rigid bodies etc. (p. 126).

Pace Routley, however, there are substantial differences between discourse about (non-existent) scientific objects and discourse about fictional objects. Consider one of Routley's examples:

(S) Phlogiston is a heat substance.

(S) may be uttered in (at least) three different contexts:

(i) the context of assumption, where (S) is uttered as a hypothesis that is yet to be proved;

(ii) the context of assertion, i.e. the context of proponents of the phlogiston theory;

(iii) the context of the history of science, in which the phlogiston theory is reported, where it is understood that this theory is false.

In the assumption context, S is short for "Let us assume/hypothesize that phlogiston is a heat substance". In this context, the speaker abstains from any commitment to either the existence or nonexistence of phlogiston, and, related to this, also from any commitment either to the truth or falsity of S. This is just the essence of hypothesizing. Moreover, in this context, the speaker is striving to verify or falsify S. He is, at any rate, interested in the truth value of S. In the assertion context of the phlogiston theory, the speaker is apt to commit himself both to the existence of phlogiston and to the truth of S. In the report context of history of science, in turn, the speaker is apt to commit himself to the nonexistence of phlogiston and to the falsity of S.

Now consider, by contrast:

(F_i) Sherlock Holmes is a detective.

Is there anything like the assumption context for utterances of this sort? – Well, one might construct a queer scenario in which there is: Imagine an immigrant from a foreign culture, not familiar with crime fiction, who incidentally picks up a fragment of discourse about Sherlock Holmes, wondering whether the incredible story can be true. Such a person might indeed utter F_i as a hypothesis. However, this is not a typical context of utterance of a sentence like F_i.

Is there anything like the phlogiston theory context for utterances like F_i? Again, there might be. Someone might listen to a reading of a Holmes story, take it at face value and thus acquires the belief that there is a detective named Sherlock Holmes. But again, this is a very atypical context. In fact, probably it does not exist outside philosophical thought experiments.

What about the history context? Is that perhaps the typical (or at least a common) context for utterances like F_i? I do not think so. In standard contexts, F_i is not uttered as a report about another person's false belief.

In fact, there are two distinct standard contexts for utterances like F_i. The first is the storytelling context, the second is the literary criticism context. In the storytelling context, the speaker does not claim truth for what he says, and he is not even interested in the truth value of his utterances; and the competent audience knows that no truth claims are made and is not interested in the truth values either. The question of truth or falsity is simply out of place in this context. By contrast, in none of the three contexts of the phlogiston sentence is the question of truth and falsity out of place!

Things are quite different in the literary criticism context. In this context, the speaker attributes a (internal) feature to a fictional character. Here, the question of truth and falsity is not out of place. The speaker claims truth, and the audience expects the speaker to tell the truth. Thus, the literary criticism context is clearly distinct both from the assumption context and from the context of the history of science. However, the literary criticism context is also distinct from the assertion context of the phlogiston theory, despite the parallelism that in both contexts the speaker claims truth. The difference is that the truth claim in the phlogiston theory context is not sufficiently justified and sentence S is in fact false, while the assertion in the literary criticism context is well justified and F_i is true. While S (in the assertion context) rests on a false scientific theory, F_i (in the literary criticism context) rests on a correct theory of literary criticism.

That sentences about fictional characters can be well justified and true is even more evident in the case of external attributions to fictional characters such as "Holmes is one of the most famous characters of western crime fiction", "Holmes has been created by Arthur Conan Doyle", "Holmes is embodied by Robert Downey Jr. in a recent movie series" etc.

Thus, the logic of fictional discourse is distinctly different from the logic of discourse about ancient (and nowadays dismissed) scientific theories. Routley's repeated parallelization of fictional characters with objects of (mistaken) scientific theories blurs these important differences.

There are two apparent motives for Routley's equalizing fiction and false scientific theories. Underlying both of them is the obvious fact that Routley felt a strong need to propagate his noneism against an overwhelmingly predominant philosophical establishment with a strongly sceptic or even hostile attitude against basic tenets of noneism. In fact, large portions of *Exploring Meinong's Jungle and Beyond* are devoted to a defence of noneism almost in the style of a dissident pamphlet, telling a tale of unsympathetic prejudices, misrepresentation and disregard to be fended offby a kind of philosophical Robin Hood.

The levelling down of fiction and science helps in two ways to make noneism a reputable position: First of all, it is generally considered an advantage of a theory if it can be shown that it accounts for a number of distinct phenomena in a uniform way, not just for one isolated class of phenomena. Second, and perhaps more importantly, within the "philosophical establishment", the domain of science counts as much more relevant and respectable than the domain of fiction. Thus, a theory that accounts exclusively or primarily for phenomena of fiction is much more likely to be dismissed as marginal and unworthy of serious consideration than a theory that (also) accounts for phenomena of science.

For the sake of fairness, however, it should not be concealed that at least at one point, at the very end of his chapter on fictions, Routley admits differences between objects of fiction and objects of science. Even here, however, Routley first emphasises the (alleged) kinships:

> A theoretical object is, crudely, one of the "characters" of a (scientific) theory, of a theory that, in live cases, purports to elucidate empirical reality. Objects of a dead or failed theory behave very like a fictional object, like the objects of myth. Most obviously, theoretical objects belong to some theory while fictional objects are supplied by a work of fiction or the like, not a very hard and fast boundary. (p. 310)

Only with respect to "live" scientific theories (i.e., theories that are not yet dismissed as erroneous by the scientific community), Routley finally acknowledges some differences:

> Then theoretical objects are, according to insiders, always deductively closed, fictional objects are sometimes not; theoretical objects stand in entire relations to their own kind and to entities, fictional objects do not. So in particular, with statements concerning theoretical objects, passive conversion is always admissible and frames are commonly extensional and so admit replacement of identicals: many of the contextual intensional aspects of genuine fiction are removed.

> Insiders operate in these logical ways with the theory and its objects: outsiders are more cautious. The key distinguishing features derive from the intended role of theoretical objects in giving a description and explanation of the empirico-historical world (i.e. $\underline{\underline{G}}$). So too do other distinctive constraints on theoretical objects: that the theories to which they belong are subject to empirical falsification, etc. In marked contrast, the source books of fictional objects are not so subject. Holmes does not have to answer to historical facts. With respect to Holmes we are all (well, almost all) outsiders. (p. 310f.)

There is a tension between the passage just quoted and the repeated emphasis on similarities between theoretical and fictional objects, which may be a symptom of a certain unease Routley himself felt about that matter. Moreover, as I have argued, the differences between theoretical and fictional objects concern not only "live" scientific theories, but also "dead" ones.

Contextualism as an alternative to elliptical paraphrase strategies?

Large portions of chapter 7 consist not of a construction of Routley's own theory of fictions but of critical assessments of rival theories. In particular, Routley critically discusses and rejects what he calls "paraphrastic and elliptical theories of fiction". (This is the subtitle of section 7.3.)

According to the "Reference Theory" (which Routley rejects as the core of the misguided philosophical establishment which he sets out to dismount), sentences that seem to be about fictional characters (e.g. "Holmes is a detective") carry an ontological commitment to fictional characters: sentences of this kind cannot be true unless fictional characters exist (or at least have some kind of being). According to the paraphrase theory of fiction, such ontological commitments are undesired, and they are to be avoided by not taking sentences of the relevant sort at face value. Rather, such sentences should be interpreted as improper renderings of something else, which, if formulated in the proper way, do not carry the unwanted ontological commitments.

The challenge for proponents of a paraphrase theory is to give an account of what, exactly, such sentences are supposed to express, how the proper rendering exactly is supposed to look like. The (alleged) proper rendering of a problematic sentence is called a *paraphrase* of this sentence. Thus, the task of a paraphrase theory of fiction is to provide paraphrases (or a method of paraphrasing) for all sentences that would, if taken at face value, carry an ontological commitment to fictional characters. Adequate paraphrases must fulfil two conditions: First, they should express in an unambiguous and proper way what the original sentences are meant to express. Second, they should not carry an ontological commitment to fictional characters.

There are a large variety of paraphrase theories of fiction. Here are two examples of paraphrases explicitly discussed by Routley: 1. "'Fergus' is a fabulous-entity-name" as a paraphrase of "Fergus is a fabulous entity"; 2. "Pegasus must be treated as it would be treated if it were a horse" as a paraphrase of "Pegasus is a horse". (p. 247f.)

More important than these examples, however, is a particular subclass of paraphrase theories, namely those that make use of a sentence operator ϕ, such that one gets a paraphrase of a problematic sentence simply by prefixing ϕ to the sentence in question. Theories of this sort are called "elliptical" by Routley, since the basic idea is that sentences like "Holmes is a detective" are interpreted as elliptical renderings for "ϕ: Holmes is a detective".

Elliptical theories differ with respect to the way in which they flesh out the operator ϕ. Routley provides, if not a complete, at least quite a comprehensive collection of examples of what ϕ is taken to or might stand for: "It is written that", "It is told that", "It is said that"; "It is pretended that", "It is fictionalised that", "It is imagined that"; "Doyle wrote that" (p. 250); "In the story" (p. 254).

Routley provides objections against a number of particular paraphrases (elliptic and others). More importantly, however, he rejects paraphrase strategies in general, for the following reason:

> [N]o purely elliptical theory is faithful to ordinary discourse about fictional objects [...]. For ordinarily we can say truly, at least sometimes, such things as "Holmes was a detective". But what happens to this statement A on the elliptical theory: it is replaced by another A^* (e.g. OA [where "O" stands for a sentence operator of one of the sorts listed above, or something akin]) which is supposed adequately to replace A, and thus be in some sense equivalent to A (or what A

> was intended to say). What we can say truly with A is said to be properly said with A^*. But now what about the original A [...]? A is not meaningless [...]. Nor is A true, as follows from the referential assumptions underlying elliptical theories; [...]. So it is false, which is what Ryle says and what most other elliptical theorists tend to admit at least when pressed – but which is *not* what we always say ordinarily. And if A is false, then it is not equivalent to A^*[,] which is true. Likewise if A is not truth-valued. So the elliptical theory does not provide a satisfactory paraphrase – unless A is in some way ambiguous. But to insist upon ambiguity of fictional statements such as A is again to diverge from what we ordinarily say. (p. 254.)

This line of reasoning is reminiscent of William P. Alston's objection against paraphrase strategies in Alston 1958 and 1967. To sum up Alston's argument: If A^* is an adequate paraphrase of A and if A implies a certain ontological assumption, then A^* has to imply the same ontological assumption; for adequacy requires sameness of meaning, and if two sentences share the same meaning, they also share the same implications. Therefore, either A^* is not adequate (i.e., it does not express what A expresses) or A^* does not help to avoid an undesired ontological commitment. (Cf. Alston 1958, pp. 8-10 and Alston 1967, p. 288f.)

The proper reply to Alston's argument is as follows: It is misguided to stipulate sameness of meaning as an adequacy condition for paraphrases. The paraphrase A^* should not express what A already expresses but rather what A *was intended to express* but does not express (at least not in an unambiguous and proper way). What a speaker intends to express with the utterance of a statement A is not always what A literally means. Sometimes, speakers make use of metaphors, ellipses and other improper ways of speaking, and sometimes they do so without full consciousness. (Think of examples like "The average Austrian family has 1.6 children"; "The moon was hiding behind a cloud" etc.) The task of an adequate paraphrase A^* is not to provide an alternative rendering of what A already literally and unambiguously expresses but to bring out clearly and unambiguously what a speaker (or a group of speakers) intends to express with A but fails to express, if A is taken at face value.

If A^* does not have the same meaning as A (if taken at face value), it may well be that A implies ontological commitments that A^* does not imply. Thus, Alston's argument does not show that paraphrase strategies are always unsuccessful as a means to avoid undesired ontological commitments.

That sameness of meaning (in a sense that implies equivalence) is not the proper adequacy condition for paraphrases is hinted at by Routley in the above quotation. Consider again: "[A] is replaced by [...] A^* [...] which is supposed adequately to replace A, and thus be *in some sense* equivalent to A (*or what A was intended to say*)". (p. 254; my italics.) But Routley does not give due attention to this point, for he goes on to argue as if A^* had to be equivalent to A ("[...] if A is false, then it is not equivalent to A^* [...]"; *ibid.*).

However, Routley touches upon this issue again in the same passage. Consider again: "So the elliptical theory does not provide a satisfactory paraphrase – *unless A is in some way ambiguous*". (p. 254; my italics.) Here, Routley implicitly admits that a paraphrase strategy might be a viable option if the original sentence A could be considered to be "in some sense ambiguous". Routley hastens to dismiss the hypothesis that at least some "fictional statements" might be considered to be ambiguous by arguing that this "is again to diverge from what we ordinarily say". (*ibid.*)

The latter claim is doubtful, however. Consider, for instance: "Zeus resides at Mount Olympus". Imagine two utterances of this type of statement, the first one uttered in 460 BC in Athens by a Zeus believer, making his child familiar with his religion, the second one uttered in 2018 by a classics scholar in a lecture on ancient Greek religion and mythology in an adult education centre. To me it sounds very natural to say that these two utterances do not have the same meaning, that the utterance type is thus ambiguous and that the first utterance is false, while the second one is true. Obviously, to Routley, this does not sound natural. But it is at least an open question whether the ambiguity view really diverges "from what we ordinarily say", i.e., whether it runs counter to common sense; and even if it could be shown that it runs counter to common sense, this would not yet show that it is false. For though common sense may be often a guide to truth, the mere fact that a majority of speakers is apt to agree to a certain statement is not, by itself, a sufficient reason for a philosopher to hold the statement to be true.

Whether appeal to common sense provides a strong reason to integrate a certain belief into a (philosophical) theory depends, upon other things, on how deeply and firmly the belief is entrenched in our common sense views. I doubt that the belief that "Zeus resides at Mount Olympus" is an unambiguous (and true) sentence is deeply and firmly entrenched in many ordinary people's belief systems, at least in comparison to beliefs like that there is a mind-independent reality or that there exists more than one sentient being.

What holds for "Zeus resides at Mount Olympus" holds, *mutatis mutandis*, also for "Holmes lives in London". Routley disagrees: "*The statement hil* [i.e., "Holmes lives in London"] *is not ambiguous*: that is a guiding principle that will be adopted and defended in what follows". (p. 259; emphasis taken from the original.)

Perhaps it is noteworthy that at other places Routley expresses himself more cautiously about this matter: "'Holmes lived in London' *does not appear to be* ambiguous in the sense of having more than one sense, and there *do not appear to be* any structural features which would account for the alleged ambiguity". (p. 258f.; my emphasis) "Structural features which would account for the alleged ambiguity" would be that at least one component of the statement "Holmes lived in London" is ambiguous, i.e., either "Holmes" or "London" or "lived in". But this is not the case, according to Routley. On the other hand, however, Routley admits: "Something *like ambiguity* in many statements of fiction – by no means all – is forced by any theory which allows [a] work of fiction N to ratify statements as true, which incorporates a (qualified) *sayso condition of truth* [...]". (p. 257; emphases taken from the original.)

Why does Routley hold the assumption of "something like ambiguity" for some fictional statements to be a necessary ingredient of a theory of fiction? The reason is that, as Routley notes, a statement like "Holmes lived in London" obviously contradicts empirical facts (at least if "London" is understood in its usual meaning, i.e. as the name of a city in the existing world). (*ibid.*)

Routley presents as a solution to this problem what he calls a "contextual theory" of fiction:

> There is no riddle, then, on the contextual theory, as to how the consistency problem for statements such as "Holmes lived in London" is to be resolved without resort to ambiguity in the more ordinary sense: the missing factor, which makes the difference, is context. "Holmes lived in London" is true in one context, c say, and not true in another, a referential context s say. (p. 261)

Routley does not provide an explication of the notion of context, at least not close to the above quoted passage. Only in the following section, the reader gets at least some hints as to what is meant by "context":

> There are, of course, sometimes problems as to which context one is operating in. One can be confused as to what the context is, e.g. as to which work a given character appears in, not knowing what the context is, or be entirely mistaken as to what it is. Such indeed provides part of the evidence for the contextual theory. There is much circumstantial evidence of this sort supporting the contextual account of fiction. For example, there is the experience of not knowing the context, familiar to people who have not read a book, been to a film, or seen a television show, that is under discussion. Such people are left out of discussions by not knowing enough of the context: they cannot play the fictional game properly, a little like someone who does not know a language well. (p. 266)

So the relevant context can be a work. But the explanation just quoted raises more questions than it answers. First, it seems that Routley's "contextual" solution comes very close to a story operator strategy, i.e. to prefixing fictional sentences with a "story operator", like "According to Dickens' *Pickwick Papers*", "According to Shakespeare's *Hamlet*" etc. However, the story operator strategy is one of the "elliptical theories" which Routley explicitly rejects. An explication of the difference between these dismissed theories and contextualism is not given.

Second, the passage quoted above suggests that the relevant context for utterances of fictional sentences is always a particular work. But this cannot be. For, as Routley acknowledges at several other places, it makes sense to compare fictional characters of distinct works within a single sentence. Consider, for instance: "Holmes is smarter than Poirot". In cases like this, the context cannot be restricted to a single work. It remains unclear what, then, the context is supposed to be, according to Routley's theory. The problem is even more pressing for "transfictional sentences" like "Holmes is the most famous character of crime fiction in the history of literature", "Goethe's Faust character is based upon several older Faust characters" or sentences like "Holmes does not really exist; he is a fictional character".

Third, it is unclear why epistemic uncertainties as to the context should support evidence for Routley's "contextual theory", as Routley claims. Sure, people who do not know the work in which a character originates often cannot join in a discussion about the character in question. This certainly shows that knowledge about a work in which a character originates is an important source for knowledge about the character. (It is not the only source: one could learn a lot about Shakespeare's Hamlet character without ever reading the play itself, just by studying secondary literature.) From this, however, it does not follow that the "contextual theory" is superior to its rivals, for instance to the abstractness theory, i.e., the view that fictional characters are abstract objects. (See p. 256f. for Routley's rejection of this theory.)

Fourth, it is curious that Routley uses the term "fictional game" here. None of Routley's claims about fictional works and their objects motivates the use of this term. To the contrary: it is often used in the context of "pretence theories" of fiction, but these belong to the theories of fiction which Routley explicitly rejects:

> Most authors are not engaged in an elaborate process of pretence. They tell a story; they do not pretend that it is literally true. Dickens did not pretend that Mr. Pickwick was fat or that he existed. On the contrary, authors (like their audiences) are well aware that fictional characters do not exist; recall, for instance, the notices that it used to be common for authors of fiction to insert at the beginning of their works, which implied, among other things, but not always honestly, that none of the characters ever existed. [...] The core thesis of the pretence theory is false: what is right about the thesis that fiction is pretence is that fiction is fiction. (p. 251)

Fifth, even more curious is the suggestion at the end of the above quotation: Routley seems to maintain that a person who cannot join in a discussion about a fictional character because she lacks acquaintance with the relevant work or works in fact lacks linguistic competence (is "like someone who does not know a language well"). This does not only suggest that there is a particular language of "fictional discourse" but even the stronger claim that there is a particular language for discourse about each single fictional work. Not only is this highly implausible; it also seems to run counter to Routley's denial of any ambiguity in fictional sentences.

Apart from these minor difficulties with the explication of the notion of context, there is a principled objection against the contextual theory: It seems that it amounts to the claim that fictional sentences are ambiguous. For Routley claims that "Holmes lived in London" is true in some contexts and false in others. This sounds plausible. If this is so, however, there must be some reason for it. It suggests itself that the reason is that "Holmes lived in London" has different meanings in different utterance contexts. But this is one of the views that Routley fervently rejects, as has been noted above.

Routley himself takes notice of this problem: "One objection is that the contextual theory still treats fictional claims as ambiguous; it merely relocates the ambiguity, as in the context". (p. 263) He replies as follows:

> If "ambiguity" is stretched in this way, beyond its dictionary senses (such as "double meaning"), if one insists upon calling different contextual resolutions those of ambiguity, then of course the theory is a further "ambiguity" position – any position which resolves the inconsistencies is. So is any statement involving "egocentric" particulars; e.g. "I am hot" is (potentially) infinitely ambiguous! But if due account is taken of what ambiguity *is* like, if different senses are required for ambiguity, then the contextual theory is very different from an ambiguity position. (p. 263f.)

This reply is unconvincing. If a sentence's truth value differs in different contexts of utterance, this contextual truth-value variability needs to have some basis. The basis may consist in an ambiguity of the sentence. Admittedly, however, this is not the only possible basis for contextual truth-value variability. I agree with Routley that in the case of "I am hot" the basis for the contextual truth-value variability is not an ambiguity. Rather, "I am hot" has different truth values in different contexts because it contains an indexical expression ("I"). Indexical expressions (like "I", "here", "now", "this", "yesterday", "5 minutes ago", "my sister" etc.) involve an implicit reference to the context of utterance. This is an essential semantic feature of this particular type of expressions. Therefore, the contextual truth-value variability of sentences that contain indexical expressions is an effect of an essential feature of indexicals. This feature cannot, however, be transferred arbitrarily to other types of expressions – which seems to be what Routley attempts to do.

Are there any other possible bases for contextual truth-value variability, besides indexicality and ambiguity? To examine this question, consider some examples of sentences that might be conceived as context sensitive with respect to their truth values:

(1) Berlin is the capital of Germany.

The truth value of this sentence is context sensitive, because it depends on the date of utterance. How is this to be explained? As I see it, the explanation is as follows: Although the sentence does not contain an indexical expression, there is an implicit reference to the date of utterance, and this reference is given through the *tense* of the verb. "Berlin is the capital of Germany" is (in standard contexts at least) synonymous with "Berlin is *now* the capital of Germany".

In some special cases, the grammatical present tense is used as atemporal or historical present tense. These are clearly non-standard uses, chosen for stylistic reasons. If "is" is used in the atemporal sense, it is to be interpreted as "was, is now and will be"; if it is used in the historical sense, it is to be interpreted as "was". If there is an unclarity as to whether the present tense is used in the standard sense or in the atemporal or in the historical sense, this seems to be a case of ambiguity.

(2) Action A is legal.

This sentence, too, seems to involve an implicit indexical reference to the context of utterance. To say that A is legal is the same as to say that A conforms to the laws that are in force *here and now*. Analogously, to say that A *was* legal is to say that A conformed to the laws that *were* in force *back then*.

If one does not hold moral objectivism, one might hold similar considerations for

(3) An action A is morally acceptable.

One might argue that to say that A is morally acceptable is to say that A conforms with the moral norms and values that are generally accepted today/within the speaker's cultural environment etc.

(4) S knows that p.

According to epistemic contextualism, knowledge attributions are context sensitive with respect to their truth value. The basic idea of epistemic contextualism is that certain contexts may require more demanding standards of justification than others. Let us assume, for the sake of argument, that epistemic contextualism is right. Does this imply that knowledge attributions are ambiguous?

I think that this is a possible (not outrightly implausible) explanation. It may be that in some contexts, knowledge implies certainty and in other contexts it doesn't. If this is the case, "to know" is ambiguous.

Perhaps there is, however, another possible explanation for the contextual truth-value variability of knowledge attributions. One might argue that knowledge attributions involve an implicit reference to the context of utterance: To say that S knows that p is to say that S is *sufficiently well* justified to believe that p; and "su ffi ciently well" means "good enough *for the present context*". According to this explanation, there is again an implicit indexical reference to the context of utterance.

To sum up the discussion: in all the examples so far considered, the contextual truth-value variability is based either on ambiguity or on (explicit or implicit) indexicality. In fictional sentences like "Holmes lived in London" I cannot find any hidden indexical reference. Therefore, it seems to me to be reasonable to assume that in this case the contextual truth-value variability is based on an ambiguity.

Of course, I did not show that there are no other possible bases (besides indexicality and ambiguity) for contextual truth-value variability. Furthermore, I did not show that there is no hidden indexicality in "Holmes lived in London". However, in this case, the burden of proof is on Routley's side: for he claims that "Holmes lived in London" is not ambiguous, although it has different truth values in different contexts of utterance, without giving any account of this context sensitivity. Thus, there is (to say the least) an explanatory gap in his contextual theory.

Some further difficulties

(i) Remember that, according to the Characterisation Postulate (CP), "nonentities have their characterising properties" ($EMJB_2$, Vol. 1, p. 32), but that properties concerning an item's ontological status are not counted among the item's characterising properties (in order to avoid "clashes with empirical reality"). In the case of fictional objects, however, the exclusion of "ontic properties" from the domain of characterising properties is implausible. For we can (and do) distinguish – within the context of a fictional work

– between existent and non-existent items. For instance, within Shakespeare's *Hamlet*, Hamlet clearly exists, but – at least according to one interpretation of *Hamlet* – the ghost of Hamlet's father does not exist (but is just a figment of Hamlet's and his friends' imagination).

Thus, it is a desideratum of an adequate theory of fiction to allow for the distinction between existent and non-existent fictional objects. It is unclear, however, how one can do this without accepting that ontic properties may be characterising properties of fictional objects.

(ii) On p. 80, Routley writes: "Some identity claims concerning nonentities are indeterminate, e.g. which of the various Faustus' of the literature are the same". Thus, Routley admits a serious indeterminacy for at least some fictional objects. By "serious indeterminacy" I mean indeterminacy that results in truth-value gaps for identity statements. The feature of incompleteness explained in the first section of this paper does not result in "serious indeterminacy". It is curious that Routley admits indeterminacy of identity, for the so-called "extensional identity criterion" which Routley postulates for nonentities in general does not admit of such an indeterminacy – or so it seems.

According to the extensional identity criterion, an object x and an object y are identical if, and only if, x and y share exactly the same extensional properties. (pp. 71-75) As far as I can see, this should suffice to settle the question of "which of the various Faustus' of the literature are the same". A Faust character who enters a pact with the devil cannot be identical with a Faust character who refrains from doing so; a Faust character that finally goes to hell cannot be identical with a Faust character who finally receives salvation by God's grace; etc. Thus, it is unclear as to how identity indeterminacy can arise within Routley's theory of fiction.

(iii) Another difficult remark concerning identity conditions for fictional objects is to be found on p. 154: "Venus is the same as Aphrodite because their parts (histories, etc.) coincide; Holmes is not the same as Moriarty because their parts go their separate ways". Routley does not explicate his notion of a "part" of a fictional object, which makes this remark hard to understand. Furthermore, it is left unclear what is meant with "histories" in this context: Is it the story or stories in which Venus/Aphrodite are "native", or is it the history of creation, passing on and reception? If the latter, in what sense can this be a "part" of a character? Or is, in the above quotation, "histories" not to be understood as an *explication* of "parts" but rather as an *addition* (such that the whole passage reads "because their parts *and* histories etc. coincide")? At any rate, this remark is not really embedded in an elucidating context, and the notions of a fictional object's part as well as a fictional object's history remain unexplicated.

(iv) Consider the following passage, in which Routley sets out to argue against reductionist strategies (i.e., strategies to reduce discourse allegedly about nonentities to discourse about entities such as literary works and scientific theories):

> The Meinongian approach does not end – cannot satisfactorily end – with a statement of simple facts such as that a nonexistent object, Sherlock Holmes, has certain properties, e.g. he smoked a pipe but not cigars. There is always an account to be given of how nonentities acquire the properties that they have. In certain cases this will involve details of actual stories[, in] other cases theories, and so on. *In some cases it may involve details of merely possible or conjectural stories.* For the theory should be able to take account of the deeds of unsung heroes, just as it does of unstated propositions and unformulated theories – matters that make for very serious trouble for "realist" accounts. Indeed for this sort of reason a full reduction to actual source books is impossible. (p. 161; my emphasis)

Once more, Routley here equates fictional discourse with scientific discourse. However, this passage is worthy of attention for still another reason: Routley here lumps together (actual) fictional objects with merely possible fictional objects. This strikes me as a serious weakness. As I see it, there is a robust sense in which one may (truthfully) say that there is a fictitious object that is a pink panther. There is also, as I learned recently, a fictitious pink rhinoceros, a fictitious pink kangaroo and a fictitious pink porcupine – but (to my knowledge, at least) so far there is no fictitious pink bedbug. It is a desideratum of a theory of fiction to us allow to distinguish between the ontic status of the fictitious pink panther (rhinoceros, kangaroo etc.) on the hand and the fictitious pink bedbug on the other. It seems that Routley's theory does not allow for this distinction, for, according to him, fictitious objects can be native both in nonexistent stories and in existent stories.

(v) Remember that Routley admits (correctly) that what *is true* according to a (literary) fictional work does not have to (and typically does not) coincide with what *is told* in the work. Routley rightly points out that there may be (and typically are) sentences that are true according to a fictional work but do not occur in the work. (See section "Fictions and fictional objects in the framework of noneism" above.) However, Routley does not take notice of the fact that the reverse holds as well: There may be sentences that occur within a work and yet are not true according to the work. Routley does not notice that the narrator (who is not to be identified with the empirical author!) may be an unreliable figure. He (the narrator) may be a liar, intending to deceive the reader, or he may be epistemically unreliable (for instance, because prejudices, animadversions or sympathies distort his perception). In such cases, it is usually the author's intention that the readers realise the narrator's unreliability, such that the readers realise that some of the sentences that the work consists of are not true according to the work.

Summary and short assessment

Routley's theory of fiction is embedded in his much more comprehensive and extremely complex theory of nonentities, his noneism, of which I could provide in this paper only a rough sketch. Yet, it seems that the topic of fiction and fictitious objects provides a good deal of Routley's motivation for noneism.

I have tried to give a concise (and, as I hope, basically adequate and fair) presentation of his theory of fiction. I refrained from criticising his theory of fiction directly for its dependency on certain tenets of noneism. Rather, I tried to grant, for the sake of this paper, the basics of noneism but rather pointed out a number of more or less internal difficulties (though I did not intend to conceal my general "referential" outlook, which Routley would, of course, reject as fundamentally misguided).

Some of these difficulties might be overcome without much effort in a slightly revised version of Routley's theory; others might be more obstinate, but still resolvable. As I see it, Routley's theory of fiction in *Exploring Meinong's Jungle and Beyond* is not a fully developed mature theory, but partly tentative; at some places, the theory suffers from an undue alignment of fictitious objects with other "nonentities". However, for those who are willing to accept noneism in general, Routley's theory of fiction surely is still a viable option – as is evident from a number of publications of contemporary noneists (see, e.g., Priest 2005 and Priest 2011, Berto 2008).

References

$EMJB_1$ Routley, R., *Exploring Meinong's Jungle and Beyond*, Research School of Social Sciences, Australian National University, 1980.

$EMJB_2$ Routley, R. et al., *The Sylvan Jungle*, edited by M. Eckert et al., vol. 1–4, Springer Verlag, Berlin, 2018+.

Alston, William P., "Ontological commitments", *Philosophical Studies* **9** (1958) 8–17.

Alston, William P., "The Ontological Argument revisited", in *Descartes*, edited by W. Doney, Notre Dame Press, Notre Dame, Ind., 1967, 278–302.

Berto, Francesco, "Modal Meinongianism for fictional objects", *Metaphysica* **9** (2008) 205–218.

Priest, Graham, *Towards Non-Being: The Logic and Metaphysics of Intentionality*, Clarendon Press, Oxford, 2005.

Priest, Graham, "Creating non-existents", in *Truth in Fiction*, edited by F. Lihoreau, Ontos, Heusenstamm, 2011, 278–302.

Routley's second thoughts

Fred Kroon

Routley[1] died in 1996, but his work is still throwing up surprises. The present paper is devoted to a topic for which he is perhaps best known: his rediscovery and development of Meinong's theory of objects. In this paper I first discuss Routley's canonical version of Meinongianism in *Meinong's Jungle* ($EMJB_1$; reprinted here as $EMJB_2$) and some criticisms of views like Routley's that have led Graham Priest and Franz Berto to suggest an alternative Meinongian view, Modal Meinongianism. Notably, this new view rejects a pivotal part of Routley's version of Meinongianism: the nuclear/extranuclear distinction among properties. I then turn to Routley's attempt to liberalise his earlier views in a paper published shortly before he died, which has recently led to speculation on the part of Priest that Routley had come round to a view that is rather close to Modal Meinongianism despite retaining the nuclear/extranuclear distinction. After briefly outlining Routley's revisions of his earlier view, I compare the new account to Modal Meinongianism and argue that there is a good reason why Routley revised his account in the way he did rather than rejecting the nuclear/extranuclear distinction.

Meinong's Jungle: map of the terrain

Like Meinong, Routley repudiates the assimilation of objects to objects that have some kind of being or ontological status. What is vastly more important is the fact that objects present themselves to us in some way: they are given in terms of properties, and objects can be given in this way without having any sort of being or *Sein*. The golden mountain,

[1] Richard Routley changed his name to Richard Sylvan, of course, but his work on Meinong began when he was Routley and that is what I will call him in this paper.

for example, is given in terms of the properties of *being a mountain* and *being golden*, the round square in terms of *being round* and *being square*, Holmes in terms of the properties he is credited with in the *Holmes* stories, and so on. None of these objects exist. More precisely, none of them have any form of being or *Sein*, where for Meinong the category of being includes, and is exhausted by, two modes of being: what he called existence (*existenz*), the mode of being appropriate to all objects that are genuinely located in space/time, as well as subsistence, the mode of being appropriate to all abstractions from the actual relationships that hold among the former. Routley himself parts with Meinong on this point. Roughly speaking, the only kind of being there is for Routley is temporal being. All objects not located in time – mathematical objects, universals, ideal objects, ... – lack any kind of genuine ontological status. He calls this view *noneism*, and his magnum opus *Meinong's Jungle* is the undisputed bible of noneism.

But what objects are there? Early Meinong seems to have thought that *any* combination of properties determines an object, that the realm of objects is governed by Unrestricted Freedom of Assumption. Routley agrees (pp. 216-219). In short,

UFA For any set S of properties, there is an object that is determined by (the properties in) S.

But this threatens contradiction, as Russell saw with the case of the round square. Consider the following intuitive principle, the unrestricted *characterisation principle*: an object that is determined by a combination S of properties has each of the properties in set S, i.e.

UCP Given a set S of properties, if O is the object determined by S and if P is a member of S, then O has P.

If this principle is correct, the round square would have to be both round and square, and so would be a contradictory object. To this formulation of the problem Routley accepts the following solution: being square, the round square is indeed both round and non-round (being non-round is what Meinong calls a consecutive property, one that is entailed by an object's constitutive properties). But we can't then infer that the round square is not round. Having a negative property doesn't entail not having the corresponding positive property (in this case the object possesses both).[1]

[1] Meinong himself didn't see this as a deep problem, and responded that there is no reason to think that the law of non-contradiction extends beyond the actual and the possible. (This didn't satisfy Russell,

But of course it is now only a short step away to a more serious problem, also posed by Russell, the famous problem of the existent round square. By Meinong's reasoning (and assuming both UFA and UCP) the existent round square should not only be round and square but also existent (Russell 1907). And this is absurd, since nothing that is both round and square can possibly exist. (Russell used this objection to press home his worry about impossible objects, but the point is independent of the notion of internally inconsistent objects like the round square; it applies equally to objects whose nature is internally consistent, like the existent golden mountain, and that is the example I will work with.)

Crucially, Routley takes the properties of being golden, round, or a detective living in London to be very different sorts of properties from the properties of existence and non-existence. The former are nuclear properties, the latter extranuclear (or, in Routley's preferred terminology: the former are characterising properties, the latter non-characterising). The distinction is one that Meinong himself made in Meinong 1915, using a suggestion made by his student Ernst Mally, and is also defended in the work of Terence Parsons and Dale Jacquette (Parsons 1980, Jacquette 2015). Existence and non-existence share their extranuclear status with such properties as being possible, impossible, complete, and incomplete. The rough idea is that a nuclear property is the kind of property that helps to make an object what it is, while an extranuclear property captures something of the relationship between the real world and the object, given the properties that determine the object. Take the golden mountain. It lacks existence since nothing in the world is a golden mountain. Or take the *possible* round square (that's the way the object is presented to us). It has the nuclear properties of being round and square, but whether or not it is possible depends on whether there *can* be something with the properties of being round and square in the world – the fact that *possible* occurs in the characterisation is entirely irrelevant. So "possible", like "exists" is extranuclear.

To solve the problem of the existent golden mountain, Routley replaces UCP with RCP, a restricted form of the characterisation principle:

RCP Given a set S of properties, the object O determined by S possesses each *nuclear* property in S.

---

of course, because he understood the contradiction as follows: the round square is round and square, so both round and not round; and that is a straight-out *propositional* contradiction of the kind that the law of non-contradiction rules out. This is an inference Routley rejects.)

Given that being round and being square are nuclear properties, it follows that the round square is round and square. Knowledge of the incompatibility of these two properties then tells us that this object is *not* possible and so lacks the extranuclear property of being possible. We can similarly conclude that the golden mountain is both golden and a mountain, but lacks the extranuclear property of existence. There is no need to worry about the existent golden mountain, for while the existent golden mountain is indeed an object (that much is guaranteed by UFA) and is both golden and a mountain (by RCP), the restriction imposed by RCP means that we can't deduce that it exists.

Much of this will be familiar. Those acquainted with the neo-Meinongian literature will know that there are two traditional ways of dealing with the problem of the existent golden mountain. Routley, like Parsons and Jacquette, follows the two-kinds-of-property or dual-property strategy (2P), while Zalta and some others follow the two-modes-of-predication or dual-copula strategy (2C). (Zalta calls the two modes *exemplification* and *encoding*; see Zalta 1983, Zalta 1988). Actual mountains like Mt Ruapehu exemplify being a mountain, while the existent golden mountain is an abstract object that encodes it. And Mt Ruapehu exemplifies existence, while the existent golden mountain encodes existence but exemplifies nonexistence. So Zalta and other 2C-ers need only one kind of property but two modes of predication, while Routley and fellow 2P-ers only need one mode of predication but two kinds of properties. Note, however, that Zalta and Routley agree on one thing. They both subscribe to a version of (UFA): for any set S of properties, there is an object determined by S. (For Zalta, this is given by a comprehension principle.) By contrast, Parsons and Jacquette accept only a restricted version of (UFA): for any set S of *nuclear* properties, there is an object determined by S. The difference will be important later in the paper.

Fiction and the problem of existence

Routley offers his brand of Meinongianism as a viable theory of the semantics of fiction, and not just the semantics of talk involving (apparently) non-referring descriptions like "the golden mountain". Other neo-Meinongians share this goal for their own versions of Meinongianism. But fiction presents 2P-ers with special problems. To see this, consider Conan Doyle's *Holmes* stories. These stories make it true that Holmes lived in London, at 221B Baker Street, and that he was a detective. But they also make it true that

Holmes *existed.* The stories don't need to say this, because the point is utterly obvious: if Watson had jokingly told one of Holmes's clients "Holmes doesn't exist", Holmes would have regarded this as the most elementary of falsehoods, far more basic a falsehood than the claim that he wasn't a clever detective. Note that the property of existence at play here is the extranuclear property of existence, not some weakened notion of existence. (In fact, it is hard to imagine what it would be like for Holmes to have only a weakened property of existence; would he appear a ghostly figure to Watson, perhaps?)

The evident falsity of the sentence "Holmes doesn't exist" in the mouth of the person who knew Holmes best (Watson) might suggest that it is true that Holmes has the property of existence, just as it is true that he has the property of being a detective. But that of course would be the wrong conclusion to draw, since we also think that it is categorically false that Holmes ever existed. What has gone wrong?

One way to deal with this problem is to make use of Meinong's doctrine of the modal moment and the ensuing view that there are weakened nuclear versions of standard extranuclear properties like "existence": there is not just existence, there is also the nuclear property of *being existent,* and what Holmes has is the latter while he lacks the former. In *Meinong's Jungle* Routley claimed that his theory didn't need Meinong's doctrine of the modal moment. (He didn't, in fact, consider the problem we are presently considering.) Another way out, explicitly adopted by Jacquette, is to say that the relevant nuclear property that properly characterises Holmes is not existence but something like *being described in the story as existing.* Holmes has this property, so does Hamlet (relative to the story in which he appears), and so does every other fictional character, relative to the works in which they appear.

It is hard to see how either way out solves the problem, however. The main difficulty for both remedies is that it is unclear why we shouldn't apply notions like weakening or story-relativising to all the other properties that are ascribed to fictional objects, leaving us with a Holmes who is weakly a detective or described as a detective in the Holmes stories, without being a detective in the full extranuclear sense. There is in fact a good reason why we should take this route, one that will be familiar from the work of Graham Priest. Most properties we think of as clearly nuclear seem to be existence-entailing. Try as we might, it is hard to think of anyone being a detective or a prince or an unhappy house-wife, or, simply, a *person,* without also thinking them as existing. To be a person requires an individual to be located in space and time, to be flesh and blood (so *concrete*, not abstract), and that requires existence. That is precisely the intuition we exploited when presenting the scenario of Watson's declaring Holmes not to exist. That claim struck as ludicrous because it is obvious that to be a person Holmes had to exist. If we now want to say, as Routley and other 2P-ers do, that full-blooded existence is *not* one of Holmes's properties, even though Holmes's existence is implied by his having properties that 2P-ers regard as paradigm instances of nuclear properties, that suggests we should say the same about the other properties standardly ascribed to Holmes. If the claim is that Holmes has full-blooded existence in the story, but that *apart from the story* he is existent without existing we should similarly say that he has the full-blooded extranuclear property of being a detective from the point of view of the story, but apart from the story he has only a weakened nuclear version of this property, not the full-blooded version. Or if, like Jacquette, we want to say that Holmes only has the property of *being described in the story* as existing, we should say that he similarly only has the properties of *being described in the story* as being a detective. In reality he is not a detective, since in reality he doesn't exist.

Call properties denoted by predicates of the form "being described / characterised / represented, etc., as P (in work W/ by individual X)" P*-representational* properties. I take it that the strategy of characterising Meinongian objects in terms of either *weakened* or P*-representational* versions, of ordinary (extensional) properties P is entirely inimical to the idea of a robust distinction among nuclear and extranuclear properties.[1] If one further suspects, as many will, that the notion of a weakened version of a property suffers from clear disadvantages relative to the notion of a representational version of a property, the observations above might suggest that 2P-ers should replace the nuclear/extra-nuclear distinction with a distinction between ordinary properties and representational properties. And that gets us close to an increasingly popular version of Meinongianism known as Modal Meinongianism.

[1] Some may think that the only way to make sense of the idea of a weak nuclear correlate of an extranuclear property like being a detective is to think of the former as really being the property of being *described* or *represented* as being a detective. I think there is at least a logical difference between the views, but discussing this will take us too far afield.

Modal Meinongianism

Graham Priest and Francesco Berto espouse a version of Meinongianism that, unlike Zalta's object theory, rejects 2P's distinction among properties while retaining a single mode of predication (Priest 2016, Berto 2012). It achieves this by direct appeal to representational properties. Objects are characterised in terms of certain properties – for example, an object may be represented as being a unique golden mountain or a unique round square or as the person who has all the properties that Holmes is credited with in the *Holmes* stories. Unless the object exists, it doesn't have these properties in the actual world but only in worlds in which the object is the way it is characterised to be. (Formally, this is secured by construing "It is represented in work W that p" as a modal operator, subject to a version of [possible-] world semantics.) Where the properties are existence-entailing, the object also exists in those worlds. Holmes exists in worlds that realise the *Holmes* stories, and the golden mountain exists in worlds in which it is golden and a mountain (and so, of course, does the existent golden mountain). The round square, in turn, is round and square but only in impossible worlds.[1] Following now-common practice, we'll call the Priest-Berto view *Modal Meinongianism.*

Non-existent objects also have *non*-representational properties in the actual world. In particular, they have properties that reflect their relationship to things that actually exist. The golden mountain is an object that Meinong often thought about, for example, just as Holmes is one of the best-loved characters in fiction. As on Zalta's view, these properties are not regarded as in any way different in kind from ordinary properties, except in so far as they are not existence-entailing. (They may even form part of the way an object is characterised. Take the object x such that x is both a golden mountain and my favourite object. This object may indeed be my favourite object; in that case it is my favourite object in the actual world, where it is an object that doesn't exist, as well as my favourite object in worlds in which it does exist and is a golden mountain.)

It is not hard to see why we might consider Modal Meinongianism the natural development of a Meinongianism that is impressed by Meinong's basic ideas but is equally impressed by the kind of considerations adduced in the last section. For Priest in partic-

[1] Because Priest takes as his criterion of existence having the potential to interact causally, he denies that abstract objects have existence. (Like Routley, he takes himself to be a noneist.) So for Priest the round square has these properties in certain (impossible) worlds without *existing* in those worlds. Berto is less sure.

ular, sharing as he does so many of Routley's logical and metaphysical commitments, this is where Routley should have ended up had he seen the error of insisting on a distinction among nuclear and extranuclear properties.

Routley's second thoughts

In the new 2016 edition of Priest's *Towards Non-Being* (Priest 2016), Priest declares that Routley came closer to such a view than anyone had so far realised. He notes that recent consultations of the Routley/Sylvan Archive at the University of Queensland brought to light an article Routley published just before he died: "Re-exploring item-theory" (Sylvan 1995). The article reviews elements of *Meinong's Jungle*, and explores ways to liberalise them. The most thorough engagement with the new view occurs in the evocatively entitled "The future perfect of *Exploring Meinong's Jungle and Beyond*" by Filippo Casati, based on notes in the Routley/Sylvan archives (Sylvan, Box 291). (Casati 2018 in $EMJB_2$, Vol. 1). In what follows, my account of Routley's liberalisation will rely both on the 1995 article and on Casati's work.

Routley's concerns about his earlier view are centred on one aspect in particular. Recall that Routley advocated a version of Meinongianism on which every description determines an object: according the principle of Unlimited Freedom of Assumption, for any set S of properties, there is an object – the thing described as being $F_1 - F_n$ (where F_i are the properties in S). But there are *a vast number of these objects* whose nature we can't access completely on this account and where we only have the illusion that we have complete access to their nature. The existent golden mountain is one such object; distinct from the golden mountain, it presents itself to us as existent as well as golden and a mountain, even though it is in fact a golden mountain that *doesn't* exist. Routley thinks we ought to be able to do better.

In fact, even the scope of Meinongianism is up for review, according to Routley. He now thinks that it is wrong simply to acknowledge a broad class of objects, many non-existent. Item-theory trades in items, and these extend far beyond the realm of objects as we usually conceive of them (Sylvan 1995, pp. 554ff.): *every* "well-formed subject signifies an item", including even a term like "sake" as it occurs in "for my sake" (pp. 550-1). But not every well-formed subject signifies *faithfully*, that is, has the properties listed in the subject. Despite this, Routley now wants to insist that there is an important sense in which an item has all those characterising properties used to characterise it. And he thinks that by focusing on such bare objects as the golden mountain, which are presented without much setting – just a bare description – theorists have distorted the possibilities open to item-theory. It is, he thinks, "far more revealing to begin with items with rich sources such as those of intricate mathematics, challenging theoretical science, and high fiction", and to treat the other cases as default cases when other normal sources fail (*ibid.*, p. 564).

The point is well put in his archival notes:

> Every item has a source of some sort ... For fictional characters, for instance, the source is a text or a set of texts. For theoretical objects, the sources are the appropriate theories. For empirical items, a prime source is experience or, more generally, aspects of a world of experience.

On Routley's new view, the source of an object determines what the object is like, without restriction:

> [An item's] source ... supplies all the information as to what it is like. Sources are many and various. They include such items as books, poems, myths, theories, films, theatre, rituals, dreams, perceptions, sketches, scenarios, descriptions, postulate systems and so on. ... Every item has a source of some sort which supplies its characterizing properties and perhaps other properties or supposed properties as well.

But, of course, this sounds for all the world like a version of UCP: "other properties or supposed properties" includes extranuclear properties like existence, being possible, and so on. Routley makes it abundantly clear that he would now accept that in the *Holmes* stories Holmes exists, just as the golden mountain exists in "The King of the Golden Mountain" (one of the fairly tales collected by the brothers Grimm) and just as the existent golden mountain exists simply in virtue of our describing it that way. So something has to give in the theory presented in *Meinong's Jungle*, else the ensuing account would have us class Holmes, the existent golden mountain, and the golden mountain in "The King of the Golden Mountain" as existing.

Routley's strategy, one that is not explored in great detail in either his archival notes or in Sylvan 1995, appeals to a metaphor of a city centre and suburbs (or sometimes a capital region and provinces):

> Item-theory can be initially separated into two interconnected, sometimes overlapping but far from exhaustive, areas: old down-town and suburban. Pictorially, the ancient city coincides approximately with older object-theory, as postulationally constrained by characterisation postulates of type CP. ... In the suburbs items are characterised or axiomatised through different postulates, appropriate to the suburb. ... [T]he items of the region (or the theory depicting it) have the features their assumptive principles ascribe. (Sylvan 1995, pp. 564-5)

This description is somewhat confusing. It may suggest that different sorts of items occupy different strata of the realm of items. That is not Routley's meaning. The distinction is rather one between what holds where and how. But despite now admitting other worlds or regions in which items possess the properties in terms of which they are presented, Routley thinks that how matters stand at the actual world(s) often holds a kind of pride of place. As he puts it, "it is often important, and sometimes essential, to be able to report on what holds of items not merely in other worlds, but in actual worlds" (*ibid.*, p. 571). And this is not always easy:

> [O]ut in its region, an item has all those features, properties and relations, it is ascribed by its relevant source. If its source reliably says it is thus and so then it is thus and so. Like what is told in a story, believed in a religion, or explained within science, this may comprise a bizarre or unlikely collection of features ... Down, however, where truth is evaluated, as in an actual world, affairs are different, and more demanding, as (real-world) constraints operate. (*ibid.*, p. 572)

To help us see what is involved, Routley talks of "preparation" of an item's prominent features before claims about the thing can be accepted as factually correct (*ibid.*, pp. 571-2). Thus:

> *One* way that preparation may proceed in the transfer from country to evaluation market, consists in deletion or reduction of non-

> characterizing (or non-primary) features and compensating adjustment of deleted or reduced features to presentational features. For instance, an item Doddy, a dodo, that is portrayed as existing in Mauritius in Prof Somersault's work, a bird which Somersault takes to be as real as quarks, is reassigned presentational existence when entered in the domain of an accommodating actual world. (*ibid.*, p. 572-3)

(Routley doesn't define *presentational existence*; I surmise it is a weakened form of existence that something has in virtue of existing according to the source in which it is presented.) Similarly, in the case of a mathematical theory existential claims have to be neutralized by using a neutral quantifier before we can regard its claims as factually correct, while in the case of traditional Meinongian objects Routley suggests we might appeal to methods like the watering-down of extra-nuclear predicates, as suggested by Terence Parsons in Parsons 1980. If we do, we can say that what is genuinely true about the existent golden mountain is that it has a weak, watered-down kind of existence rather than full extra-nuclear existence. What is not in contention is that it is at the same time golden and a mountain (Routley tellingly affirms that "nuclear predicates stand" (Sylvan 1995, p. 575).

It seems, then, that Routley wants to retain the thesis UFA, highlighting its strengths by stressing that we can assume items of the most attenuated kinds (even "sakes") and in terms of the most bizarre and seemingly incoherent combination of properties. Second, he wants a thesis that makes sense of the claim that an item possesses each property in its nature:

RCP_{SU} Given a specification by a source of an item O in terms of a set S of properties, there is a suburb in which O possesses each property in S.

Finally, he wants a thesis pertaining to what is true about the nature of items from the point of view of the centre:

RCP_{C} Given a specification in a source of an item O in terms of a set S of properties, the properties that O possesses at the centre are the nuclear properties in S as well as appropriate nuclear weakenings of the extra-nuclear properties in S (say, as watered-down properties).

Talk of city centre and suburbs is metaphorical, and we will never, I suspect, be entirely sure why Routley chose the metaphor. For the most part, it suffices to take the city centre to be the actual world, and suburbs alternative (possible and impossible) worlds. But towards the end Routley may have grown suspicious of the idea of worlds, since he explicitly claimed that there can be a plurality of actual worlds. That may explain his preference for metaphor. Suburbs other than the centre may be actual. (The thought that there can be a plurality of actual worlds was further explored in Sylvan 1988.)

Was Routley a proto–Modal Meinongian?

We needn't worry about the idea that there may be more than one actual world, a view that has largely been dismissed by even close followers. Priest, for example, comments that "For my part, I take it that there is only one actual world. For better or for worse, this is it" (Priest 2016). But commentators have been more complementary about another aspect of Routley's view, his liberalising attitude that makes him want to return to an unrestricted version of CP. And they have been quick to suggest that Modal Meinongianism seems to do what Routley was trying to do, but without retaining the problematic nuclear/extranuclear distinction. Here is Priest 2016 again:

> This is clearly a move in the general direction that [Priest's *Towards Non-Being*] was to take. For the city centre, take the actual world; for the suburbs, take other worlds, and one is pretty close. Indeed, Richard does sometimes speak of worlds in this connection ... (p. xxxiii)

But Priest also notes that the parallel is not exact, not only because Routley thinks there is more than one actual world but also because he still maintains that instances of CP are true at the actual world whenever they feature specifications of objects in terms of nuclear predicates. That, Priest thinks, is not necessary, for

> once one can accommodate a completely unrestricted CP in the way that [*Towards Non-Being*] does, it is not clear why one should worry about or employ the notion of a nuclear predicate at all. (p. xxxiii)

Reflecting on similar passages, Casati comes to a more guarded conclusion:

> ... a lot of problems remain open. For instance, how shall we formally deal with this new version of item-theory? Can we have a formal semantics that is able to faithfully represent Routley's new intuitions? Moreover, Graham Priest has already developed a semantics of possible and impossible world in order to make justice to Meinong's ontology. Can Priest's Modal Meinongianism reflect all the ideas sketched by Routley? If not, why? All these questions need to be addressed in future works. And this is the real inheritance from Routley, namely, living thoughts that need our contribution to be developed. (Casati 2018, p. 599)

My own view is that Casati was right to come to a more guarded conclusion. There is a reason why Routley wanted to retain the role of nuclear properties in his two-fold version of UCP. It concerns a problematic feature of any account that places overriding emphasis on representational properties as a way of eliminating effective appeal to nuclear properties, which is what Priest and Berto do in their Modal Meinongianism.

But before we discuss this problematic feature, let me briefly summarise what I think Routley was trying to do. Routley thought that any Meinongian object had a nature – its way of being "what it is like" – given in terms of the way the object was specified in its source. Apprehending its nature is what allows us to grasp the object, and the properties make the object the object that it is. Armed with the nuclear/extranuclear distinction, Routley thought that the idea of an object's nature is relatively unproblematic if this nature is given in terms of nuclear properties: an object characterised a certain way in terms of nuclear properties just has those properties, so an object given in terms of that kind of nature unproblematically has or possesses that nature. The golden mountain just is golden and a mountain, just as the round square is round as well as square. But what, then, should we do about all the other objects (or, more generally, items) that are specified in part or in whole in terms of properties that are extranuclear? We grasp such an object, say the existent golden mountain or Holmes, in terms of these properties, but what sense can we make of the thought that this is their nature if we can't even deduce that have these properties? This, I think, was one of the conundrums facing Routley at the time he decided to liberalise his item-theory.

A problem for Modal Meinongianism

The last paragraph stressed the issue of how we have access to, or grasp, Meinongian objects. I now want to argue that this issue presents Modal Meinongianism with a looming problem. The problem is this. Consider an object like Holmes. Let ϕ be the big conjunctive predicate in terms of which Holmes (h) is characterised in the *Holmes* stories. So the Holmes stories represent Holmes in the following way: $\phi(h)$. According to Modal Meinongianism, $\phi(h)$ is indeed the case in worlds that realise the *Holmes* stories, while the actual world is not like this: in the actual world Holmes is just *represented* as being like this. Now ask: what, in that case, is Holmes? How do we pick "him" out when declaring that "he" is the subject of this representation? (Because being male is existence-entailing, the neutral "it" would be more precise, but let us stay with "him".) The obvious answer is that we defer to Conan Doyle. Whoever Conan Doyle picked out, he is the one we have in mind. But this just postpones the problem. How does Conan Doyle do it? Not by stipulating that Holmes is the unique entity who bears the name "Holmes", is extremely clever, became a detective, went to live with someone called Watson in 221B Baker St., London, and so on. That is simply how Holmes is *represented.*[1]

Note that we have no such problem where ordinary existents are concerned. Tolstoy knew of Napoleon, and represented him as being a certain kind of person when writing *War and Peace*. Unsurprisingly, perhaps, as Modal Meinongians Berto and Priest disagree that the cases are different. Here is their Characterisation Principle for Modal Meinongianism:[2]

> If ϕ is a choice function on sets of objects, then the denotation of $\epsilon x A(x)$ is:
> $\phi(\{x$: (α and $A(x)$) or ($\sim\alpha$, x is a non-existent object, and the envisaged world is one where $A(x)\})$ [where α abbreviates "At @, some y is such that $A(y)$"] (Berto and Priest 2014, p. 196)

[1] Incidentally, this worry is substantially independent of the modal element in Modal Meinongianism. The worry arises as soon as you think that non-existents like Holmes or the round square only have representational properties as part of their nature, whether or not you also think that they inhabit other possible or impossible worlds. In short, it is a worry for all representationalist versions of Meinongianism, I think, the kind of view of non-existent objects that Tim Crane has defended in various places (see especially Crane 2013).

[2] This is a revision of their earlier accounts of CP for Modal Meinongianism, in response to an objection (cf. Kroon 2012, Berto and Priest 2014, Priest 2016).

This suggests that the idea of selection-by-intention of an object as the denotation of a description $\epsilon x A(x)$ is indifferent to whether there is something that satisfies the predicate A. If there is, an existent object is selected; if not a non-existent object is selected, one that satisfies A in some other world(s).

But how does one intend a non-existent object? Berto and Priest think there is no in-principle difficulty:

> One answer is that it is the very nature of intentionality to single out an object of a certain kind, and given a bunch of objects, one can just point mentally to one, in the same way that, given a bunch of physical objects, one can point physically to one of them (indeed, an act of physical pointing presupposes the intentional act that goes along with it: otherwise, one could be pointing at many things). And if one imagines an object with certain properties then, *ex hypothesi*, what one does is imagine such a thing. (Berto and Priest 2014, pp. 197-8)

This may look promising. It claims that to intend an object $\epsilon x A(x)$ is to imagine an object as satisfying the predicate $A(x)$. If nothing *actually* satisfies $A(x)$, one has thereby intended a non-existent object, one that only satisfies $A(x)$ in worlds other than the actual world. Call this the Intending-As-Imagining thesis.

But this Intending-As-Imagining thesis should strike one as troubling. Remember that we are to take the predicate $A(x)$ in the description $\epsilon x A(x)$ as representational: an object is *represented* as satisfying $A(x)$. Now consider the sentence "if one imagines an object with certain properties then, *ex hypothesi*, what one does is imagining such a thing". Presumably the phrase "such a thing" stands for the intended object, the object that in the actual world is *represented* as having certain properties. Call this the *de re* reading of the term. For this object to be the appropriate target for our imaginative exercise, it must also be the object that features in the antecedent; it must be the focus of our imagining when we "imagine an object with certain properties". But what is it to imagine an object as satisfying $A(x)$, say?

The most natural reading of this process is *de dicto*. One simply imagines a situation in which there *is* an object that satisfies $A(x)$, and then one focuses on this object, all within the scope of the imagining. It is not *de re* to apprehend something which one then imagines as satisfying $A(x)$, presumably by transporting oneself imaginatively to a world where it does satisfy $A(x)$. Indeed, how would one do the *de re* apprehending of the object in the first place?[1]

The alternative way to understand the argument is by adding the needed *de re* element at the end. Suppose one goes through the imaginative exercise under its *de dicto* construal, imagining that there is an object that has certain properties. By doing this, so the present construal of the argument goes, one thereby intends an object. But this object need not have the properties in question in the actual world, since there may be nothing in the actual world that has these properties: there is no golden mountain in the actual world, for example, so in imagining a golden mountain, or the golden mountain, we are imagining something that doesn't exist. All we can say, therefore, is that the object is one that is imaginatively *represented* as having the properties.

On this account of the argument, the imagining and the intending can't really be separated. But there is an obvious problem with this way of construing the argument: how can one reasonably go beyond a premise about the composition of the world that is being imagined to a conclusion about the actual world? Why see this imaginative exercise – imagining a world in which there is something that is golden and a mountain, say – as testament to there being something in the actual world that is being represented in the imagination as being golden and a mountain? And why one thing rather than another? It is hard to avoid the impression that here Modal Meinongians are guilty of a curious kind of modal fallacy.

Back to the centre and suburbs

To avoid committing such a fallacy, it seems that the best hope for the Modal Meinongian is to show how to make sense of a *de re* grasp of an object that can then be identified, on principled grounds, as being a non-existent object. This is something that Modal Meinongianism has troubles with. All it can say about the allegedly non-existent objects in question is that they are *represented* as having certain properties, properties that are

[1] One reason for thinking that we do have such a *de re* grasp on a non-existent object is that the invitation to imagine typically involves names or definite descriptions: we are asked to imagine *Holmes*, for example, or *the* golden mountain. But nothing can be concluded from such formulations. They only show, I think, that we are invited to imagine that there is a unique object of a certain kind, as in a game of make-believe where we invite fellow game-players to imagine that "that lion hiding behind that tree has noticed us", when there is nothing at all behind the tree.

then identified as being uninstantiated in the actual world. But this is no identification at all – we need to know *what* is being represented.[1] Call the problem of specifying appropriate identifying conditions of this kind the *identification problem*.

Whatever the theory's other problems, no such problem attends the way we identify Meinongian objects in the framework of Zalta's object theory (cf. Bueno and Zalta 2017). And arguably no such problem attends the way Routley thinks of our access to the items of his item-theory. As I understand the theory, for Routley a Meinongian non-existent item is constituted by its nature, and we grasp the item by apprehending its nature. This nature is the way the item is specified in its source. The item possesses the properties mentioned in the specification, although often only at the suburbs (a fictional character like Holmes, for example, possesses the extra-nuclear property of existence in the suburbs, not at the centre). But from the point of view of *grasping* the item, what properties it has at the suburbs, or at other worlds, is of little moment. We grasp the item *at the centre*, or the actual world, and to do so we need to apply relevant *actual world* criteria of identity that tell us what makes the item what it is. As far as is known, Routley never changed his mind about this after he wrote "On what there isn't". There he had this to say:

> [T]he very same notions of identity – most importantly, extensional identity – and distinctness that apply to entities apply likewise to nonentities, objects that do not exist. The criterion for identity for nonentities is, as for entities, coincidence in extensional properties. Thus, for instance, Hercules and Heracles are identical, though some people did not and do not know this. The criterion for distinctness is that of positively differing on extensional properties. Thus, for example, Pegasus is distinct from Thunderhead because Pegasus has the (extensional) property of being winged and Thunderhead does not. (Routley 1982, p. 156; see also, this volume p. 71)

Routley makes it clear that he means extensional *characterising* or *nuclear* properties. In line with this view, when we first grasp the golden mountain and then the

[1] The mistake here is akin to someone's thinking they have identified an object when they think of it as the object they are referring to with the name "Smith", say. That is no identification at all, as Kripke reminds us when articulating his non-circularity condition on descriptivist accounts of reference (Kripke 1980, pp. 68-70). Mistakes of this kind are surprisingly common in attempts to resuscitate a reasonable form of Meinongianism. Thus cf. Paoletti's appeal to properties like *having S as a source* in Paoletti 2013.

existent golden mountain we grasp distinct objects because we are focusing on different sets of constituting properties. While they share the nuclear properties of being golden and a mountain, the golden mountain is distinct from the existent golden mountain because only the second has the nuclear property of weak existence. Similarly, we grasp Holmes in terms of the properties he is assigned in the source where he is first found. These properties include full-blooded extra-nuclear existence, but to apprehend Holmes we need to prepare such features for proper inclusion at the actual world, the centre. That means we grasp Holmes in terms of the nuclear properties assigned to him in the source as well as appropriate weakenings of other properties like existence (recall Routley's *presentational* existence). This is a very different view from Modal Meinongianism, not simply because it continues to rely on the nuclear/extranuclear distinction among properties but, more importantly, because it is motivated in part by the need to specify what it is that we grasp when we grasp a non-existent object or item. If I am right, this problem of identification is one that stymies Modal Meinongianism.

Conclusion

Earlier, in section 2, I argued that the case of fiction presents Routley's object theory in *Meinong's Jungle* with a serious problem. One way in which Routley could have responded would have been to weaken UFA, restricting the class of *bona fide* objects to objects specified in terms of nuclear properties alone. Routley was resolutely set against this path. As we saw earlier, in liberalising the theory he presents in *Meinong's Jungle* he was concerned to broaden, not tighten, our understanding of UFA.

This puzzles Priest, who complains that "once one can accommodate a completely unrestricted CP in the way that [*Towards Non-Being*] does, it is not clear why one should worry about or employ the notion of a nuclear predicate at all." The answer, I think, is that Routley had as background a different project from Priest and Berto, one that had the issue of the identification of items as its core. (This is no less true of other 2P-ers like Parsons and Jacquette.) Assuming the nuclear/extranuclear distinction, RCP seemed well placed to say how we identify non-existent objects when these are specified in terms of bare descriptions. With this successful theory in place, Routley was in no mood to give it up: he wanted to achieve something no less convincing for all the other items permitted by UFA, using related resources. Given this, I think it is scarcely a surprise that Routley stayed with the nuclear/extranuclear distinction.

This is not to say that Routley was right to maintain the distinction. As I have been at pains to point out, it is hard to disagree with the criticisms of commentators like Zalta and Priest and Berto. We should acknowledge, however, that theories that reject the distinction incur a serious debt – if I am right, a debt that Modal Meinongianism in particular has grave difficulty paying.

References

$EMJB_1$ Routley, R., *Exploring Meinong's Jungle and Beyond*, Research School of Social Sciences, Australian National University, 1980.

$EMJB_2$ Routley, R. et al., *The Sylvan Jungle*, edited by M. Eckert et al., vol. 1–4, Springer, 2018+.

Berto, F., *Existence as a Real Property: the Ontology of Meinongianism*, Springer, 2012.

Berto, F. and Priest, G., "Modal Meinongianism and characterization: reply to Kroon", *Grazer Philosophische Studien* **90** (2014) 183–200.

Bueno, O. and Zalta, E., "Object theory and Modal Meinongianism", *Australasian Journal of Philosophy* **95** (2017) 761–78.

Casati, F., "The future perfect of *Exploring Meinong's Jungle and Beyond*", in *$EMJB_2$*, vol. 1, 2018, 583–601.

Crane, T., *The Objects of Thought*, Oxford University Press, Oxford, 2013.

Jacquette, D., *Alexius Meinong: the Shepherd of Non-Being*, Springer, Dordrecht, 2015.

Kripke, S., *Naming and Necessity*, Harvard University Press, Cambridge Mass., 1980.

Kroon, F., "Characterization and existence in Modal Meinongianism", *Grazer Philosophische Studien* **86** (2012) 23–34.

Meinong, A., *Über Möglichkeit und Wahrscheinlichtkeit*, Verlag von Johann Abrosius Barth, Leipzig, 1915.

Meinong, A., "The theory of objects [Über Gegenstandstheorie]", in *Realism and the Background of Phenomenology*, edited by R.M. Chisholm, trans. by I. Levi, D.B. Terrell and R.M. Chisholm, The Free Press, Glencoe, 1960.

Paoletti, M.P., "*Exploring Meinong's Jungle and Beyond*: an investigation of noneism and the theory of items", *Humana Mente Journal of Philosophical Studies* **25** (2013) 275–92.

Parsons, T., *Nonexistent Objects*, Yale University Press, New Haven, 1980.

Priest, G., *Towards Non-Being*, 2nd ed., Oxford University Press, Oxford, 2016.

Quine, W.V.O., "On what there is", *The Review of Metaphysics* **2** (1948) 21–38.

Routley, R., "On what there isn't", *Philosophy and Phenomenological Research* **43** (1982) 151–77. Reprinted in this volume – chapter 3.

Russell, B., "On denoting", *Mind* **14** (1905a) 479–93.

Russell, B., "Review of: A. Meinong, *Untersuchungen zur Gegenstandstheorie und Psychologie*", *Mind* **14** (1905b) 530–38.

Russell, B., "Review of: A. Meinong, *Über die Stellung der Gegenstandstheorie im System der Wissenschaften*", *Mind* **16** (1907) 436–39.

Sylvan, R., "Radical pluralism – an alternative to realism, anti-realism, and relativism", in *Relativism and Realism in Science*, edited by R. Nola, Kluwer, Dordrecht, 1988, 253–91.

Sylvan, R., "Re-exploring item-theory: object-theory liberalized, pluralized and simplified but comprehensivized", *Grazer Philosophische Studien* **50** (1995) 47–85. Reprinted in $EMJB_2$, Vol. 1, pp. 546-81. All page references are to the reprint.

Zalta, E., *Abstract Objects: an Introduction to Axiomatic Metaphysics*, Dordrecht Reidel, Dordrecht, 1983.

Zalta, E., *Intensional Logic and the Metaphysics of Intentionality*, MIT Press, Cambridge, MA., 1988.

Index

D. Hyde (ed.), *Noneist Explorations I*, Synthese Library 415,
https://doi.org/10.1007/978-3-030-26309-6

Kolibri 360 GmbH
Ferdinand-Jühlke-Straße 7
99095 Erfurt, Deutschland
produktsicherheit@kolibri360.de